Wildfire Hazard and Risk Assessment

Wildfire Hazard and Risk Assessment

Editor

James R. Meldrum

MDPI • Basel • Beijing • Wuhan • Barcelona • Belgrade • Manchester • Tokyo • Cluj • Tianjin

Editor
James R. Meldrum
Fort Collins Science Center
U.S. Geological Survey
Fort Collins
United States

Editorial Office
MDPI
St. Alban-Anlage 66
4052 Basel, Switzerland

This is a reprint of articles from the Special Issue published online in the open access journal *Fire* (ISSN 2571-6255) (available at: www.mdpi.com/journal/fire/special_issues/wildfire_assessment_fire).

For citation purposes, cite each article independently as indicated on the article page online and as indicated below:

LastName, A.A.; LastName, B.B.; LastName, C.C. Article Title. *Journal Name* **Year**, *Volume Number*, Page Range.

ISBN 978-3-0365-3582-1 (Hbk)
ISBN 978-3-0365-3581-4 (PDF)

© 2022 by the authors. Articles in this book are Open Access and distributed under the Creative Commons Attribution (CC BY) license, which allows users to download, copy and build upon published articles, as long as the author and publisher are properly credited, which ensures maximum dissemination and a wider impact of our publications.

The book as a whole is distributed by MDPI under the terms and conditions of the Creative Commons license CC BY-NC-ND.

Contents

Preface to "Wildfire Hazard and Risk Assessment" vii

Michael D. Caggiano, Todd J. Hawbaker, Benjamin M. Gannon and Chad M. Hoffman
Building Loss in WUI Disasters: Evaluating the Core Components of the Wildland–Urban Interface Definition
Reprinted from: *Fire* **2020**, *3*, 73, doi:10.3390/fire3040073 1

Nathan Mietkiewicz, Jennifer K. Balch, Tania Schoennagel, Stefan Leyk, Lise A. St. Denis and Bethany A. Bradley
In the Line of Fire: Consequences of Human-Ignited Wildfires to Homes in the U.S. (1992–2015)
Reprinted from: *Fire* **2020**, *3*, 50, doi:10.3390/fire3030050 19

James R. Meldrum, Christopher M. Barth, Julia B. Goolsby, Schelly K. Olson, Adam C. Gosey and James (Brad) White et al.
Parcel-Level Risk Affects Wildfire Outcomes: Insights from Pre-Fire Rapid Assessment Data for Homes Destroyed in 2020 East Troublesome Fire
Reprinted from: *Fire* **2022**, *5*, 24, doi:10.3390/fire5010024 39

Benjamin M. Gannon, Yu Wei and Matthew P. Thompson
Mitigating Source Water Risks with Improved Wildfire Containment
Reprinted from: *Fire* **2020**, *3*, 45, doi:10.3390/fire3030045 61

Michael J. Campbell, Philip E. Dennison, Matthew P. Thompson and Bret W. Butler
Assessing Potential Safety Zone Suitability Using a New Online Mapping Tool
Reprinted from: *Fire* **2022**, *5*, 5, doi:10.3390/fire5010005 87

Johannes Heisig, Edward Olson and Edzer Pebesma
Predicting Wildfire Fuels and Hazard in a Central European Temperate Forest Using Active and Passive Remote Sensing
Reprinted from: *Fire* **2022**, *5*, 29, doi:10.3390/fire5010029 111

Den Boychuk, Colin B. McFayden, Douglas G. Woolford, Mike Wotton, Aaron Stacey and Jordan Evens et al.
Considerations for Categorizing and Visualizing Numerical Information: A Case Study of Fire Occurrence Prediction Models in the Province of Ontario, Canada
Reprinted from: *Fire* **2021**, *4*, 50, doi:10.3390/fire4030050 135

Ujjwal KC, Jagannath Aryal, James Hilton and Saurabh Garg
A Surrogate Model for Rapidly Assessing the Size of a Wildfire over Time
Reprinted from: *Fire* **2021**, *4*, 20, doi:10.3390/fire4020020 155

Cory W. Ott, Bishrant Adhikari, Simon P. Alexander, Paddington Hodza, Chen Xu and Thomas A. Minckley
Predicting Fire Propagation across Heterogeneous Landscapes Using WyoFire: A Monte Carlo-Driven Wildfire Model
Reprinted from: *Fire* **2020**, *3*, 71, doi:10.3390/fire3040071 173

Panteleimon Xofis, Pavlos Konstantinidis, Iakovos Papadopoulos and Georgios Tsiourlis
Integrating Remote Sensing Methods and Fire Simulation Models to Estimate Fire Hazard in a South-East Mediterranean Protected Area
Reprinted from: *Fire* **2020**, *3*, 31, doi:10.3390/fire3030031 189

Preface to "Wildfire Hazard and Risk Assessment"

Wildfire risk can be perceived as the combination of wildfire hazards (often described by likelihood and intensity) with the susceptibility of people, property, or other valued resources to that hazard. Reflecting the seriousness of wildfire risk to communities around the world, substantial resources are devoted to assessing wildfire hazards and risks. Wildfire hazard and risk assessments are conducted at a wide range of scales, from localized to nationwide, and are often intended to communicate and support decision making about risks, including the prioritization of scarce resources. Improvements in the underlying science of wildfire hazard and risk assessment and in the development, communication, and application of these assessments support effective decisions made on all aspects of societal adaptations to wildfire, including decisions about the prevention, mitigation, and suppression of wildfire risks. To support such efforts, this Special Issue of the journal *Fire* compiles articles on the understanding, modeling, and addressing of wildfire risks to homes, water resources, firefighters, and landscapes.

James R. Meldrum
Editor

Article

Building Loss in WUI Disasters: Evaluating the Core Components of the Wildland–Urban Interface Definition

Michael D. Caggiano [1,*], **Todd J. Hawbaker** [2], **Benjamin M. Gannon** [1] **and Chad M. Hoffman** [1]

[1] Department of Forest & Rangeland Stewardship, Colorado State University, 1472 Campus Delivery, Fort Collins, CO 80523, USA; Benjamin.Gannon@colostate.edu (B.M.G.); C.Hoffman@colostate.edu (C.M.H.)
[2] Geosciences and Environmental Change Science Center, U.S. Geological Survey, Denver, CO 80225, USA; tjhawbaker@usgs.gov
* Correspondence: michael.caggiano@colostate.edu

Received: 30 October 2020; Accepted: 17 December 2020; Published: 20 December 2020

Abstract: Accurate maps of the wildland–urban interface (WUI) are critical for the development of effective land management policies, conducting risk assessments, and the mitigation of wildfire risk. Most WUI maps identify areas at risk from wildfire by overlaying coarse-scale housing data with land cover or vegetation data. However, it is unclear how well the current WUI mapping methods capture the patterns of building loss. We quantified the building loss in WUI disasters, and then compared how well census-based and point-based WUI maps captured the building loss. We examined the building loss in both WUI and non-WUI land-use types, and in relation to the core components of the United States Federal Register WUI definition: housing density, vegetation cover, and proximity to large patches of wildland vegetation. We used building location data from 70 large fires in the conterminous United States, which cumulatively destroyed 54,000 buildings from 2000 through to 2018. We found that: (1) 86% and 97% of the building loss occurred in areas designated as WUI using the census-based and point-based methods, respectively; (2) 95% and 100% of all of the losses occurred within 100 m and 850 m of wildland vegetation, respectively; and (3) WUI components were the most predictive of building loss when measured at fine scales.

Keywords: wildfire; risk assessment; structure loss; wildland–urban interface; mitigation; mapping; land use; disaster

---

1. Introduction

The rapid development and expansion of the wildland–urban interface (WUI) into areas with highly-flammable vegetation has significantly increased the potential for building loss during wildland fires [1–5]. Recent wildfire losses in the United States, Australia, and Spain have highlighted the global nature of this phenomenon [6]. Globally, wildfires have occurred across the full spectrum of rural to urban communities with a range of housing densities and arrangements, and have even spread into urban areas that were considered low risk [7]. In the United States, which has seen extensive losses in the 2020 fire season, there are currently an estimated 1.7 million residences in areas at high or extreme risk of wildfire [8], the majority of which are in the WUI.

Wildfires that cause substantial building loss are known as WUI disasters [9]. Many WUI disasters follow the same sequence of events. First, a wildfire ignites in an extreme environment (i.e., fuel, weather, and topography) that allows it to escape initial attack. Then, as the fire grows, it threatens a large number of homes and buildings relative to firefighting resources, which reduces the efficacy of the fire suppression, which leads to numerous losses [9,10]. The increased fuel loading in fire-adapted ecosystems due to historical fire suppression practices [9], increased residential development and increased anthropogenic ignitions [11] due to an expanding WUI [12], and climate change [13] is believed to exacerbate the WUI problem. Although the exact number of WUI disasters is unknown, these events likely represent only a small subset of the total number of wildfires that ignite annually [14]. Understanding when and where WUI disasters are likely to occur, and how buildings and community characteristics influence the likelihood of building loss is critical for the development of strategies for the reduction of wildfire risk and increasing community resiliency [9,10,15–20]. Although wildfires can impact WUI communities in multiple ways (e.g., loss of life, smoke, evacuation, post-fire economic recovery), we focused on building loss because, at its core, community wildfire risk has been described as a home ignition problem, with building loss being responsible for the majority of economic loss during WUI disasters [9,10]. Accordingly, land managers tasked with reducing wildfire-caused building loss risk require accurate maps of at-risk buildings within the WUI, calculated at appropriate scales [17,18,20].

WUI mapping efforts in the United States are based on the Federal Register definition of WUI communities [21], which focuses on buildings within or adjacent to wildland vegetation. The Federal Register definition further differentiates two subtypes within the WUI based on housing density and the proximity of buildings relative to wildland vegetation: (1) intermix, in which housing units are dispersed among wildland vegetation; and (2) interface, in which housing units are adjacent to wildland vegetation. Several different WUI mapping efforts have operationalized these definitions in order to estimate the number of at-risk housing units and the total extent of the WUI [1,5,12,22–24]. The SILVIS WUI maps made available by the University of Wisconsin SILVIS Lab have emerged as the most widely-used WUI maps in the United States [5,22,23]. For example, a presidential executive order in 2016 directed federal wildfire mitigation and planning efforts to use SILVIS WUI maps or their equivalent when determining wildfire risk [25]. SILVIS WUI maps rely on the U.S. Census Bureau (Census) block level accounting of housing units and wildland vegetation extents extracted from land cover data in order to classify WUI based on three specific components: housing unit density, vegetation cover, and proximity to large patches of contiguous wildland vegetation [5,22,23]. Although SILVIS WUI maps are often used to estimate the number of at-risk housing units and communities at risk in the WUI, they may be insufficient for wildfire planning and response due to the relatively coarse spatial scale of the census blocks, the decadal update interval, and the lack of consideration of fire's behavior [17,18,20].

The increasing availability and use of individual building location data in WUI mapping overcomes the uncertainty associated with housing unit locations and other critical factors that influence the fire behavior and response in census block data [1,18]. Although fine-scale data have historically only been available for limited spatial extents [17,18], recent developments in remote sensing technology have greatly increased the ability to collect fine-scale building locations [26,27]. Point-based building location and development data are now available across large extents through commercial vendors and open data archives [28,29]. With these advances in technology, a single dataset of at-risk buildings, the finest spatial unit of the WUI problem, can be used in various applications, and provide consistency across multiple scales.

Wildland–urban interface categories can be assigned to building points using the spatial analysis of WUI components (housing unit density, vegetation cover, and proximity to large patches of contiguous wildland vegetation) calculated at appropriate scales [1,5,17,24]. Regardless of whether WUI maps are developed based on individual building locations or census blocks as inputs, WUI maps should align with our understanding of what and where the WUI is [24]. Although numerous concerns exist in the WUI (e.g., loss of life, evacuation, smoke impacts), of particular interest to managers is the ability of WUI maps to identify buildings that are at risk of igniting [20,30]. However, there is a paucity of information on how well WUI maps capture the observed patterns of building loss during WUI disasters. A systematic assessment of building loss alignment with point-based and census-based WUI maps is needed [30,31].

In order to advance our understanding of where building loss occurs and how we spatially conceptualize the WUI problem [9,21], this study sought to: (1) characterize the occurrence of WUI disasters; (2) assess building loss relative to the three core components of the WUI (building density, vegetation cover, and proximity to large patches of contiguous wildland vegetation) at multiple scales, hereinafter referred to as WUI component-scale combinations; and (3) evaluate how well the SILVIS and point-based WUI maps capture the patterns of building loss in WUI disasters. We anticipated that building loss would be positively-correlated with housing density, vegetation cover, and proximity to vegetation, as all three of these factors increase the potential for heat and fire brand exposure [9]. We also expected that both WUI mapping methods would capture a majority of the building loss in WUI disasters, but that point-based maps would perform better due to their finer spatial resolution.

2. Methods and Data

2.1. Overview of Methods

We utilized a combination of spatial and non-spatial datasets to identify WUI disasters and examine the patterns of building loss within those disasters. We start with a summary of our analyses, followed by detailed description of the data sources and methods. First, we queried the National Wildfire Coordinating Group Wildfire Incident Status Summary Reports, which are referred to as ICS-209 [32], in order to identify individual wildfires and the associated building loss. WUI disasters were identified through the inclusion of only those wildfires that reported more than 50 destroyed buildings, which is similar to the thresholds used by others [9,10]. Next, we examined how well the SILVIS WUI maps and point-based WUI maps captured the patterns of building loss. This involved the use of geospatial datasets of fire perimeters and building location points affected by and adjacent to WUI disasters. Using a series of spatial overlays, we attributed each building location to a specific fire, burn status (burned or unburned), census block housing unit density, and SILVIS WUI type [23]. The building density, percent vegetation cover, and distance to vegetation were then calculated at multiple scales, following the framework proposed by Bar-Massada et al. [18], using the National Land Cover Database (NLCD) [33]. Figure 1 provides an overview of the spatial datasets used in this analysis for the 2012 Waldo Canyon Fire in Colorado.

2.2. Identifying WUI Disasters

In order to identify WUI disasters, we queried the ICS-209 databases from 2000 to 2018. Individual ICS-209 reports capture daily information on the incident type, fire size, percent containment, significant events, resource needs, and cumulative number of threatened and destroyed structures [32]. The ICS-209 reports changed the ways in which they recorded structure types over time. Before 2014, they recorded structures as either primary buildings, outbuildings, or commercial buildings, whereas in 2014 and after, they included several additional categories. We used the total destroyed primary structures before 2014, and the total destroyed single residences for 2014 onwards as proxies for building loss. Primary structures made up 63% of all of the destroyed structures from 2000 to 2013. This change in the data reporting introduces a small bias, but the focus on homes and residential structures aligns well with prior conceptualizations of the WUI problem [9,10], and better matches the spatial building location datasets described below. We then identified 108 wildfire incidents with more than 50 destroyed buildings. We chose the 50-building threshold, which is slightly lower than the thresholds used by others to identify WUI disasters [9,10], in order to account for potential errors in the ICS-209 reported structure loss [34].

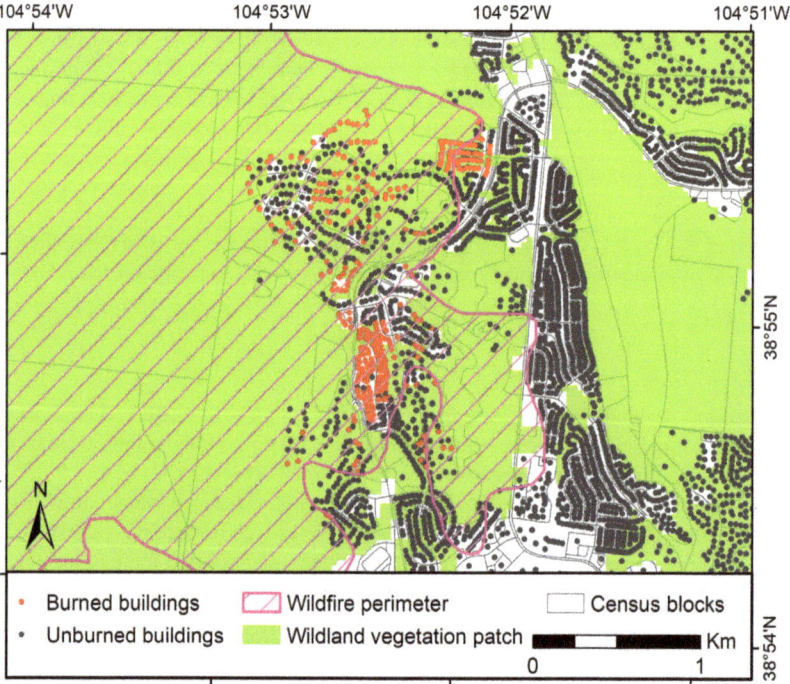

Figure 1. An example of the spatial datasets used in our analysis, including fire perimeters, wildland vegetation land cover, census blocks, and building locations using the 2012 Waldo Canyon Fire in Colorado. Each building location was assigned a burn status and attributed with information from layers derived from the above data and the SILVIS wildland–urban interface data, including building density, percent vegetation cover, and distance to vegetation.

2.3. Building Location Data

We were able to acquire the building location data for 70 of the fires we identified as WUI disasters in the ICS-209. We achieved this by querying and merging four distinct point-based building location datasets. Henceforth, we refer to buildings, but also acknowledge the differences in terminology and potential bias between buildings and the housing units referenced in the Census-based data, and between buildings and the destroyed structures referenced in ICS-209 reports. For the subsequent analysis and discussion, we equate the three terms. First, the building location data were acquired for 44 fires from a spatial dataset of burned and unburned buildings created by visually digitizing the buildings and evaluating the building burn status from high-resolution pre- and post-fire Google Earth imagery from 2000 through 2013, which was produced by Alexandre et al. [35]. Second, the building location data were acquired for 15 fires from the California Department of Forestry and Fire Protection (CalFire) as part of their Damage Inspection Report Program (DINS) [36]. This dataset was produced from door-to-door damage inspections collected by CalFire staff with GPS-enabled tablets during and after wildfire incidents. However, the CalFire dataset is only available for a limited set of fires in California, starting in 2013. The CalFire data collection process, data attributes, and limitations are further described in Syphard and Keeley [7]. Third, the building location data were acquired for 11 fires identified in the ICS-209 analysis that were missing the building location data from other sources. For these fires, the affected buildings were digitized, and the burn status (burned or unburned) was determined using pre- and post-fire Google Earth imagery using similar methods to those described in Alexandre et al. [35]. The fourth building location dataset used in this analysis was the Microsoft structure footprint dataset [29]. This dataset allowed us to augment the CalFire data by identifying missing undamaged building locations within the fire perimeters. It also allowed us to identify buildings within 2400 m of each fire perimeter, and to reduce edge effects when calculating the building density. The Microsoft dataset was generated using the RefineNet convolutional neural network for semantic segmentation, followed by a polygonization process to identify the building footprints in high-resolution satellite images [37] collected roughly around 2015. Although it is impossible to determine the exact build date of a specific structure in the Microsoft data, the building locations have a reported accuracy of 99.3%, with commission and omission error rates of 0.7% and 6.5%, respectively [29]. For this study, the footprint polygon centroid locations were used in order to match the other building datasets.

The remaining 38 WUI disasters were excluded because of data limitations. These limitations included: limited pre- and post-fire image availability regarding the evaluation of the fate of the buildings (6 fires), insufficient building location data (15 fires), missing fire perimeters (5 fires), or errors in the ICS-209 reports (12 fires). Fires which were missing high-resolution imagery included the 2002 Hayman Fire in Colorado and the 2004 Bear Fire in California, and recent fires such as the 2018 Roosevelt Fire in Wyoming and the 2018 Spring Creek Fire in Colorado, which both lacked publicly-available post-fire imagery as of 1st January 2020. Other fires, such as the 2003 Padua Fire and the 2007 Ham Lake Fire, were excluded due to insufficient building location data. For example, the ICS-209 reported that the 2003 Padua Fire destroyed 60 buildings, but the corresponding spatial dataset of the building locations only included 22 burned building location points. Fires for which the ICS-209 reported more than 50 burned buildings, but the spatial data had less than 50 burned buildings, were included in the initial ICS-209 analysis, but were excluded from subsequent building location analyses evaluating the census-based and point-based WUI map accuracy. Fires with missing or insufficient building location data occurred in states that were generally well-represented by other fires, allowing us to assume that the missing fires would not bias the results. The final fire perimeter for each WUI disaster identified in the ICS-209 database was identified using data from either the Monitoring Trends in Burn Severity (MTBS) [38] or the Geospatial Multi-Agency Coordination program (GeoMAC) [39]. Because the perimeters for the 2008 Parker Road Fire, the 2013 Carolina Forest Condo Fire, the 2016 Glendale fire, the 2017 NEU Wind Complex, and the 2017 County Road 630 E were not included in these databases, they were removed from all of the subsequent analyses.

2.4. Census-Based WUI Map Building Loss

We assigned a SILVIS WUI type and census block-based housing unit density to each building location prior to the assessment of the proportion of the building loss in the different WUI categories [5,23]. For each WUI disaster, we calculated the total number of exposed buildings (total buildings within the fire perimeter), the total number of burned buildings, the proportion of burned buildings to total exposed buildings, and the total number of unburned buildings adjacent to the fire (within 2400 m) in each SILVIS WUI type. We used a buffer distance of 2400 m because this is the buffer distance commonly used in Census-based WUI maps, and it represents the expected maximum travel distance for fire brands [5,23]. The total number of buildings and proportion of loss between the WUI and non-WUI buildings, as well as among the WUI types (interface, intermix, low density vegetated, high density non-vegetated), were assessed for both the individual fires and the dataset as a whole.

2.5. Point-Based WUI Map and Building Loss

We calculated the values for each WUI component (housing unit density, vegetation cover, and proximity to large patches of contiguous wildland vegetation) at multiple scales, referred to as WUI component-scale combinations, using the framework proposed by Bar-Massada et al. [18], and then assigned values to each individual building point. The building density was calculated using the locations of the building points measured at different neighborhood sizes, ranging from 100-m to

1000-m radii in 100-m increments. In order to calculate the percentage of vegetation cover around each building, we first reclassified the 2011 NLCD data [33] in order to delineate the wildland vegetation using forest, shrub/scrub, grassland, wetland, and open space categories. We then estimated the percentage of the vegetation cover surrounding each building using circular neighborhoods with radii increasing in 100-m increments from 100 to 1000 m. In order to identify the proximity to contiguous patches of wildland vegetation at multiple scales, we varied the patch size. A single binary morphology shrink and expand algorithm was applied to the national NLCD wildland vegetation data in order to remove isolated pixels of vegetation and generalize the edges [40]. The resulting data were converted into polygons, from which the patch size was measured. Five different layers of wildland vegetation were created, with minimum patch sizes of 0.2, 0.4, 1.25, 2.5, and 5.0 km^2, and then the distance to the vegetation was measured for each building location using the approach outlined by Chen and McAneney [41].

We used logistic regression on the entire dataset in order to assess the relationship between the building burn status and the WUI components at different scales after first conducting indicator kriging with semivariogram analysis in order to determine the distance at which the burn status was correlated between adjacent building locations [42,43]. This resulted in correlative models, the fit of which was then assessed using the area under the curve (AUC) of the receiver operating characteristic plots [44] and p-values to assess the predictor significance. WUI component-scale models and AUC results were used in order to determine which scale had the highest predictive performance for each WUI component.

For each building location, we measured the value of each of the three WUI components at the scale with the highest predictive performance based on the AUC. Point-based WUI classifications were then determined for each building location using these three component values and the framework presented by Bar-Massada et al. [18]. The classification of the WUI type using both point-based and census-based WUI mapping approaches allowed us to compare the ways in which each approach aligned with the observed building loss in WUI disasters, and to categorize the burned buildings as either WUI or non-WUI. Using a heuristic approach, we assessed how the burned buildings were distributed across the range of building density, vegetation cover, and distance to contiguous patches of wildland vegetation using histograms and cumulative distribution functions. This facilitated the examination of the building loss in relation to both the distribution extremes and the WUI component thresholds operationalized in the WUI definitions; a minimum building density of 6.17 buildings per km^2, a maximum distance to wildland vegetation of 2400 m, and a vegetation cover percentage threshold of 50% in order to differentiate the intermix from the interface [5,23,24].

3. Results

3.1. Identifying WUI Disasters

Between 2000 and 2018, the ICS-209 reports documented 2777 wildfires that burned at least one building, and cumulatively burned 58,705 buildings (Table 1). Of these 2777 fires, 108 fires burned more than 50 buildings, and were classified as WUI disasters. These 108 fires were responsible for 83% of all of the building loss reported in the ICS-209 database. Geographically, the WUI disasters were distributed primarily across the western United States (Figure 2). Over half of the WUI disasters occurred in California (51%), followed by Texas (10%), Colorado (7%), Arizona (7%), and New Mexico (6%). On average, there were five WUI disasters and 2433 buildings lost per year. We observed a slight increase in both the number of buildings lost and the number of WUI disasters per year through time, but the building loss was heavily influenced by fires in 2018, and the number of WUI disasters had a high year to year variability (Figure 3).

Table 1. The number and percentage of wildfires and building loss reported in the Incident Status Summary Reports (ICS-209) between 2000 and 2018.

Buildings Lost per Wildfire	Number of Wildfires		Number of Buildings Lost	
	Total	%	Total	%
>1000	9	0.3%	31,321	53.4%
401–1000	8	0.3%	4849	8.3%
101–400	46	1.7%	9283	15.8%
51–100	45	1.6%	3053	5.2%
21–50	103	3.7%	3439	5.9%
6–20	286	10.3%	3024	5.2%
1–5	2280	82.1%	3736	6.4%
Total	2777	100.0%	58,705	100.0%

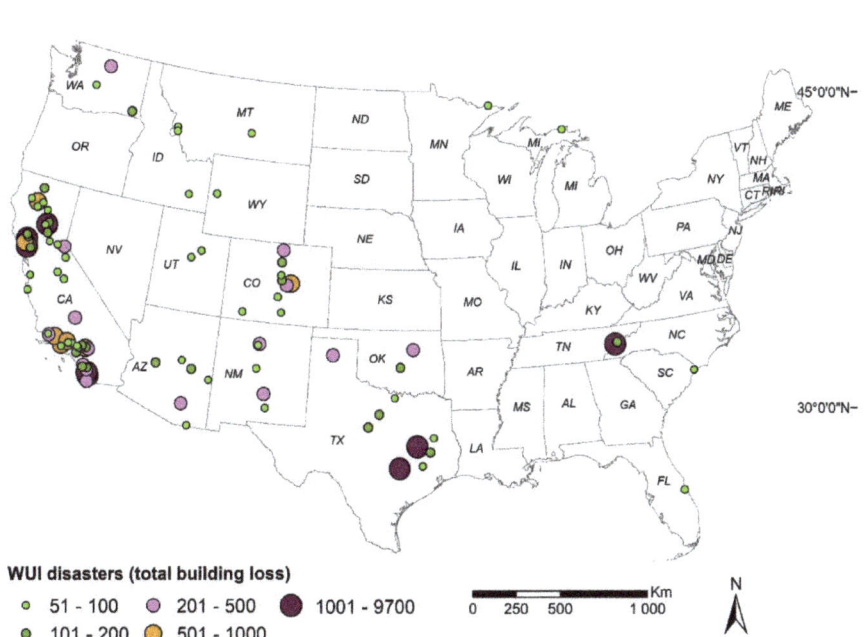

Figure 2. The spatial distribution and magnitude of wildland–urban interface (WUI) disasters in the conterminous United States from 2000 to 2018.

3.2. Census-Based WUI Map Building Loss

The composite building location dataset created by merging the four point-based building location datasets contained 987,430 total buildings within and adjacent to fire perimeters. This included 135,416 buildings that were within fire perimeters, of which 53,786 were burned. The minimum, median, and maximum number of burned buildings per fire were 51, 194, and 18,831, respectively. California's 2018 Camp Fire was responsible for 35% of the total building loss in our dataset. Eighty-six percent of all of the burned buildings were in areas classified as WUI interface (16,009 buildings) and intermix (30,441 buildings) based on the SILVIS WUI map (Table 2). The remaining 14% of the burned buildings were located in areas classified as non-WUI, either areas with vegetation and low building density, or areas with no vegetation and medium or high building density. Most of these buildings

(11% of total) were in areas with building densities of less than 6.17 buildings per km^2, which is the threshold used by SILVIS to distinguish WUI from non-WUI. There was considerable variability among the individual WUI disasters in terms of how the building loss was distributed between the SILVIS WUI categories (Figure 4). In total, 42 of 70 WUI disasters experienced 80% or more of their losses in the WUI interface or intermix categories, but 12 WUI disasters had more than 50% of their loss occur in non-WUI areas. While most WUI disasters (90%) engendered some building loss in low-density non-WUI areas, only 9% caused building loss in high-density non-WUI areas.

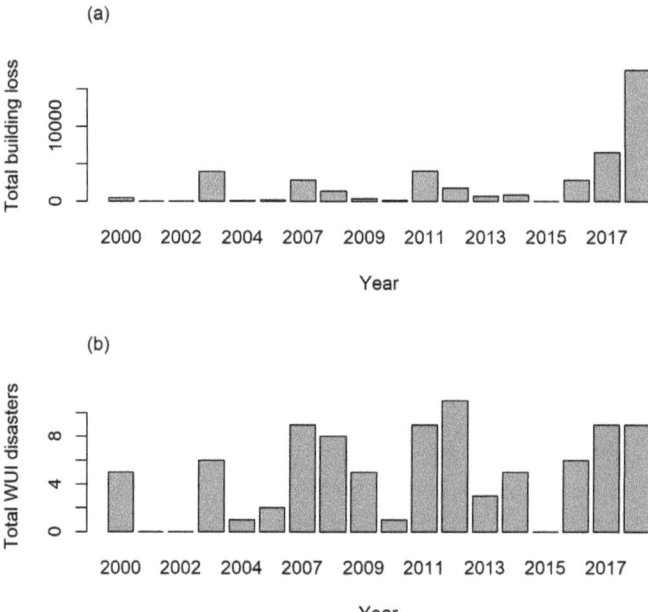

Figure 3. Trends in (**a**) wildland–urban interface (WUI) disaster annual building loss from 2000 to 2018, and (**b**) the number of WUI disasters from 2000 to 2018.

Table 2. Total number of buildings and burned buildings by SILVIS wildland–urban interface (WIU) type for WUI disasters from 2000 to 2018.

Wildland–Urban Interface Category	Buildings within Fire Perimeters				Buildings Outside Fire Perimeters (≤2400 m)
	Count	Percent of Grand Total	Burned	Percent of Total Burned	
WUI					
Interface	44,913	33%	16,009	30%	553,847
Intermix	68,170	50%	30,441	56%	225,205
WUI total	113,083	83%	46,450	86%	779,052
Non-WUI					
Vegetation and no housing	4089	3%	819	2%	12,532
Vegetation and very-low housing density	12,107	9%	3880	7%	39,009
No vegetation and low to very-low housing density	2719	2%	986	2%	27,347
No vegetation and medium to high housing density	3418	3%	1651	3%	129,490
Non-WUI total	22,333	17%	7336	14%	208,378
Grand total	135,416	100%	53,786	100%	987,430

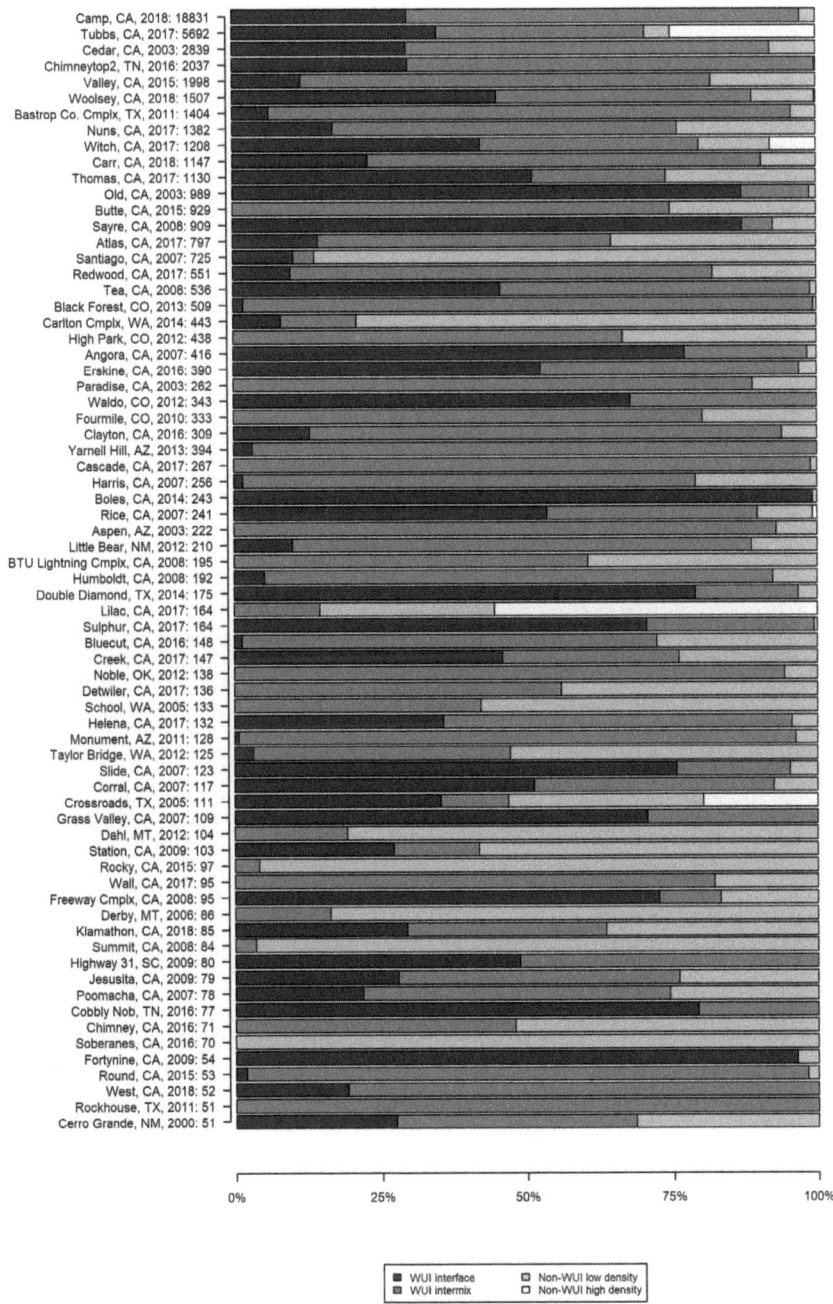

Figure 4. Percentage of building loss in wildland–urban interface (WUI) and non-WUI land use types for the most destructive WUI disasters, using SILVIS WUI categories. The data were sorted by most to least destructive, with the fire name, state, year, and total building loss.

3.3. Determining the Scale to Measure WUI Components for Point-Based WUI Maps

Indicator kriging and semivariogram analysis documented the sample independence between the point-based building locations, with a range, sill, and nugget of <0.01 m^2, 0.002 m, and 0.14 m^2, respectively. The small values for the range, sill, and nugget indicated no spatial autocorrelation in the building location burn status at meaningful scales. After measuring and assigning values for each WUI component at multiple scales for all 135,416 affected buildings, we assessed their relationship to burn status using bivariate logistic regression models (Table 3). Based on the AUC, the best performing scales were a 1000-m neighborhood for building density and a 100-m neighborhood for vegetation cover. The multiple WUI component-scale combinations for distance to wildland vegetation patch size all resulted in the same AUC value, indicating that most buildings were lost near a 5 km^2 or larger patch of wildland vegetation. Overall, low AUC values suggest poor predictive power for most individual WUI component-scale combinations. Building density AUC values tended to slowly increase and then level off with neighborhood sizes up to 1000 m. We observed building density AUC values dropping off and becoming erratic after 1000 m. Vegetation cover AUC values were greatest at 100 m, and declined with increasing neighborhood size. The relationships between building loss and proximity to the various vegetation patch sizes we tested were all similar to one another, indicating no difference in the predictive performance between the various patch sizes. The p-values did not vary and were significant (<0.0001) for all of the models, likely due to the large sample size.

Table 3. Logistic regression results testing WUI component-scale combinations for building density, vegetation cover, and distance to wildland vegetation.

Building Density			Vegetation Cover			Distance to Wildland Vegetation		
Radius (m)	AUC	R (Pseudo)	Radius (m)	AUC	R (Pseudo)	Patch Size (km^2)	AUC	R (Pseudo)
100	0.583	0.017	100	0.550	0.006	0.20	0.475	0.001
200	0.591	0.023	200	0.542	0.005	0.40	0.475	0.001
300	0.597	0.027	300	0.538	0.004	1.25	0.475	0.001
400	0.600	0.029	400	0.536	0.004	2.50	0.475	0.001
500	0.602	0.031	500	0.534	0.004	5.00	0.475	0.001
600	0.603	0.033	600	0.533	0.004			
700	0.603	0.034	700	0.533	0.003			
800	0.604	0.035	800	0.532	0.003			
900	0.604	0.036	900	0.531	0.003			
1000	0.604	0.036	1000	0.529	0.003			

The wildland–urban interface component values for the burned buildings were distributed across a wide range of building densities, vegetation cover, and proximities to wildland vegetation when calculated at the highest performing WUI component-scale combinations. However, the majority of the building loss occurred in areas with low building density, high vegetation cover, and within close proximity to large patches of wildland vegetation (Figure 5). Only 2.7% of the burned buildings occurred at densities of <6.17 buildings per km^2 (the minimum housing density threshold used to differentiate WUI from non-WUI areas in the census-based SILVIS WUI maps). Ten percent of the burned buildings occurred in areas with less than 25% vegetation cover. In comparison, 80% of the building loss occurred in areas with 50% or more vegetation cover (the threshold used in the SILVIS WUI data to distinguish WUI interface from WUI intermix). Eighty percent of all of the burned buildings were located within wildland vegetation (0 m distance), 95% of the burned buildings occurred within 100 m, and 100% of the burned buildings occurred within 850 m, which is substantially less than the 2400-m threshold used in the SILVIS WUI data to distinguish WUI from non-WUI areas. Only seven WUI disasters resulted in building loss at distances greater than 300 m from wildland vegetation, and only California's 2017 Tubbs Fire caused building loss further than 500 m from large patches of wildland vegetation.

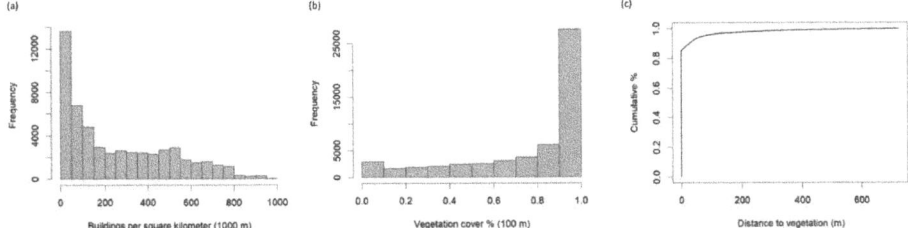

Figure 5. Building loss by wildland–urban interface components: (**a**) histogram of the building density calculated with a 1000-m neighborhood size, (**b**) histogram of the vegetation cover within 100 m, and (**c**) cumulative percentage of the building loss by distance to 5 km² or larger patches of wildland vegetation.

3.4. Building Loss by Point-Based WUI Type

Each building's point-based WUI type was determined using the point-based WUI framework and the WUI component values calculated at the best-performing predictive scale. We used a minimum building density threshold of 6.17 buildings per km², calculated using a 1000-m neighborhood in order to distinguish WUI from low-density non-WUI; a vegetation cover threshold of 50%, calculated using a 100-m neighborhood in order to differentiate the WUI interface from WUI intermix; and a maximum distance of 850 m from wildland vegetation patches of 5 km² or larger. The point-based WUI map classified 97% of the burned buildings as either WUI interface (20%) or WUI intermix (77%) (Table 4). The remaining 3% of the burned buildings were located in areas classified as non-WUI, primarily in areas with vegetation and low to very-low building density (less than 6.17 buildings per km²). The percentage of building loss was 22% to 31% in non-WUI areas, and 33% and 42% for interface and intermix areas. The WUI intermix areas experienced the highest rates of loss, at 42%.

Table 4. Total buildings and building loss by wildland–urban interface (WUI) type, created using a point-based WUI mapping method in which the WUI component values were calculated using the best-performing scale for each. The building density and percentage of vegetation cover were measured in 1000-m and 100-m neighborhoods, respectively, and the distance to wildland vegetation was measured using a 5 km² minimum patch size and a maximum distance of 850 m.

Wildland–Urban Interface Category	Buildings within Fire Perimeters				Buildings Outside Fire Perimeters (≤2400 m)
	Count	Percent of Grand Total	Burned	Percent of Total Burned	
WUI					
Interface	32,727	24%	10,815	20%	514,241
Intermix	97,909	72%	41,492	77%	393,682
WUI total	130,636	96%	52,307	97%	907,923
Non-WUI					
Vegetation and no housing	0	0%	0	0%	
Vegetation and very-low housing density	4534	3%	1424	3%	16,795
No vegetation and low to very-low housing density	246	<1%	55	0%	2769
No vegetation and medium to high housing density	0	0%	0	0%	59,943
Non-WUI total	4780	4%	1479	3%	79,507
Grand total	135,416	100%	53,786	100%	987,430

4. Discussion

This study characterized the occurrence of WUI disasters and assessed building loss relative to housing density, vegetation cover, and proximity to large patches of contiguous wildland vegetation, and evaluated how well census-based and point-based WUI maps captured patterns of building loss in WUI disasters. We identified 108 wildfires in the United States between 2000 and 2018 that burned more than 50 buildings. These WUI disasters were the most destructive wildfires in the United States in terms of building loss. However, they represented only a small fraction of the estimated 1.4 million wildfire ignitions estimated to have occurred during our study period [14]. The WUI disasters tended to be concentrated in California, Colorado, Texas, Arizona, and New Mexico. The number of WUI disasters and the associated building loss has been increasing since 2000. However, there was considerable interannual variation, with extensive losses in 2017 and 2018, and with the Camp Fire in Paradise, California heavily influencing the overall trend. Widespread drought conditions in the western United States in 2020, along with climactic and demographic trends, resulted in extensive areas burned and building losses in many western states [9,11–13,23,45], with over 17,663 buildings destroyed as of 30 November 2020 [46].

We defined WUI disasters as those wildfires that burned more than 50 buildings, but there is no formally agreed-upon definition. Cohen suggested 100 burned buildings was an appropriate threshold [9], but we lowered that number in order to capture more loss and account for potential errors in the ICS-209 data. Had we used Cohen's threshold of 100 burned buildings, the proportion of the WUI interface loss relative to WUI intermix loss would have likely been slightly higher. If we had further lowered the threshold or examined all of the fires with building loss, the proportion of WUI interface loss would have likely been slightly lower. Future research investigating the sensitivity of WUI disasters' building-loss thresholds would be useful in advancing our understanding of the spatial nature of building loss, and could help to develop a framework to synthesize findings from disparate post-wildfire building-loss case studies [5,18,24].

Census-based and point-based WUI maps, respectively, classified 86% and 97% of all of the buildings lost in WUI disasters as being in either the WUI interface or intermix. Despite the coarse resolution of the SILVIS WUI data, these maps effectively identify the majority of structures that are at risk of loss during WUI disasters. The SILVIS WUI mapping methods are transparent and easy to replicate with publicly available data [23]. Census and land cover data collection efforts are both well-documented and routinely updated, allowing managers and researchers to account for temporal changes in WUI development [12,33,47]. Both the demonstrated utility of the SILVIS WUI maps and their acceptance as a standard through Executive Order #13728 [25] suggest that they will continue to serve an essential role at a national scale, and for many broad-scale WUI mapping efforts that do not require specific building locations. That said, point-based WUI mapping methods classified more burned buildings as either WUI interface or intermix. Their ability to include the specific locations of at-risk building locations and exclude low-risk buildings offers additional advantages for certain applications at multiple scales, and more closely aligns with our conceptual understanding of the WUI problem as one for which the fundamental unit is the individual building [9,10,17,18,20].

Our finding, that building loss mostly occurs in areas with low building densities and high vegetation cover, supports previous findings [35,48–50]. Higher losses in rural intermix environments with more wildland vegetation may be due to multiple factors, such as: continuous vegetation providing more potential for fire spread and building ignition, lower fire response capacity, the relative inaccessibility of dispersed buildings, or incident managers' choices to direct resources to areas with higher-density clusters of buildings [51]. However, 20.5% of the building loss occurred in areas with > 500 buildings per km^2, suggesting that the building loss in dense environments may be influenced by a set of unique environmental factors related to the buildings themselves, such as construction materials, landscaping, and proximity to neighboring buildings, all of which may facilitate building-to-building ignition [35,48,52]. Our results indicate that the building loss declined rapidly with the increasing distance from vegetation. All building loss in the study occurred within 850 m from large patches of wildland vegetation, and less than 5% of the building loss occurred at distances exceeding 100 m. The fires' spread is likely hindered in environments with fragmented vegetation, and likely depends on embers, which rarely travel and ignite buildings farther than several hundred meters from the nearest patch of wildland vegetation [52,53].

Point-based WUI maps identified a higher percentage of burned buildings as WUI compared to the census-based methods. This suggests that, if the goal is to identify at-risk buildings, map developers can adjust WUI mapping methods to align with the observed losses. This could involve the selection of specific scales and thresholds for the calculation of WUI component values (minimum housing density, vegetation cover, and maximum buffer distance). The fact that all of the building loss occurs at distances substantially less than the 2400 m used in other WUI mapping efforts [5,22,23] suggests that this threshold could potentially be altered for WUI maps that seek to identify at-risk buildings. The 2400-m distance threshold used in the SILVIS WUI maps may be appropriate for census-based WUI maps, but our results suggest that a more conservative distance threshold may be appropriate for point-based WUI maps. The reduction of the buffer distance threshold to 850 m or shorter would focus attention on buildings with the highest risk of ignition, whilst also removing buildings at lower risk.

Detailed WUI maps of individual at-risk structures can also improve land use and wildfire mitigation planning at both the community and national scales [10,15,19,30]. Detailed point-based WUI maps overcome a key limitation of Census-based maps, which are sufficient at a national scale, but often insufficient for local applications such as wildfire incident response or neighborhood evacuation planning [17,18]. Furthermore, the documentation of losses across a range of building densities and vegetation cover percentages underscores the need to customize wildfire mitigation techniques for different environments [16,52]. In areas between 100 and 850 m from wildland vegetation, buildings may still be at risk of igniting, but the exposure in this zone is likely due to embers [53], and ignition becomes more influenced by the built environment. In areas where wildland hazardous fuel reduction projects are not feasible, there may be a benefit to the removal of flammable landscaping and the implementation of building hardening techniques. Our results, which describe the pattern of loss in different environments, underscore the need to tailor distinct mitigation techniques across a range of building density, vegetation cover, and community types, including some areas that are not traditionally considered at risk [16,48–50]. Fine-scale WUI maps of at-risk buildings also provide the potential to improve quantitative wildfire risk assessments, refine estimates of values at risk, and better identify community wildfire exposure zones [17,18,20,52].

Detailed point-based maps can help local governments assess strategies to limit residential development in fire-prone areas, and enact building codes that require defensible space, [10,15,16,18]. The use of detailed WUI maps focused on at-risk buildings enables more in-depth discussions about the ways in which communities and property owners can mitigate risk, and facilitates more nuanced mitigation efforts that consider local knowledge about the building's immediate surroundings, property access, construction materials, and vegetation cover on the property. The knowledge from detailed WUI maps reduces the uncertainties that WUI residents and communities need to navigate, including the ways in which buildings and vegetation are co-located within a census block. Point-based maps also have additional benefits because they offer a nationally-consistent approach that can be scaled up and summarized at multiple scales (e.g., counties, firesheds, states, regions) [30,54,55]. Their improved spatial resolution relative to census blocks allows for improved estimates of community wildfire exposure, helps to refine estimates of values at risk, and can inform local fire planning and response strategies, as well as regional land management budget allocations tied to building exposure [20,55–57].

Our findings are particularly informative in light of previous WUI sensitivity analyses. Previous work has examined the ways in which the adjustment of WUI component parameters would change the spatial extent of the WUI and the number of buildings categorized as intermix or interface areas [5,18,23,24]. For example, shrinking the WUI buffer distance from 2400 m to 850 m would exclude low-risk buildings while reducing the total number of buildings and the spatial extent of the WUI interface. The reduction of the building density threshold or the calculation of the building density using a smaller neighborhood size would increase the number of buildings in the WUI intermix. Many building points have building densities >6.17 buildings per km^2 when calculated using a 1000-m neighborhood, but <6.17 buildings per km^2 when calculated at the census-block scale. These adjustments could have significant implications for informing land management decisions. Expanding the WUI intermix and shrinking the WUI interface could influence funding allocations tied to the number of at-risk WUI buildings present in a landscape. The expansion of how and where the intermix is represented in WUI maps could also inform land management agencies whose hazardous fuel reduction efforts are often tied to WUI community protection objectives [57]. Additional research is needed in order to investigate the potential implications these adjustments may have for different states and regions, depending on their unique residential development patterns and mix of WUI interface and WUI intermix.

In conclusion, our results indicate that point-based WUI maps can improve our understanding of where building loss occurs during WUI disasters, and can more accurately identify at-risk buildings relative to existing census-based mapping. By documenting the pattern of home loss relative to the standard WUI mapping components (building density, vegetation cover, distance from vegetation), this study also improves our understanding of the WUI problem as a complex spatial phenomenon. Although the WUI is the fastest-growing land-use type in the United States [23], our results suggest that building loss is variable, localized, and is not equally distributed throughout the interface, intermix, and WUI-adjacent areas. Although this variation needs to be explored further, the findings herein have important implications for land-use planning, risk assessments, and wildfire mitigation. Our findings improve our understanding of historical patterns of loss and the potential for loss in the future, which will become increasingly important as the WUI expands, and the number of at-risk buildings increases.

Author Contributions: Conceptualization, M.D.C., T.J.H., B.M.G., C.M.H.; methodology, M.D.C., T.J.H., B.M.G., C.M.H.; formal analysis, M.D.C., T.J.H., B.M.G., C.M.H.; writing—original draft M.D.C., T.J.H., B.M.G., C.M.H. All authors have read and agreed to the published version of the manuscript.

Funding: Funding for T.J. Hawbaker was provided by the U.S. Department of the Interior, U.S. Geological Survey (USGS), Land Change Science Program managed by the Core Science Systems Mission Area, and the U.S. Department of the Interior North Central Climate Adaptation Science Center, which is managed by the USGS National Climate Adaptation Science Center.

Conflicts of Interest: The authors declare no conflict of interest.

Disclaimers: Any use of trade, firm, or product names is for descriptive purposes only and does not imply endorsement by the U.S. Government.

References

1. Theobald, D.M.; Romme, W.H. Expansion of the US wildland–urban interface. *Landsc. Urban Plan.* **2007**, *83*, 340–354. [CrossRef]
2. Platt, R.V.; Schoennagel, T.; Veblen, T.T.; Sherriff, R.L. Modeling wildfire potential in residential parcels: A case study of the north-central Colorado Front Range. *Landsc. Urban Plan.* **2011**, *102*, 117–126. [CrossRef]
3. Syphard, A.D.; Massada, A.B.; Butsic, V.; Keeley, J.E. Land use planning and wildfire: Development policies influence future probability of housing loss. *PLoS ONE* **2013**, *8*, e71708. [CrossRef] [PubMed]
4. Haas, J.R.; Calkin, D.E.; Thompson, M.P. Wildfire risk transmission in the Colorado Front Range, USA. *Risk Anal.* **2015**, *35*, 226–240. [CrossRef]
5. Radeloff, V.C.; Hammer, R.B.; Stewart, S.I.; Fried, J.S.; Holcomb, S.S.; McKeefry, J.F. The Wildland–Urban Interface in the United States. *Ecol. Appl.* **2005**, *15*, 799–805. [CrossRef]
6. Gill, A.M.; Stephens, S.L. Scientific and social challenges for the management of fire-prone wildland–urban interfaces. *Environ. Res. Lett.* **2009**, *4*, 034014. [CrossRef]
7. Syphard, A.D.; Keeley, J.E. Factors associated with structure loss in the 2013–2018 California wildfires. *Fire* **2019**, *2*, 49. [CrossRef]
8. Corelogic. 2019 Wildfire Risk Report. Available online: https://www.corelogic.com/insights-download/wildfire-risk-report.aspx (accessed on 1 December 2020).

9. Cohen, J.D. The wildland-urban interface fire problem: A consequence of the fire exclusion paradigm. *For. Hist. Today* **2008**, *Fall*, 20–26.
10. Calkin, D.E.; Cohen, J.D.; Finney, M.A.; Thompson, M.P. How risk management can prevent future wildfire disasters in the wildland-urban interface. *Proc. Natl. Acad. Sci. USA* **2014**, *111*, 746–751. [CrossRef]
11. Balch, J.K.; Bradley, B.A.; Abatzoglou, J.T.; Nagy, R.C.; Fusco, E.J.; Mahood, A.L. Human-started wildfires expand the fire niche across the United States. *Proc. Natl. Acad. Sci. USA* **2017**, *114*, 2946–2951. [CrossRef]
12. Hammer, R.B.; Stewart, S.I.; Radeloff, V.C. Demographic trends, the wildland–urban interface, and wildfire management. *Soc. Nat. Resour.* **2009**, *22*, 777–782. [CrossRef]
13. Westerling, A.L.; Hidalgo, H.G.; Cayan, D.R.; Swetnam, T.W. Warming and earlier spring increase western U.S. forest wildfire activity. *Science* **2006**, *313*, 940–943. [CrossRef] [PubMed]
14. Short, K.C. A spatial database of wildfires in the United States, 1992–2011. *Earth Syst. Sci. Data* **2014**, *6*, 1–27. [CrossRef]
15. Smith, A.M.S.; Kolden, C.A.; Paveglio, T.B.; Cochrane, M.A.; Bowman, D.M.; Moritz, M.A.; Kliskey, A.D.; Alessa, L.; Hudak, A.T.; Hoffman, C.M.; et al. The science of firescapes: Achieving fire-resilient communities. *BioScience* **2016**, *66*, 130–146. [CrossRef]
16. Paveglio, T.B.; Moseley, C.; Carroll, M.S.; Williams, D.R.; Davis, E.J.; Fischer, A.P. Categorizing the cocial context of the wildland urban interface: Adaptive capacity for wildfire and community "archetypes". *For. Sci.* **2015**, *61*, 298–310. [CrossRef]
17. Calkin, D.E.; Rieck, J.D.; Hyde, K.D.; Kaiden, J.D. Built structure identification in wildland fire decision support. *Int. J. Wildland Fire* **2011**, *20*, 78–90. [CrossRef]
18. Bar-Massada, A.; Stewart, S.I.; Hammer, R.B.; Mockrin, M.H.; Radeloff, V.C. Using structure locations as a basis for mapping the wildland urban interface. *J. Environ. Manag.* **2013**, *128*, 540–547. [CrossRef]
19. Jakes, P.; Burns, S.; Cheng, A.; Saeli, E.; Brummel, K.N.R.; Grayzeck, S.; Sturtevant, V.; Williams, D. Critical elements in the development and implementation of Community Wildfire Protection Plans (CWPPs). In *The Fire Environment—Innovations, Management, and Policy, Proceedings of the RMRS-P-46CD, Destin, FL, USA, 26–30 March 2007*; CD-ROM; Butler, B.W., Cook, W.C., Eds.; U.S. Department of Agriculture, Forest Service, Rocky Mountain Research Station: Fort Collins, CO, USA, 2007; pp. 613–624.
20. Scott, J.H.; Thompson, M.P.; Calkin, D.E. A wildfire risk assessment framework for land and resource management. In *Gen. Tech. Rep. RMRS-GTR-315*; U.S. Department of Agriculture, Forest Service, Rocky Mountain Research Station: Station, 2013; Volume 315, p. 83. [CrossRef]
21. Federal Register. Urban Wildland Interface Communities within the Vicinity of Federal Lands That Are at High Risk from Wildfire. Available online: https://www.federalregister.gov/documents/2001/01/04/01-52/urban-wildland-interface-communities-within-the-vicinity-of-federal-lands-that-are-at-high-risk-from (accessed on 2 December 2020).
22. Martinuzzi, S.; Stewart, S.I.; Helmers, D.P.; Mockrin, M.H.; Hammer, R.B.; Radeloff, V.C. *The 2010 Wildland-Urban Interface of the Conterminous United States*; U.S. Department of Agriculture, Forest Service, Northern Research Station: Newtown Square, PA, USA, 2015. [CrossRef]
23. Radeloff, V.C.; Helmers, D.P.; Kramer, H.A.; Mockrin, M.H.; Alexandre, P.M.; Bar-Massada, A.; Butsic, V.; Hawbaker, T.J.; Martinuzzi, S.; Syphard, A.D.; et al. Rapid growth of the US wildland-urban interface raises wildfire risk. *Proc. Natl. Acad. Sci. USA* **2018**, *115*, 3314. [CrossRef]
24. Stewart, S.I.; Radeloff, V.C.; Hammer, R.B.; Hawbaker, T.J. Defining the wildland–urban interface. *J. For.* **2007**, *105*, 201–207. [CrossRef]
25. Federal Register. Wildland-Urban Interface Federal Risk Mitigation. Available online: https://www.federalregister.gov/documents/2016/05/20/2016-12155/wildland-urban-interface-federal-risk-mitigation (accessed on 2 December 2020).
26. Blaschke, T. Object based image analysis for remote sensing. *ISPRS J. Photogramm. Remote Sens.* **2010**, *65*, 2–16. [CrossRef]
27. Caggiano, M.D.; Tinkham, W.T.; Hoffman, C.; Cheng, A.S.; Hawbaker, T.J. High resolution mapping of development in the wildland-urban interface using object based image extraction. *Heliyon* **2016**, *2*, e00174. [CrossRef] [PubMed]
28. Leyk, S.; Uhl, J.H. HISDAC-US, historical settlement data compilation for the conterminous United States over 200 years. *Sci. Data* **2018**, *5*, 180175. [CrossRef] [PubMed]

29. Microsoft Building Footprints-Bing Maps. Available online: https://www.microsoft.com/en-us/maps/building-footprints (accessed on 2 December 2020).
30. Scott, J.H.; Thompson, M.P.; Gilbertson-Day, J.W. Exploring how alternative mapping approaches influence fireshed assessment and human community exposure to wildfire. *GeoJournal* **2017**, *82*, 201–215. [CrossRef]
31. Stewart, S.I.; Wilmer, B.; Hammer, R.; Aplet, G.; Hawbaker, T.; Miller, C.; Radeloff, V. Wildland-urban interface maps vary with purpose and context. *J. For.* **2009**, *107*, 78–83.
32. National Wildfire Coordinating Group. Wildfire Incident Status Summary Reports (ICS-209). Available online: https://famit.nwcg.gov/applications/SIT209/NIMSICS209 (accessed on 2 December 2020).
33. Yang, L.; Jin, S.; Danielson, P.; Homer, C.; Gass, L.; Bender, S.M.; Case, A.; Costello, C.; Dewitz, J.; Fry, J.; et al. A new generation of the United States National Land Cover Database: Requirements, research priorities, design, and implementation strategies. *ISPRS J. Photogramm. Remote Sens.* **2018**, *146*, 108–123. [CrossRef]
34. Katuwal, H.; Dunn, C.J.; Calkin, D.E. Characterising resource use and potential inefficiencies during large-fire suppression in the western US. *Int. J. Wildland Fire* **2017**, *26*, 604–614. [CrossRef]
35. Alexandre, P.M.; Stewart, S.I.; Keuler, N.S.; Clayton, M.K.; Mockrin, M.H.; Bar-Massada, A.; Syphard, A.D.; Radeloff, V.C. Factors related to building loss due to wildfires in the conterminous United States. *Ecol. Appl.* **2016**, *26*, 2323–2338. [CrossRef]
36. California Department of Forestry and Fire Protection CAL FIRE Damage Inspection Program. Available online: https://www.fire.ca.gov/ (accessed on 2 December 2020).
37. Lin, G.; Milan, A.; Shen, C.; Reid, I. RefineNet: Multi-path refinement networks for high-resolution semantic segmentation. In Proceedings of the 2017 IEEE Conference on Computer Vision and Pattern Recognition (CVPR), Honolulu, HI, USA, 21–26 July 2017; pp. 5168–5177.
38. Eidenshink, J.; Schwind, B.; Brewer, K.; Zhu, Z.-L.; Quayle, B.; Howard, S. A project for monitoring trends in burn severity. *Fire Ecol.* **2007**, *3*, 3–21. [CrossRef]
39. U.S. Geological Survey. GeoMAC Wildfire Application. Available online: https://www.geomac.gov/ (accessed on 2 December 2020).
40. Environmental Systems Research Institute. *ArcGIS Desktop: Release 10*; Environmental Systems Research Institute: Redlands, CA, USA, 2011.
41. Chen, K.; McAneney, J. Quantifying bushfire penetration into urban areas in Australia. *Geophys. Res. Lett.* **2004**, *31*. [CrossRef]
42. Solow, A.R. Mapping by simple indicator kriging. *Math. Geosci.* **1986**, *18*, 335–352. [CrossRef]
43. Curran, P.J. The semivariogram in remote sensing: An introduction. *Remote Sens. Environ.* **1988**, *24*, 493–507. [CrossRef]
44. Fawcett, T. An introduction to ROC analysis. *Pattern Recognit. Lett.* **2006**, *27*, 861–874. [CrossRef]
45. Mueller, S.E.; Thode, A.E.; Margolis, E.Q.; Yocom, L.L.; Young, J.D.; Iniguez, J.M. Climate relationships with increasing wildfire in the southwestern US from 1984 to 2015. *For. Ecol. Manag.* **2020**, *460*, 117861. [CrossRef]
46. Headwaters Economics Wildfires Destroy Thousands of Structures Each Year. Available online: https://headwaterseconomics.org/natural-hazards/structures-destroyed-by-wildfire/ (accessed on 2 December 2020).
47. Wickham, J.; Stehman, S.V.; Gass, L.; Dewitz, J.A.; Sorenson, D.G.; Granneman, B.J.; Poss, R.V.; Baer, L.A. Thematic accuracy assessment of the 2011 National Land Cover Database (NLCD). *Remote Sens. Environ.* **2017**, *191*, 328–341. [CrossRef]
48. Syphard, A.D.; Keeley, J.E.; Massada, A.B.; Brennan, T.J.; Radeloff, V.C. Housing arrangement and location determine the likelihood of housing loss due to wildfire. *PLoS ONE* **2012**, *7*, e33954. [CrossRef]
49. Kramer, H.A.; Mockrin, M.H.; Alexandre, P.M.; Stewart, S.I.; Radeloff, V.C. Where wildfires destroy buildings in the US relative to the wildland–urban interface and national fire outreach programs. *Int. J. Wildland Fire* **2018**, *27*, 329–341. [CrossRef]
50. Kramer, H.A.; Mockrin, M.H.; Alexandre, P.M.; Radeloff, V.C. High wildfire damage in interface communities in California. *Int. J. Wildland Fire* **2019**, *28*, 641–650. [CrossRef]
51. Clark, A.M.; Rashford, B.S.; McLeod, D.M.; Lieske, S.N.; Coupal, R.H.; Albeke, S.E. The impact of residential development pattern on wildland fire suppression expenditures. *Land Econ.* **2016**, *92*, 656–678. [CrossRef]
52. Maranghides, A.; Mell, W. *Framework for Addressing the National Wildland Urban Interface Fire Problem—Determining Fire and Ember Exposure Zones Using a WUI Hazard Scale*; U.S. Department of Commerce, National Institute of Standards and Technology: Newtown Square, PA, USA, 2012; p. 25.

53. Koo, E.; Pagni, P.J.; Weise, D.R.; Woycheese, J.P. Firebrands and spotting ignition in large-scale fires. *Int. J. Wildland Fire* **2010**, *19*, 818–843. [CrossRef]
54. Ager, A.A.; Day, M.A.; Palaiologou, P.; Houtman, R.M.; Ringo, C.; Evers, C.R. Cross-boundary wildfire and community exposure: A framework and application in the western U.S. In *Gen. Tech. Rep. RMRS-GTR-392*; U.S. Department of Agriculture, Forest Service, Rocky Mountain Research Station: Fort Collins, CO, USA, 2019; Volume 392, p. 36.
55. Evers, C.R.; Ager, A.A.; Nielsen-Pincus, M.; Palaiologou, P.; Bunzel, K. Archetypes of community wildfire exposure from national forests of the western US. *Landsc. Urban Plan.* **2019**, *182*, 55–66. [CrossRef]
56. Miller, S.R.; Wuerzer, T.; Vos, J.; Lindquist, E.; Mowery, M.; Holfeltz, T.; Stephens, B.; Grad, A. *Planning for Wildfire in the Wildland-Urban Interface: A Resource Guide for Idaho Communities*; Social Science Research Network: Rochester, NY, USA, 2016.
57. Schoennagel, T.; Nelson, C.R.; Theobald, D.M.; Carnwath, G.C.; Chapman, T.B. Implementation of National Fire Plan treatments near the wildland-urban interface in the western United States. *Proc. Natl. Acad. Sci. USA* **2009**, *106*, 10706–10711. [CrossRef] [PubMed]

Publisher's Note: MDPI stays neutral with regard to jurisdictional claims in published maps and institutional affiliations.

© 2020 by the authors. Licensee MDPI, Basel, Switzerland. This article is an open access article distributed under the terms and conditions of the Creative Commons Attribution (CC BY) license (http://creativecommons.org/licenses/by/4.0/).

Article

In the Line of Fire: Consequences of Human-Ignited Wildfires to Homes in the U.S. (1992–2015)

Nathan Mietkiewicz [1,2,3,*], **Jennifer K. Balch** [1,2,4], **Tania Schoennagel** [4], **Stefan Leyk** [1,4], **Lise A. St. Denis** [1,2] **and Bethany A. Bradley** [5]

1. Earth Lab, 4001 Discovery Drive Suite S348—UCB 611, University of Colorado-Boulder, Boulder, CO 80309, USA; jennifer.balch@colorado.edu (J.K.B.); stefan.leyk@colorado.edu (S.L.); Lise.St.Denis@colorado.edu (L.A.S.D.)
2. Cooperative Institute for Research in Environmental Sciences, Boulder, CO 80309, USA
3. National Ecological Observation Network, Battelle, 1685 38th Street, Boulder, CO 80301, USA
4. Department of Geography, GUGG 110, 260 UCB, University of Colorado-Boulder, Boulder, CO 80309, USA; tania.schoennagel@colorado.edu
5. Department of Environmental Conservation, 160 Holdsworth Way, University of Massachusetts-Amherst, Amherst, MA 01003, USA; bbradley@eco.umass.edu
* Correspondence: mietkiewicz@battelleecology.org

Received: 21 August 2020; Accepted: 5 September 2020; Published: 7 September 2020

Abstract: With climate-driven increases in wildfires in the western U.S., it is imperative to understand how the risk to homes is also changing nationwide. Here, we quantify the number of homes threatened, suppression costs, and ignition sources for 1.6 million wildfires in the United States (U.S.; 1992–2015). Human-caused wildfires accounted for 97% of the residential homes threatened (within 1 km of a wildfire) and nearly a third of suppression costs. This study illustrates how the wildland-urban interface (WUI), which accounts for only a small portion of U.S. land area (10%), acts as a major source of fires, almost exclusively human-started. Cumulatively (1992–2015), just over one million homes were within human-caused wildfire perimeters in the WUI, where communities are built within flammable vegetation. An additional 58.8 million homes were within one kilometer across the 24-year record. On an annual basis in the WUI (1999–2014), an average of 2.5 million homes (2.2–2.8 million, 95% confidence interval) were threatened by human-started wildfires (within the perimeter and up to 1-km away). The number of residential homes in the WUI grew by 32 million from 1990–2015. The convergence of warmer, drier conditions and greater development into flammable landscapes is leaving many communities vulnerable to human-caused wildfires. These areas are a high priority for policy and management efforts that aim to reduce human ignitions and promote resilience to future fires, particularly as the number of residential homes in the WUI grew across this record and are expected to continue to grow in coming years.

Keywords: WUI; fire; defensible space; prescribed fire; community vulnerability; fire suppression costs; Zillow

1. Introduction

Wildfire poses a direct threat to communities and people living in fire-prone areas [1,2]. Communities that meet or intermingle with undeveloped wildland vegetation, creating zones known as the wildland-urban interface (WUI; <10% of the U.S. land area), are at the greatest risk of wildfire across the U.S. [3–5]. Humans' relationship with wildfire in the WUI is a complex interaction between an increasing number of people living in flammable landscapes (increased number of at-risk communities and ignition sources) [6–9], a warmer and drier climate that is more conducive to fire [10–12], and increased fuels accumulated in some places due to years of fire suppression [13,14].

Elements of these interactions have been the focus of contemporary scientific debate and public discourse [15], but we still lack fine-scale estimates of homes threatened from wildfire and how WUI communities contribute to wildfire at a national scale.

The WUI has expanded in the U.S. by a third between 1990 and 2010 [7] and is projected to double by 2030 [3]. Currently, there are an estimated 2.5 million residential structures in fire-prone areas of the WUI in the coterminous U.S., equivalent to $1.4 trillion in property value at risk [5]. Importantly, only ~15% of the WUI in the west is developed while the remaining 85% is available for development and deemed potential WUI expansion areas [16], highlighting the pressing need to implement strategies now that will reduce future wildfire risk in an expanding WUI. Moreover, exposure to wildfire smoke can have profound negative health effects [17]. An estimated 17.5 million people living in these fire-prone WUI areas [5] in the western US are expected to experience longer and more intense smoke waves, in which unhealthy particulate matter from wildfire smoke persists for more than two days [18]. Displacement from wildfires due to home loss or poor air quality are expected to increase as wildfire occurrence continues to rise [15,19].

Despite the call to adapt to increasing wildfires [15], we lack yearly, fine-scale estimates at a national scale of the number of homes actually threatened by wildfire and the role of human-related ignitions in starting fires near these communities. Previous work estimated that 286,000 homes were threatened by large wildfires (small wildfires, <400 ha, were not accounted for) between 2000–2010, based on decade-interval data within census blocks [7]. Also, it has been documented that 84% of the nation's wildfires were caused by people between 1992 and 2013 [8], but this work does not quantify the role of human-caused wildfires where people actually live. Further, it is problematic that there are no estimates of the wildfire threat to homes past 2010, as four of the top ten largest wildfire years (>32,000 km^2) occurred in 2011, 2012, 2015, and 2017 [20].

Human influence on the landscape is an important predictor of wildfire ignition likelihood [4,21–23]. For example, proximity to roads at local scales is positively correlated with wildfire ignition density [4,22], while fire density has a hump-shaped relationship with population density, peaking around ~10 people/km2 [24]. At regional to national scales, it has been shown that proximity to WUI areas influences the occurrence of large wildfires (>40 ha) [25]. Recent work by Syphard et al. [9] has further shown that human presence is more important in wildfire ignition than climate across the conterminous US, but burned area is primarily driven by climate conditions [10,26]. Despite the prominent role of humans in igniting fires, it remains unknown how the WUI acts as a source of human ignitions as well as how much fire damage the WUI potentially sustains at fine scales and on an annual basis. Here, we investigate how human ignitions have altered fire frequency, burned area, and seasonality in the WUI compared to wildland areas. We further assess the number of residential structures threatened and the associated suppression costs of human-started wildfires within the WUI, providing a comprehensive assessment of the consequences and costs of human-started wildfire in our most vulnerable communities.

To address these questions, we utilized over 1.6 million wildfire events (1992–2015) designated as human or lightning caused [27] and a novel spatio-temporal housing dataset of over 200 million housing records derived from Zillow's ZTRAX dataset [28] to provide a direct and improved estimate of the number of residential units threatened (defined based on being located within 1 km of the wildfire perimeter) by wildfire by year. We quantified wildfire costs from ~150,000 wildfire situation reports (1999–2014) that provide daily estimates of fire suppression costs [29]. Combining these three datasets we assessed the role of human-caused wildfires in the WUI (769,087 km^2; Interface WUI = 162,368 km^2; Intermix WUI = 606,719 km^2), very low-density housing (VLDH; 2,283,410 km^2), and wildland areas (2,565,320 km^2) [7]. We also provide an estimate of the increase in the number of homes in the WUI across the study time period.

Hypotheses

We hypothesized that human-started wildfires would be the dominant type near the WUI, while the VLDH would contain a mix of human- and lightning-started wildfires, and lightning-started wildfires would dominate the wildlands. Similarly, we expected that wildfires would be more expensive in the wildlands due to the larger wildfire footprint and the difficulty in navigating the terrain during suppression efforts. Conversely, we expected the cost of wildfires near the WUI to be less per wildfire, but more expensive proportional to wildfire size due to the more proximate threat to communities and people. We further hypothesized that the distance to human settlement and the median home density were strongly related to the prevalence of human versus lightning ignitions. Fire frequency and fire season length were expected to vary with the type of ignition source (i.e., lightning versus human) and the distance from the urban core boundary. Lastly, we hypothesized that there would be important regional differences and therefore, we refined some of our analysis based on reorganized level 1 ecoregions to capture the major eastern and western U.S. patterns.

2. Datasets

2.1. The Wildland-Urban Interface

A WUI spatial database [7] was used to assess the distribution of wildfire activity within the WUI, very low-density housing (VLDH), and wildland areas for 1990, 2000, and 2010. Each polygon within this dataset represents a U.S. census housing block group, which is defined as an interface, intermixed, or non-WUI category. Intermix WUI and Interface WUI both have census blocks that exceed 6.17 housing units per km^2 and have >50% wildland vegetation or <50% wildland vegetation, respectively. Interface WUI areas must also be located within 2.4 km of an area larger than 5 km^2 containing >75% wildland vegetation. We combined the interface and intermixed WUI zones to represent the total area of the WUI within the US. Very low-density housing areas were delineated based on census block groups that had <6.17 housing units per km^2 and wildland vegetation that is >50%. Wildlands were defined as census block groups housing density of 0 units per km^2 and wildland vegetation >50%.

Previous research has encompassed the WUI within a 2.4 km buffer based on the assumption that fire events outside of the WUI can produce firebrands that are likely to be transported ahead of a wildfire front and ignite a new wildfire in a WUI census block [30,31]. In this current work, we opted to not use this 2.4 km buffer and instead directly evaluate the wildfire activity in each of the main classes independently (e.g., Intermix WUI, Interface WUI, very-low density housing, and wildlands) [32]. In effect, the VLDH category represents a more spatially explicit and representative area between the WUI and "true" wildlands than the 2.4 km buffer zone. Additionally, we acknowledge that the wildland agricultural interface (WAI) and wildland industrial interface (WII) are important areas of wildfire risk near human settlement, the data product used in this study did not differentiate those locations from the WUI or VLDH [7].

2.2. U.S. Forest Service Fire Program Analysis-Fire-Occurrence Database

The U.S. Forest Service Fire Program Analysis-Fire-Occurrence Database (FPA-FOD) [27] documents the location, cause, discovery date across ~1.8 million wildfires from 1992–2015 from various sources, including U.S. federal, state, and local records of wildfires on public and private lands. We collapsed the US Forest Service-designated cause categories of equipment use, smoking, campfire, railroad, arson, debris burning, children, fireworks, power line, structure, and miscellaneous fires into a human-started wildfire class. All fires <0.001 acres or fires that had a missing/undefined cause were removed from the FPA-FOD dataset [8]. Wildfire size was evaluated by small (<4 km^2), large (4–500 km^2), and very large (>500 km^2) classes. There is acknowledged reporting bias over time (e.g., completeness issues) at the state level [33] making trend analysis difficult [8].

2.3. Monitoring Trends in Burn Severity

Annual burned area and large wildfire perimeters (1992–2015) were obtained through a combination of buffering the FPA-FOD points based on the size of the fire and from the Monitoring Trends in Burn Severity (MTBS) group [34] (see http://mtbs.gov/ for information on pre-processing, specific information on index calculation, and derived dataset available). MTBS leverages the extensive Landsat TM/ETM+/OLI record to produce spatially explicit fire perimeters for all large fires (>400 ha in the western US and >200 ha in the eastern US). These data are produced using the differenced Normalized Burn Ratio (dNBR), which uses pre- and post-fire spectral response to define full extents of the disturbance events. In addition to this automated process, and to ensure consistency and precision, manual digitization is performed at on-screen display scales between 1:24,000 and 1:50,000.

2.4. Zillow Transaction and Assessment Dataset

To measure residential settlement at very fine spatial scales and annually we used the Zillow Transaction and Assessment Dataset (ZTRAX), which contains unique data on housing transactions, home values, rental estimates, spatial location, home- and property-related information as well as built-year information, for existing homes and certain other properties across the United States. The ZTRAX data are obtained from a major large third-party provider as well as through an internal initiative, called County Direct. County Direct prioritizes counties based on different characteristics and supplements the third-party coverage by collecting data directly from county Assessor and Recorder's offices and represents a growing share of the ZTRAX dataset. The ZTRAX database over the study period includes public records and assessor data for approximately 200 million parcels in over 3100 counties in the U.S. (https://www.zillow.com/ZTRAX/). While this database is of unique nature covering most of the U.S. there are some issues related to data quality as described above, including spatial, temporal and thematic uncertainties that we assessed and measured in order to create accompanying uncertainty layers. The database is under continuous revision and will be updated regularly. The raw data and the created SQLite databases cannot be shared publicly per the established data share agreement, but there are recently published aggregations to 5-year intervals between 1810 and 2015 [28].

In this study, we measure residential land use at a given point in time using the construction year information within the ZTRAX database. Furthermore, we use the geographic information provided for each record, which represents approximate address point locations, to characterize residential land use at fine spatial and temporal resolution [28]. Using the raw ZTRAX point database, we selected only housing points that were defined as residential properties and had a defined built year to reduce temporal uncertainty. In recent research, data limitations including low spatial precision in census-based housing data [7], low classification accuracy in land cover data [35], and lack of small fires (e.g., MTBS [34]) resulted in limited accuracy of national assessments. Improved estimates of WUI-wildfire interactions are critical to assess the vulnerability of current and future housing development to wildfire risks and costs. By utilizing the novel ZTRAX data in conjunction with extensive federal fire records we were able to measure threatened homes with high spatial precision and on an annual basis, with future potential to continue to identify regions with elevated risk. These new fine-resolution data products provide unprecedented opportunities for such risk analysis at fine analytical scales, fundamentally changing the way patterns of human-fire interactions can be described.

2.5. United States National Incident Command System Historical ICS-209 Reports

We further leverage an expanded dataset mined from the United States National Incident Management System (NIMS) Incident Command System (ICS) Historical ICS-209 Reports acquired from the publicly available ICS-209 data on the Fire and Aviation Management website (https://fam.nwcg.gov/fam-web/), created by some of the co-authors. We performed systematic and reproducible data cleaning protocols to access these data in a usable form [29]. The cleaned ICS-209 database, henceforth known as ICS-209-PLUS-WF, resulted in 20,353 unique incidents compiled from over 150,000 daily records between 1999–2014, containing information on total estimated fire suppression cost, total residential and commercial structures destroyed and threatened, total fatalities, and wildfire cause. Here, we defined a community threatened by wildfire when threatened or destroyed structures for a given event were >0. These data are a spatially explicit summary of 67% of the total estimated fire suppression costs from all wildfires that required an incident management team but represents <2% of the total wildfires that actually occurred from 1999–2014. The fire suppression field is systematically underestimated [29]. Therefore, all of our cost estimates are lower than true costs, but they are proportionally similar to the total accrued costs reported by the National Interagency Fire Center (NIFC) [20].

3. Methods

The details of the WUI classes were extracted based on the point location of wildfire ignition in the FPA-FOD database. The classes within the WUI layer [7] are based on statistics and classified at the census block group level, which vary widely in size from 0.001 km^2 to 3403 km^2, with a mean of 0.843 km^2 across the conterminous U.S. The size of the census block groups from east to central to west vary from 0.0007 km^2 to 1231 km^2 (mean = 0.469 km^2), 0.001 km^2 to 1020 km^2 (mean = 0.937 km^2), 0.001 km^2 to 3403 km^2 (mean = 1.55 km^2), respectively. While the mean area of census block groups for the WUI, VLDH, and wildlands are 0.651 km^2, 5.72 km^2, and 1.57 km^2, respectively, the classes in the WUI layer are interconnected along an urban-rural continuum characterized by large homogenous clusters of class-similar census block groups that tend to be on the order of 10 s to 1000 s km^2 in size. Thus, the census block groups allow for an appropriate minimum mapping unit to satisfy the assessment from Short [33]; "The FPA FOD should provide point locations of wildfires at least as precise as a PLSS section (2.6 km^2 grid)", while retaining the finer detail in class variation across space without having to aggregate further to an arbitrary pixel unit. Given these vector mapping units of the WUI layer at fine and coarse scales, we believe that our analysis is more than appropriate for quantifying human- and lightning-caused wildfires at the WUI, VLDH, and the wildlands.

Information contained in the WUI data was spatially joined to the FPA-FOD and ICS-209-PLUS-WF point database and temporally aligned such that the discovery wildfire year was matched with the appropriate WUI delineation between 1992–2015. This resulted in an FPA-FOD + WUI and ICS-209-PLUS-WF + WUI databases totaling 1,386,595 and 23,607 wildfire ignition points, respectively. Additional ancillary data layers were subsequently intersected (e.g., level 3 ecoregions, state boundaries), allowing us to further refine boundaries into general regions (i.e., west, central, southeast, etc.). A systematic 50-km grid was imposed to quantify the relative contributions of human- vs. lightning-started wildfire ignition within Intermix WUI, Interface WUI, VLDH, and wildlands. Using these point databases, we were then able to create summary statistics across all grouping combinations (i.e., WUI classes, ignition type, seasonality, fire size, regions, 50-km grid, etc.). We calculated the frequency of wildfire ignitions, mean and 95th confidence interval of fire size, interquartile range of discovery day-of-year. Seasonality was defined as the season that fire ignition was most prevalent across all classes.

For the FPA-FOD and ICS-209-PLUS-WF database, we estimated burned area footprint by developing a circular buffer around each wildfire ignition point, equating to the total burned area associated with the fire event. For each wildfire ignition point that was not associated with an MTBS (large fires) wildfire perimeter, we used this buffering logic to estimate the burned area footprint; for all fires that did fall within the wildfire perimeter and have an association with the MTBS fire, we used the MTBS wildfire polygon. In addition, and because of known inaccuracies in the FPA-FOD point locations [33], we buffered the MTBS polygons equivalent to 2.6 km^2 as suggested by Short et al. (2015), which identified only 16 MTBS wildfires outside those fire polygons, which we subsequently buffered to the fire size (see Accuracy Assessment for more details).

To create estimates of wildfire threat, buffered rings were created around the estimated wildfire polygons at distances of 250 m, 500 m, and 1 km from the wildfire perimeter edge. Three buffer distances were selected to better understand and quantify the estimated number of homes threatened than with a single metric because there are no national standard defining distances from a wildfire's edge that determine evacuations and highest home threats. Those definitions are made at the moment and can vary across space as the wildfire event unfolds. These boundary distances were chosen to best exemplify a gradient of moderate (1 km) to high (250 m) home threat by wildfires. Homes within 1 km of a wildfire are not only at higher threat of wildfire relative to homes further than 1 km from wildfire edge, but those homes will likely fall within an evacuation zone and directly impact homeowners during the event.

We extracted all residential homes from the cleaned ZTRAX database for all years leading up to and including the year of the wildfire event that fell within the wildfire perimeter and each of the buffered FPA-FOD and ICS-209-PLUS-WF rings. We further spatially intersected the WUI census block groups with these buffered polygons to estimate the total WUI class area that was burned/threatened by wildfire. ANOVAs with a pairwise comparison of the means using the Tukey HSD function were used to test for differences between fire size, ignition type, homes threatened, and wildfire suppression costs between the WUI, VLDH, and wildlands.

To test whether wildfire ignition type, timing, and frequency is a function of the distance from human settlement and home density, we calculated the distance of all wildfires from the urban core boundary, proxied by the high-density urban class in the WUI database, and subsequently binned the data by 10-km grid cells across the contiguous US. This allowed us to more easily generalize relationships while retaining the spatial integrity of the data. We also used the 10-km grid cell to estimate the median home density. We employed generalized additive models (GAMs), a common statistical tool in wildfire science [36,37], to explore the relationship between the frequency and IQR of human versus lightning-caused wildfire against median home density and distance the WUI edge.

Accuracy Assessment

We conducted a comparative analysis to estimate a baseline accuracy metric using two datasets, (1) the MTBS polygons that were found within the FPA-FOD database and (2) the buffered burned area estimate using the FPA-FOD points that had an associated MTBS polygon. We did this assessment based on homes found in the ZTRAX database within wildfire perimeters. We found 288,846 homes threatened within buffer-estimated wildfire polygons. The raw MTBS polygons reported 156,915 homes threatened within the burned perimeter. This equates to the buffered burned area overestimating the number of homes threatened within the wildfire polygon by 84%. The major source of these differences between the buffered points and the raw MTBS polygons are due to the irregular shape of the raw wildfire polygons that tend to be more elongated and follow the topography, vegetation distribution, and wind direction while the buffered points do not take into account the underlying land structure.

Of the 17,381 wildfires found in the MTBS database from 1992–2015, 8576 were found within the FPA-FOD, which equates to 0.6% of the total wildfires (1,700,188) found in the FPA-FOD, or 60% of all fires >200 ha (14,327) in the FPA-FOD. We found that 49% of the wildfires in the MTBS product between 1992 and 2015 had an associated FPA-FOD ignition point that was located within the wildfire burned perimeter. When we buffered the perimeters equivalent to 2.6 km^2, we only found 16 additional FPA-FOD points associated with the MTBS product out of the 8864 polygons that were not found within the perimeters, suggesting that the remaining polygons that did not have an FPA-FOD ignition within the perimeter and beyond were not solely due to spatial inaccuracies but instead these fires were completely absent from the FPA-FOD product. It is known that there is a lack of overlap between the FPA-FOD when compared to satellite-based detections, which may relate to intentional burns, a lack of suppression efforts or reporting on small fires, or just the absence of actual events [38]. The additional 16 polygons that had an associated FPA-FOD ID within the 2.6 km^2 buffer had a median distance from the perimeter of 76 m, with over half of those points located with 135 m from the perimeter edge further suggesting that the 2.6 km² metric may be overestimating the inaccuracies of the FPA-FOD database. More refined assessments of this validation are needed for the wider fire community.

4. Results

4.1. Percent of All Wildfires Originated in the WUI

The WUI represents ~10% of the total conterminous U.S. land area in 2010 (ca. 7,780,091 km²), yet 32% of all wildfires (N ~ 437,233 between 1992 and 2015) originated in the WUI. The remaining 29% and 39% of wildfire events started in very low-density housing (VLDH) and wildlands, respectively. Wildfires starting in the WUI burned a total of 16,412 km², representing 4% of the total area burned in this record. Mean fire size in the WUI was significantly smaller (4 ha (3.3–4.3)) than in the VLDH (23.6 ha (21.8–25.9)) and wildland areas (58.3 ha (54.4–62.5); values in brackets indicate a 10,000 repeated bootstrapped 95th confidence interval around the mean). Further, across the record, the total number of residential homes in the WUI (both interface and intermix) increased by 145% with ~32 million new residential homes in the combined WUI between 1990 and 2015 (Figure S1).

4.2. In the WUI, Humans Caused Nearly All Wildfires, Doubled Fire Season Length, and Increased the Number of Wildfires More than 20-Fold, Compared to Lightning-Started Wildfires

Humans caused 97% of all wildfires (n = 424,700) in the wildland-urban interface, 85% of all wildfires (n = 459,054) in the very-low-density housing, and 59% of all wildfires (n = 242,047) in the wildlands between 1992 and 2015 (Figure 1A, Table 1). All fires < 0.001 acres or fires that had a missing or undefined cause were removed from the dataset prior to analysis [8]. In total, human-caused wildfires, irrespective of where they were ignited, burned 12,411 km² of the WUI (Figure 1B), 78,484 km² of the VLDH (Figure 2B,C), and 82,934 km² of the wildlands (Figure 2B,C). In the WUI and VLDH ~50% and ~30% of that area were burned by small fire events (< 4 km²), while in the wildlands ~62% of that area was burned by large fire events (4–500 km²). Human-caused large wildfires (4–500 km²) burned 3311 km² of the WUI. Notably, there were no very-large wildfires (>500 km²) that started in the WUI. Large, human-caused wildfires burned >4 times more of the WUI in the western U.S. (n = 123; 1549 km² area burned) than the eastern U.S. (n = 113; 1228 km² area burned), while small human-caused fires were frequent and burned >7 times more land area of the WUI in the eastern U.S. (n = 327,108; 7027 km² area burned) as compared to the west (n = 69,449; 1008 km² area burned) (Table S1). Wildfires in the WUI were predominantly caused by people across all regions but wildfires in the WUI, VLDH, and wildlands were most abundant in the southeast and western coastal areas of the U.S. (Figures 1A and 2A). Outside of the WUI, lightning-caused wildfires were most frequent and burned the most land area throughout the intermountain west, while human-caused wildfire dominated wildfire burned area and frequency along the Pacific coast and throughout the eastern U.S. (Figure 2). The eastern regions experienced the greatest growth in wildfire occurrence within the WUI, VLDH, and wildlands due to humans, by a factor of 5, 5, and 6, respectively, compared to the increase in wildfire occurrence due to lightning (Table S1).

Human-caused fires exceeded twice the length of the fire season in the WUI, VLDH, and wildlands, expanding it to 148, 164, and 141 days, respectively, compared to lightning wildfire season (WUI = 72 days, VLDH = 45 days, and wildlands = 42 days; Table 1, Figure 3). Human-caused wildfire ignitions were most prevalent during the spring months (March, April, May) in the east and the summer in the west (June, July, August) (Figure 4, Table S2). Human-caused wildfires in the WUI during the shoulder seasons caused the greatest area burned (spring = 4665 km²; fall = 4641 km²) and the highest number of fires (spring = 179,736; fall = 76,041) (Table S2a). While in the VLDH and wildlands, lightning-started wildfires in the wildlands during the summer months caused the greatest area burned (VLDH = 59,563 km², wildlands = 135,614 km²) and the highest number of fires in the spring for VLDH area and summer for wildland areas (VLDH = 178,206, wildlands = 134,309) (Table S2). The median discovery date for human-caused fires in the WUI was over 3-mo earlier (20

May) than lightning-caused fires (26 July), in the VLDH was 2-mo earlier (9 May) than lightning-started fires (26 July), and wildlands over 3-mo earlier (22 June) than lightning-started fires (27 July) (Table S1). Humans-caused fires in the WUI resulted in a longer wildfire season, as the median discovery date for autumn fires was 33 days later for human-caused fires than for lightning-caused fires (16 October and 13 September, respectively).

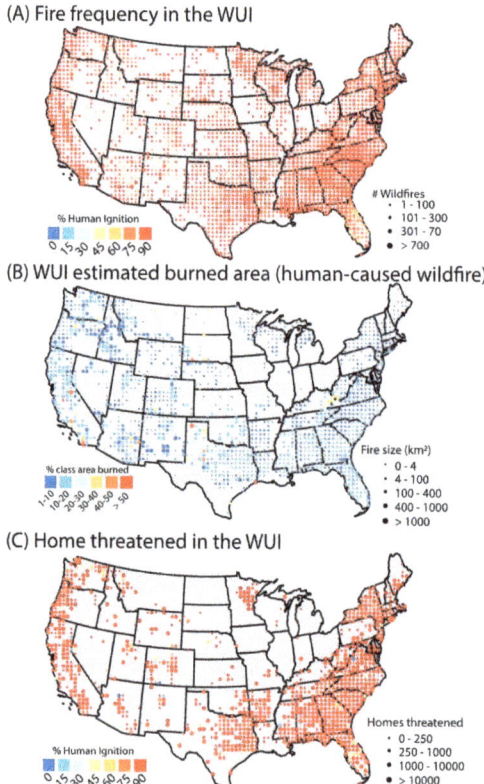

Figure 1. (**A**) The total number of wildfires (dot size) that originated in the wildland-urban interface stratified by the proportion of wildfires caused by humans, (**B**) the total fire size (dot size) as a percentage of WUI that was burned by human-caused wildfires, and (**C**) the total number of homes threatened in the WUI stratified by the proportion of wildfires caused by humans within each 50-km pixel between 1992 and 2015. Black lines represent state boundaries. For (**A**,**C**), reds indicate a greater proportion of human-caused wildfires, while blue indicates a greater proportion of lightning-caused wildfires. Note, only the buffer estimated burned area was used (**B**), not the 250/500/1000 m buffers.

Table 1. Key wildfire metrics within the wildland-urban interface (WUI), very low-density housing (VLDH), and wildlands across the conterminous U.S. (CONUS) and stratified by ignition between 1992 and 2015.

	WUI		VLDH		Wildlands	
	Human	Lightning	Human	Lightning	Human	Lightning
Cumulative Wildfire Ignitions	424,700	12,594	459,054	82,645	242,047	166,135
Cumulative Wildfire Burn Area (km^2)	15,511	1015	59,354	68,983	82,934	155,044
Cumulative Class Burn Area (km^2)	12,411	886	74,570	104,286	105,410	221,047
Average Fire Season Length (d)	148	72	164	45	141	42
Median Discovery Day	126	188	129	207	173	208
Cumulative Suppression Costs ($)	$430,272,129	$199,042,135	$1,075,374,500	$3,707,201,330	$2,684,643,910	$6,497,038,360
Cumulative Residential Structures Threatened: Within Fire	1,037,018	36,215	127,570	9,621	132,708	13,333
Cumulative Residential Structures Threatened: Edge–250 m	4,496,568	129,131	441,821	23,946	366,660	23,943
Cumulative Residential Structures Threatened: 250–500 m	11,868,411	349,220	1,165,255	61,064	1,162,470	77,845
Cumulative Residential Structures Threatened: 500–1000 m	42,448,101	1,273,532	4,811,475	277,824	6,185,076	426,951
Cumulative Residential Structures Threatened: Within Fire–1000 m	59,850,098	1,788,098	6,546,121	372,455	7,846,914	542,072

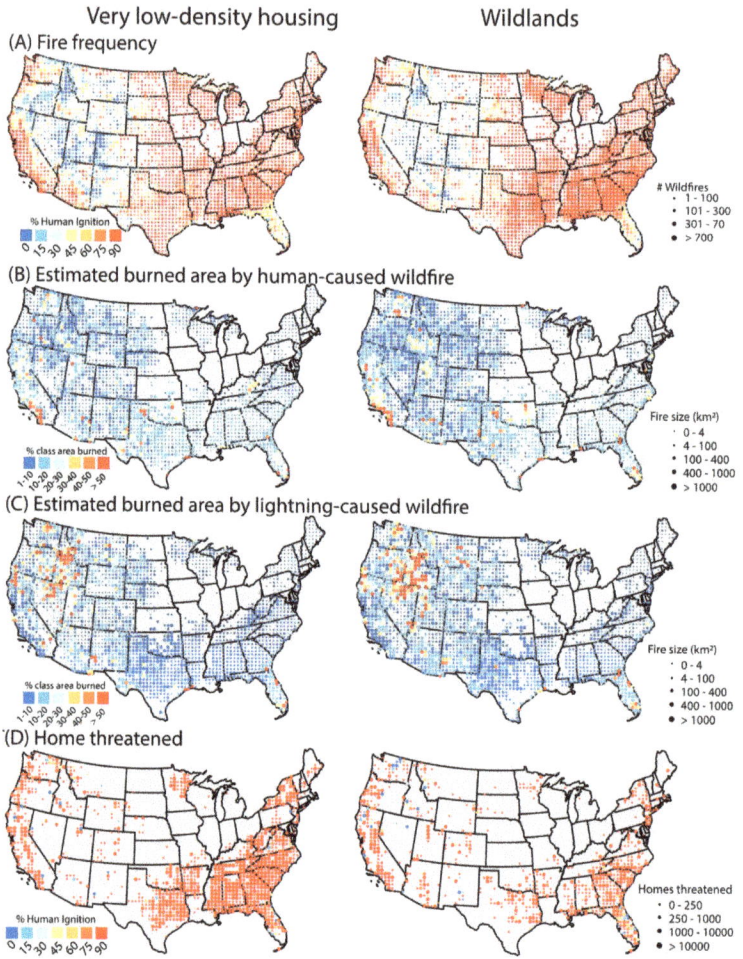

Figure 2. Spatial representation of fire effects in the very-low-density housing and wildlands. (**A**) The total number of wildfires (dot size) stratified by the proportion of wildfires started by humans, (**B**) the total fire size (dot size) as a percentage of each class that was estimated burned by human-started wildfires, (**C**) the total fire size (dot size) as a percentage of each class that was estimated burned by lightning-started wildfires, and (**D**) the total number of homes threatened stratified by the proportion of wildfires started by humans within each 25-km grid cell from 1992–2015. Black lines represent state boundaries. For (**A**) and (**D**), reds indicate a greater proportion of human-started wildfires, while blue indicates a greater proportion of lightning started wildfires.

Within the WUI, humans increased the number of wildfires, relative to the number of lightning-caused wildfires, 36-fold in the eastern (n = 318,163) and 23-fold in the western U.S. (n = 66,232). Human-caused ignitions at and near the WUI (WUI plus areas of very-low-density housing) resulted in 883,680 (64%) wildfire ignitions, 74,603 km^2 of area burned (20%). The effect of human ignitions on wildfire frequency (Figure S2) and season length (Figure S3) is clearly evident within at least 80 km from the high-density urban areas, compared to lightning ignitions. See Tables S3 and S4 for state and level 3 ecoregion analysis.

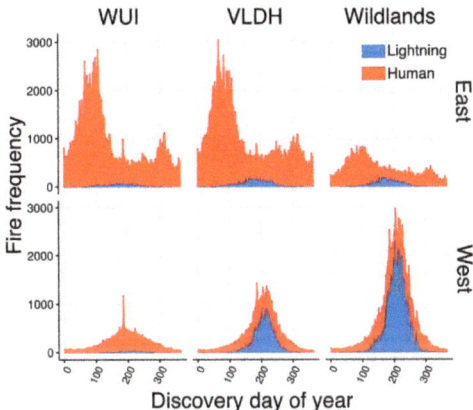

Figure 3. Frequency distributions of human- and lightning-caused wildfires by Julian discovery day of year stratified by wildfires that started either in the wildland-urban interface (WUI), Very low-density housing (VLDH), or wildlands within the eastern or western U.S. between 1992 and 2015.

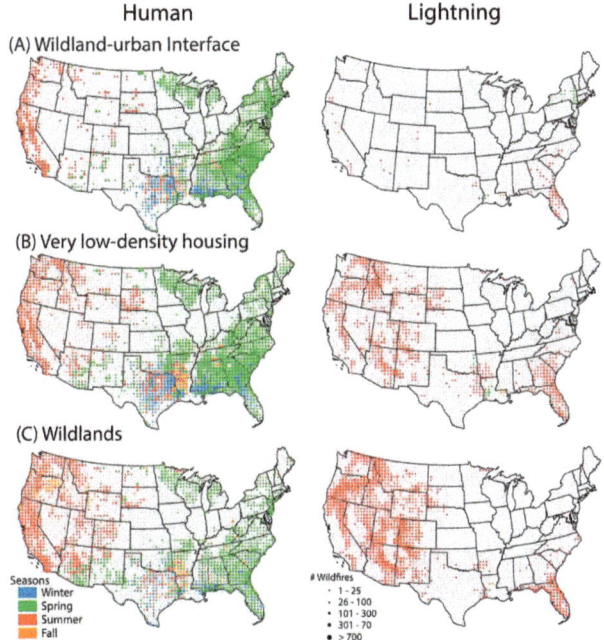

Figure 4. Spatial representation of season that fire ignition was most prevalent across all classes stratified by human and lightning started wildfire in (**A**) WUI, (**B**) very-low-density housing, and (**C**) wildlands. Winter is defined as the months of December, January, February; Spring as March, April, May; Summer as June, July, August; Fall as September, October, December. Dot sizes indicate the total number of wildfire ignitions from 1992–2015 in a 50-km grid cell.

Overall, human-caused wildfire frequency peaks at median home densities associated with the WUI in the west (40 homes/10 km^2), central (34 homes/10 km^2), and southeastern (14 homes/10 km^2) U.S. (Figure 5), and in higher median home densities in the northeastern U.S. characteristic of urban zones (3100 homes/10 km^2). There are no clear trends for lightning-caused wildfire as a function of median home density, but generally decreasing relationship with increasing home density. Across all regions, fire season length of human-caused wildfires peak at the transition between VLDH and WUI areas (Figure 5).

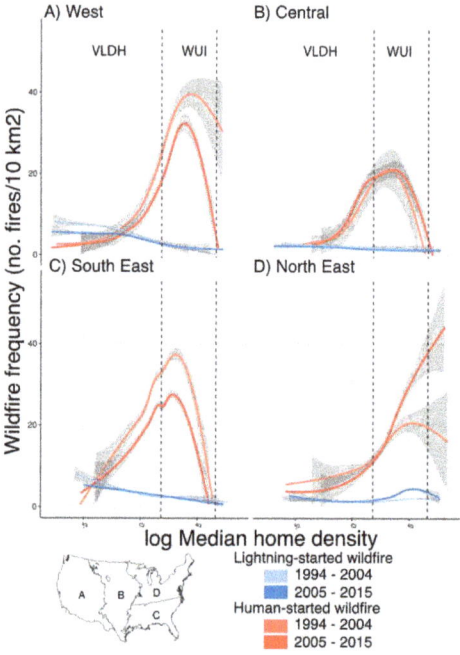

Figure 5. Regional relationships between human- and lightning-caused wildfire ignition frequency and the log median home density within 10-km pixel between two decades (1994–2004 and 2005–2015). Solid lines are based on the best fit Generalized Additive Model regressions with 95th confidence envelope. Dotted vertical lines indicate the division between the WUI and VLDH categories, assuming constant vegetation cover, where urban/WUI boundary equals log(741.3162) home density and the WUI/VLDH boundary equals log(6.17) home density.

4.3. Suppression Costs were Significantly Higher When Protecting Homes

Overall suppression costs/km^2 were more than ten times greater when any homes were at risk, regardless of ignition type, compared with when no homes were at risk ($57,670/km^2 vs. $4076/km^2). Additionally, in communities that were directly threatened by wildfires, the associated suppression costs proportionally increased with wildfire size, irrespective of ignition type (Figure 6).

The average cost of suppression in the WUI doubled ($32.5 million to $65.1 million per wildfire) and tripled in the wildlands ($559,000 to $1.7 million per wildfire) from 1999–2014. Suppression of human-caused wildfires in the WUI cost $78,105/km^2 per year ($58,202–$98,682). Similarly, the per year total suppression costs of controlling human-caused wildfires in the WUI averaged $26 million ($18,875,066 – $34,843,955) (Table S5). Human-caused wildfire costs also varied geographically, with suppression of WUI fires in the West ($220,080/km^2) costing twelve times more than in the eastern U.S. ($18,300/km^2). Similarly, VLDH and wildlands fires in the West (VLDH = $75,944/km^2 ($52,615–$102,362), wildlands = $98,105/km^2 ($72,142–$127,172)) cost six times more than the eastern U.S. (VLDH = ($9,756/km^2 ($5,852–$15,222), wildlands = $21,225/km^2 ($9,964–$38,665)). Regardless of where the fire originated, the yearly average costs to suppress wildfires nearly doubled from $809 million to $1.55 billion between 1999 and 2014. Further, over half of the costs to fight wildfires were spent on wildfires that threatened residential homes (53% or $8.8 billion; Table S6).

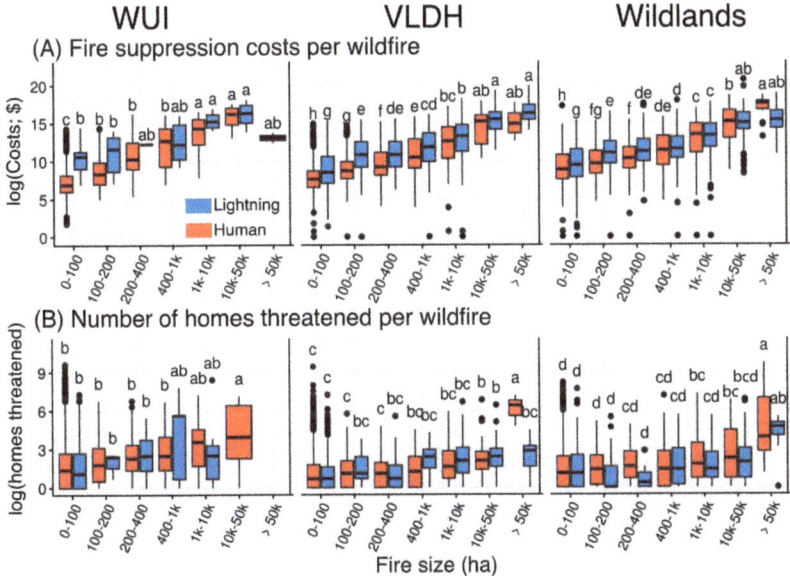

Figure 6. The log mean number of (**A**) fire suppression costs and (**B**) homes threatened per human- and lightning-started wildfire event initiated within the wildland-urban interface (WUI), Very low-density housing (VLDH), and wildlands, stratified by fire size in hectares between 1992 and 2015. Error bars indicate the 95th confidence interval around the mean of each group. Tukey's HSD pairwise comparison of the means is represented by differing letters and letter combinations indicating significant differences among groups ($p < 0.0001$).

4.4. Sixty Million Residential Homes Cumulatively were Within One km of Human-Ignited Wildfires in the WUI

During the study period (1992–2015), over one million homes cumulatively were within perimeters of human-caused wildfire in the WUI, with another 58.8 million homes within 1 km of a wildfire in the WUI (see methods for accuracy assessment). Regardless of ignition location, human-caused wildfires threatened 96% (74,243,133) of all homes threatened by wildfire between 1992 and 2015, while lightning-caused wildfire threatened 2,702,625 homes. Human-caused wildfires accounted for 97% (1,037,018), 93% (127,570), 91% (133,333) of total cumulative residential homes threatened within fire perimeters that ignited in the WUI (Figure 1C), VLDH, and wildlands (Figure 2D), respectively (Table 1).

Human-caused wildfires in the WUI threatened an annual average of 2,493,730 (2,213,141–2,787,175) homes between 1999–2014 (Table S5). Human-caused wildfires in the WUI threatened 30 times more residential homes per year (51,677 (42,692–63,871)) than lightning-caused wildfire (1640 (1021–2404)). Although human-caused fires account for less than half of the overall burned area, they were the main cause of homes threatened, with over 74 million residential structures cumulatively within 1 km of human-caused wildfires. Regardless of development class, small human-caused wildfires were more prevalent and consequently the greatest aggregate threat to homes, with 92% (73,738,356) of homes within wildfire perimeters or threatened by small fires (Table S1). The average number of homes threatened was proportional to the relative size of the wildfire area (Figure 6B). For instance, the smallest wildfires (<1 km^2) threatened on average 1.4 homes (1.3–1.5), while the largest wildfires (>500 km^2) threatened on average 551 (71.1–1350.0) homes.

5. Discussion

This study illustrates how the WUI, which accounts for only a small portion of U.S. land area (<10%), acts as a major source of wildfires, almost exclusively human-caused (Figure 1A). WUI-originated fires also account for a disproportionate number of homes threatened and suppression costs, relative to its area, and expand to burn large areas of surrounding very low-density housing areas (Table 1). Nearly a third of all wildfires between 1992 and 2015 started in the WUI, making these areas a high priority for policy and management aimed at reducing human ignitions as well as promoting resilience to future fires. Our findings directly connect two primary aims of the U.S. Forest Service's National Cohesive Wildland Fire Management Strategy [39], namely, protecting homes, communities and other values at risk; and managing human-caused ignitions. Furthermore, understanding that the WUI is a major source of human-caused ignitions, which pose significant threats to WUI communities, will be useful in designing more effective WUI-centric outreach programs to reduce the number of unwanted human-caused fires [40].

Human-caused ignitions increased the average fire season length two-fold and fire frequency more than two-fold relative to lightning ignitions within the WUI. Previous studies have found that early onset warming and drier conditions [41,42] from anthropogenic climate change [43] and human-caused wildfire [8] have increased the frequency and length of wildfire season across the U.S. This study found a more nuanced relationship: the vast majority of spring wildfires across the U.S. were almost exclusively anthropogenic and occurred within the WUI. By delineating where wildfires are occurring (i.e., the WUI) and at what time of the year the wildfire season is expanding (i.e., the shoulder seasons [8,44]), policy makers and government officials can more accurately assess budget allocations and risks associated with seasonal behaviors to reduce wildfire occurrence. We further found that wildfire in the WUI, where communities are at most risk, occurred throughout the entire year, solely by human-caused ignitions.

We documented that the presence of people enables human-related ignitions well beyond the WUI, but there are important regional differences (Figure 5 and Figures S2–S4)) in human-dominated fire frequency and fire season length. Human-caused wildfires are dominant from the WUI to at least 80 km from the urban core in the western U.S., and greater than 100 km in the northeast and central U.S. This result implies that the human imprint on wildfire extends far beyond the currently defined WUI boundary, that human ignitions may replace the role of lightning ignitions in some places, and that the footprint of WUI development may be underestimated, and more varied, than currently represented [7]. The very low-density housing class, which represent 29% of US land area, will be a critical transition zone for wildfires to burn into WUI areas due to their relatively close proximity to the WUI, yet are not accounted for in previous WUI analyses [3,5,7,31].

Changing climatic conditions have increased temperatures and promoted longer and more extreme precipitation events in some places [45,46] leading to an higher fuel loads, both live and dead, in many fire some systems [15]. Over the past several decades, greater fire activity and total burned area has occurred [10,11,19,43,47–51], which is predicted to increase in the U.S. in coming years [52–54]. This study found that the majority of human-caused wildfires were relatively small (<4 km^2) but were responsible for the vast majority of homes threatened (92%). Of critical importance, wildfires did not have to be large to threaten homes. In the future, with more extreme climatic [55] and weather conditions [49] these small wildfires, however, have a greater potential to grow into larger wildfires [47].

Our results highlight a critical overlap between human ignitions and infrastructure assets at risk from fire. Though the ZTRAX database underestimates the total number of homes across the U.S. by roughly 30% (86.7 million properties) compared to the U.S. Census (125.8 million homes) (more detailed evaluation of the ZTRAX data and the derived HISDAC-US surfaces is described in Uhl et al. [56]), the increased spatial and temporal precision allowed us to analyze the potential impact of all wildfire sizes (Figure 7), resulting in an estimated five times more homes threatened within wildfire perimeters than previously documented [7]. Moreover, from 1990 to 2015, there were an estimated ~32 million more homes built within the WUI (145% increase), demonstrating substantial growth over this period.

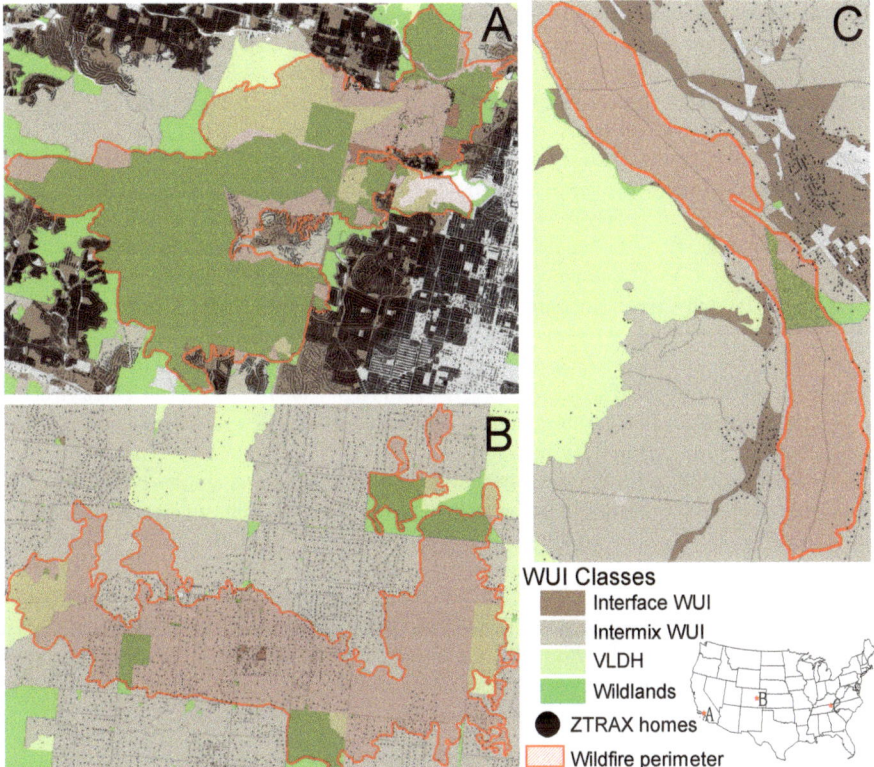

Figure 7. Examples of three wildfires that were human-caused illustrating the enhanced spatial resolution of homes threatened by wildfires using the ZTRAX database compared to the information contained within the census block groups via SILVIS. (**A**) The Topanga wildfire was an urban, human-caused wildfire near Santa Monica, California that started on 28 September 2005. The WUI dataset estimated 7595 homes, while the ZTRAX estimated 936 homes threatened within the fire perimeter. (**B**) The Black Forest wildfire was an arson-caused wildfire in the suburbs of Colorado Springs, CO that started on 11 June 2013. The WUI dataset estimated 1480 homes, while the ZTRAX estimated 859 homes threatened within the fire perimeter. (**C**) A rural, human-caused wildfire from debris burning in northeastern Tennessee that started on 2 March 2006. The WUI dataset estimated 554 homes, while the ZTRAX estimated 54 homes threatened within the fire perimeter.

We found that suppression costs are tightly linked to structures threatened, rather than simply being in the WUI, suggesting that implementing defensible spaces around homes remains a critical step in mitigating fire risk and suppression costs. While it has been suggested that the tripling in fire suppression expenses since the early 1990s has been partly due to increased wildfire risk within the WUI [16,54], we found that suppression costs from wildfires originating within the WUI were small (Table 1), likely because WUI fires tended to be small. WUI-originated fires may be likely detected, accessed, and suppressed earlier and more easily—likely resulting in smaller fires on average (mean fire size WUI = 5 ha vs. wildlands = 75 ha) without the need for an incident command response. Nonetheless, of the fires that needed an incident command response, the relatively low contribution of the WUI to fire suppression costs is counterbalanced by the significant threat of wildfires to structures in the WUI. Overall, while suppression costs within the WUI were low relative to total suppression costs across the U.S. when homes were threatened by wildfire, suppression costs were always high (Table S3).

We show that people are directly responsible for increasing wildfire risk where there are homes, highlighting the importance of policy efforts that aim to reduce ignitions in the WUI. There has been strong messaging around reducing the human-related ignitions that start forest fires, e.g., the Smokey Bear advertising campaign around campfires [57]). But it is critical to note that the primary sources of human ignitions in the WUI are debris burning, arson and equipment use; and outside the WUI the most prominent are railroads, roadways, power lines, and campsites [8,21]. There is a critical need to target messaging and efforts directed at reducing ignitions from these different and complex human activities in the WUI; we effectively need a suburban Smokey Bear. Further, given the year-round fire season—predominantly driven by human-related ignitions—it is also important to increase awareness about fire risk outside of the typical summer wildfire season, and support restrictions on debris burning in spring and fall months when fire danger is high, especially in the eastern U.S. In addition, new initiatives could focus on areas with much finer precision as an important result of employing novel settlement data products (Figure 7; [28,56]) in this study, which showed that risk across the WUI is much more varied than assumed and can be closely interlinked to the location of residential structures. Generally, more effective ways to communicate fire risk to the public and decrease ignitions near vulnerable communities is needed [58,59].

Approaches for reducing wildfire risk to communities have proposed ignition-resistant building materials and fuel-reduction treatments in and around homes in the WUI fire-prone areas of the western United States [30,31,60–62]. Community protection is a key priority in national firefighting strategies, but we should consider where fuel reduction strategies would be most effective—at the individual home level or focused within the WUI [15,63]. In conjunction with defensible space management, wildfire suppression that is more narrowly focused on lands surrounding communities that are most at-risk should be prioritized, allowing wildfire to burn in the wildlands where fire will have an ecological benefit and support future wildfire hazard mitigation and utilizing prescribed burns as a mechanism to mitigate extreme future fires [15,54,64,65].

Limitations

There are some limitations and observed biases in the utilized datasets that are important to acknowledge. First, the FPA-FOD database captures many small events; wildfires less than or equal to 100 ha account for 98% (1,675,166 out of 1,700,188 events) of the database. There is an implicit bias towards reporting of small fires near communities, as they are more likely to be reported than small events in the wildlands. These smaller events are also more likely to be human-caused, and because they are near human settlement, they are more likely to be shorter in duration and smaller because they are identified quickly and easier to access. Due to the smaller size, these fires are almost always overlooked when exploring wildfire presence on the U.S. landscape, particularly because one of the most commonly used fire perimeter datasets has a cut-off threshold for inclusion of 1000 acres in the west and 500 acres in the east (e.g., MTBS). Yet, there has been a longer selection bias in the wildfire literature excluding these smaller fires (e.g., [19,41,47,49,66,67]), which are responsible for the vast majority of homes threatened. Furthermore, smaller wildfires are much higher in frequency than large events, which may be a greater concern to WUI communities under a changing climate [43,44].

Second, while the FPA-FOD database is unique in that it contains critical information on the type and location of the wildfire ignition, it lacks final wildfire perimeter vector polygons. When perimeters were not available through the MTBS dataset, we used a circular buffer equal to the final size of the wildfire burn area. For FPA-FOD wildfire ignition points that had an associated MTBS wildfire ID we substituted that particular wildfire perimeter for the generalized buffer. These MTBS wildfire perimeters account for 83% (7693 out of 9260) of all large wildfires in the combined database (>400 ha). This addition of the 1 km (or 100 ha) buffer is large enough to capture the variation in wildfire perimeter due to topographic/weather/and or suppression efforts. When the wildfire was less than 100 ha, the 1-km buffer completely consumed the known perimeter, making the more preferable and detailed wildfire perimeter unnecessary for those calculations. Because wildfires less than or equal to 100 ha account for 98% of the wildfire database, we are confident that our estimates for homes threatened within 1 km of an event are reliable. Further, it has been estimated that embers can travel up to 1 mile (1.6 km) or further from a fireline, a mechanism generally explored only for larger wildfires [68]. Since the majority of the wildfires in our database are smaller fires (<100 ha; n = 1,675,166) rather than larger fires (>400; n = 7693), we chose a more conservative estimate of distance from the fireline of 1 km rather than 1 mile.

Third, based on comparisons with the US census data for 1990, 2000, and 2010, the ZTRAX data underestimate housing counts by 30%. Therefore, the estimates provided here are likely underestimating the true threat to homes. ZTRAX housing property counts and county stats on housing units are strong, >95% in urban counties and >75% in rural counties during this time period [56].

6. Conclusions

We demonstrate the remarkable effect that humans have had on introducing wildfire to all landscapes of the conterminous US, most importantly within the WUI, through changes in wildfire frequency, seasonality, burned area, and associated costs. We found that human-caused wildfire was responsible for the vast majority of residential homes threatened between 1992 and 2015. People are starting almost all of the wildfires that threaten our homes. Further, the costs to protect homes from wildfires account for over half of the suppression budget. These results have two implications; first, there is a great need to determine how best to reduce human-related ignitions that result in damaging wildfires; and second, this provides greater justification for implementing fuel treatments and prescribed burns, where safe, to mitigate the risk and threat of future wildfires, particularly near and within the WUI. A better understanding of our own relationship with fire, how we promote it, and how we are vulnerable to it, ultimately will help us live more sustainably with fire.

Supplementary Materials: The following are available online at http://www.mdpi.com/2571-6255/3/3/50/s1.

Author Contributions: Conceptualization, N.M., J.K.B., S.L. and B.A.B.; Data curation, N.M. and L.A.S.D.; Formal analysis, N.M.; Funding acquisition, J.K.B.; Investigation, N.M. and T.S.; Methodology, N.M., J.K.B., T.S. and S.L.; Project administration, J.K.B., T.S. and B.A.B.; Resources, J.K.B., S.L. and L.A.S.D.; Supervision, J.K.B. and T.S.; Validation, N.M.; Visualization, N.M.; Writing—original draft, N.M., J.K.B., T.S., S.L., L.A.S.D. and B.A.B.; Writing—review & editing, N.M., J.K.B., T.S. and S.L. All authors have read and agreed to the published version of the manuscript.

Funding: This research was supported by the Earth Lab through the University of Colorado, Boulder's Grand Challenge Initiative. Partial support was provided through the Humans, Disasters, and the Built Environment program of the National Science Foundation, Award Number 1924670 to the University of Colorado Boulder.

Acknowledgments: We thank A. Mahood and R. Chelsea Nagy for valuable discussions and J. Uhl for the original processing of the ZTRAX database. We also thank A. Park Williams, Todd Hawbaker, and Melanie Vanderhoof for their detailed manuscript review. Data and materials availability: The authors were provided access to the Zillow Transaction and Assessment Dataset (ZTRAX) through a data use agreement between the University of Colorado Boulder and Zillow Inc. More information on accessing the data can be found at http://www.zillow.com/ZTRAX. The results and opinions are those of the author(s) and do not reflect the position of Zillow Group. Support by Zillow Inc. is gratefully acknowledged. The code used to analyze the data can be found at https://github.com/NateMietk/human-ignitions-wui. All data used, with the exception of raw ZTRAX, is publicly available; a gridded version of the ZTRAX data can be found here: https://dataverse.harvard.edu/dataverse/hisdacus.

Conflicts of Interest: The authors declare no conflict of interest.

References

1. Calkin, D.E.; Cohen, J.D.; Finney, M.A.; Thompson, M.P. How risk management can prevent future wildfire disasters in the wildland-urban interface. *Proc. Natl. Acad. Sci. USA* **2014**, *111*, 746–751. [CrossRef] [PubMed]
2. Knorr, W.; Arneth, A.; Jiang, L. Demographic controls of future global fire risk. *Nat. Clim. Chang.* **2016**, *6*, 781–785. [CrossRef]
3. Theobald, D.M.; Romme, W.H. Expansion of the US wildland–urban interface. *Landsc. Urban Plan.* **2007**, *83*, 340–354. [CrossRef]

4. Hawbaker, T.J.; Radeloff, V.C.; Stewart, S.I.; Hammer, R.B.; Keuler, N.S.; Clayton, M.K. Human and biophysical influences on fire occurrence in the United States. *Ecol. Appl.* **2013**, *23*, 565–582. [CrossRef]
5. Thomas, D.S.; Butry, D.T. Areas of the U.S. wildland–urban interface threatened by wildfire during the 2001-2010 decade. *Nat. Hazards* **2014**, *71*, 1561–1585. [CrossRef]
6. Bowman, D.M.J.S.; Balch, J.; Artaxo, P.; Bond, W.J.; Cochrane, M.A.; D'Antonio, C.M.; DeFries, R.; Johnston, F.H.; Keeley, J.E.; Krawchuk, M.A.; et al. The human dimension of fire regimes on Earth. *J. Biogeog.* **2011**, *38*, 2223–2236. [CrossRef]
7. Radeloff, V.C.; Helmers, D.P.; Kramer, H.A.; Mockrin, M.H.; Alexandre, P.M.; Bar-Massada, A.; Butsic, V.; Hawbaker, T.J.; Martinuzzi, S.; Syphard, A.D.; et al. Rapid growth of the US wildland-urban interface raises wildfire risk. *Proc. Natl. Acad. Sci. USA* **2018**, *115*, 3314. [CrossRef]
8. Balch, J.K.; Bradley, B.A.; Abatzoglou, J.T.; Nagy, R.C.; Fusco, E.J.; Mahood, A.L. Human-started wildfires expand the fire niche across the United States. *Proc. Natl. Acad. Sci. USA* **2017**, *114*, 2946–2951. [CrossRef]
9. Syphard, A.D.; Keeley, J.E.; Pfaff, A.H.; Ferschweiler, K. Human presence diminishes the importance of climate in driving fire activity across the United States. *Proc. Natl. Acad. Sci. USA* **2017**. [CrossRef]
10. Abatzoglou, J.T.; Kolden, C.A.; Williams, A.P.; Lutz, J.A.; Smith, A.M.S. Climatic influences on interannual variability in regional burn severity across western US forests. *Int. J. Wildland Fire* **2017**, *26*, 269–275. [CrossRef]
11. Jolly, W.M.; Cochrane, M.A.; Freeborn, P.H.; Holden, Z.A.; Brown, T.J.; Williamson, G.J.; Bowman, D.M. Climate-induced variations in global wildfire danger from 1979 to 2013. *Nat. Commun.* **2015**, *6*, 1–11. [CrossRef]
12. Abatzoglou, J.T.; Williams, A.P.; Boschetti, L.; Zubkova, M.; Kolden, C.A. Global patterns of interannual climate–fire relationships. *Glob. Chang. Biol.* **2018**. [CrossRef] [PubMed]
13. Parks, S.A.; Holsinger, L.M.; Miller, C.; Nelson, C.R. Wildland fire as a self-regulating mechanism: The role of previous burns and weather in limiting fire progression. *Ecol. Appl.* **2015**, *25*, 1478–1492. [CrossRef] [PubMed]
14. Higuera, P.E.; Abatzoglou, J.T.; Littell, J.S.; Morgan, P. The Changing Strength and Nature of Fire-Climate Relationships in the Northern Rocky Mountains, U.S.A., 1902-2008. *PLoS ONE* **2015**, *10*, e0127563. [CrossRef]
15. Schoennagel, T.; Balch, J.K.; Brenkert-Smith, H.; Dennison, P.E.; Harvey, B.J.; Krawchuk, M.A.; Mietkiewicz, N.; Morgan, P.; Moritz, M.A.; Rasker, R.; et al. Adapt to more wildfire in western North American forests as climate changes. *Proc. Natl. Acad. Sci. USA* **2017**, *114*, 4582–4590. [CrossRef]
16. Gorte, R. *The Rising Cost of Wildfire Protection*; Headwaters Economics: Bozeman, MT, USA, 2013; Available online: https://headwaterseconomics.org/wp-content/uploads/fire-costs-background-report.pdf (accessed on 3 September 2020).
17. Liu, Z.; Wimberly, M.C. Direct and indirect effects of climate change on projected future fire regimes in the western United States. *Sci. Total Environ.* **2016**, *542*, 65–75. [CrossRef] [PubMed]
18. Liu, J.C.; Mickley, L.J.; Sulprizio, M.P.; Dominici, F.; Yue, X.; Ebisu, K.; Anderson, G.B.; Khan, R.F.; Bravo, M.A.; Bell, M.L. Particulate air pollution from wildfires in the Western US under climate change. *Clim. Chang.* **2016**, *138*, 655–666. [CrossRef] [PubMed]
19. Dennison, P.E.; Brewer, S.C.; Arnold, J.D.; Moritz, M.A. Large wildfire trends in the western United States, 1984–2011. *Geophys. Res. Lett.* **2014**, *41*, 2928–2933. [CrossRef]
20. NIFC. Federal Firefighting Costs (Suppression Only). Available online: https://www.nifc.gov/fireInfo/fireInfo_documents/SuppCosts.pdf (accessed on 9 June 2020).
21. Fusco, E.J.; Abatzoglou, J.T.; Balch, J.K.; Finn, J.T.; Bradley, B.A. Quantifying the human influence on fire ignition across the western USA. *Ecol. Appl.* **2016**, *26*, 2390–2401. [CrossRef]
22. Syphard, A.D.; Radeloff, V.C.; Keeley, J.E.; Hawbaker, T.J.; Clayton, M.K.; Stewart, S.I.; Hammer, R.B. Human influence on California fire regimes. *Ecol. Appl.* **2007**, *17*, 1388–1402. [CrossRef] [PubMed]
23. Syphard, A.D.; Keeley, J.E. Location, timing and extent of wildfire vary by cause of ignition. *Int. J. Wildland Fire* **2015**, *24*, 37–47. [CrossRef]
24. Braswell, A.E.; Leyk, S.; Connor, D.; Uhl, J. Historical Development of the Coastal United States: Evolution of Areas Vulnerable to Sea Level Rise. *Nat. Commun.* **2020**. Submitted.
25. Modugno, S.; Balzter, H.; Cole, B.; Borrelli, P. Mapping regional patterns of large forest fires in Wildland–Urban Interface areas in Europe. *J. Environ. Manag.* **2016**, *172*, 112–126. [CrossRef] [PubMed]

26. Malamud, B.D.; Millington, J.D.A.; Perry, G.L.W. Characterizing wildfire regimes in the United States. *Proc. Natl. Acad. Sci. USA* **2005**, *102*, 4694–4699. [CrossRef] [PubMed]
27. Short, K. *Spatial Wildfire Occurrence Data for the United States, 1992–2015 [FPA_FOD_20170508]*, 4th ed.; Forest Service Research Data Archive: Fort Collins, CO, USA, 2017. [CrossRef]
28. Leyk, S.; Uhl, J.H. HISDAC-US, historical settlement data compilation for the conterminous United States over 200 years. *Sci. Data* **2018**, *5*, 180175. [CrossRef] [PubMed]
29. St. Denis, L.A.; Mietkiewicz, N.P.; Short, K.C.; Buckland, M.; Balch, J.K. All-hazards dataset mined from the US National Incident Management System 1999–2014. *Sci. Data* **2020**, *7*, 64. [CrossRef] [PubMed]
30. Aronson, G.; Kulakowski, D. Bark beetle outbreaks, wildfires and defensible space: How much area do we need to treat to protect homes and communities? *Int. J. Wildland Fire* **2012**, *22*, 256–265. [CrossRef]
31. Schoennagel, T.; Nelson, C.R.; Theobald, D.M.; Carnwath, G.C.; Chapman, T.B. Implementation of National Fire Plan treatments near the wildland–urban interface in the western United States. *Proc. Natl. Acad. Sci. USA* **2009**, *106*, 10706–10711. [CrossRef]
32. Martinuzzi, S.; Stewart, S.I.; Helmers, D.P.; Mockrin, M.H.; Hammer, R.B.; Radeloff, V.C. The 2010 wildland-urban interface of the conterminous United States. In *Research Map NRS-8*; Department of Agriculture, Forest Service, Northern Research Station: Newtown Square, PA, USA, 2015; 124p.
33. Short, K. Sources and implications of bias and uncertainty in a century of US wildfire activity data. *Int. J. Wildland Fire* **2015**, *24*, 883–891. [CrossRef]
34. Eidenshink, J.; Schwind, B.; Brewer, K.; Zhu, Z.-L.; Quayle, B.; Howard, S. A project for monitoring trends in burn severity. *Fire Ecol.* **2007**, *3*, 3–21. [CrossRef]
35. Wickham, J.D.; Stehman, S.V.; Gass, L.; Dewitz, J.; Fry, J.A.; Wade, T.G. Accuracy assessment of NLCD 2006 land cover and impervious surface. *Remote Sens. Environ.* **2013**, *130*, 294–304. [CrossRef]
36. Krawchuk, M.A.; Moritz, M.A.; Parisien, M.A.; Van Dorn, J.; Hayhoe, K. Global pyrogeography: The current and future distribution of wildfire. *PLoS ONE* **2009**, *4*. [CrossRef] [PubMed]
37. Syphard, A.D.; Keeley, J.E.; Massada, A.B.; Brennan, T.J.; Radeloff, V.C. Housing Arrangement and Location Determine the Likelihood of Housing Loss Due to Wildfire. *PLoS ONE* **2012**, *7*, e33954. [CrossRef] [PubMed]
38. Fusco, E.J.; Finn, J.T.; Abatzoglou, J.T.; Balch, J.K.; Dadashi, S.; Bradley, B.A. Detection rates and biases of fire observations from MODIS and agency reports in the conterminous United States. *Remote Sens. Environ.* **2019**, *220*, 30–40. [CrossRef]
39. USFS. The National Strategy. 2014. Available online: https://www.forestsandrangelands.gov/strategy/documents/strategy/CSPhaseIIINationalStrategyApr2014.pdf (accessed on 3 November 2017).
40. Wildland Fire Leadership Council. Cohesive Strategy Crosswalk and Strategic Alignment. Available online: https://www.forestsandrangelands.gov/documents/strategy/reports/cohesive_strategy_crosswalk_and_strategic_alignment_report.pdf (accessed on 9 June 2020).
41. Westerling, A.L. Increasing western US forest wildfire activity: Sensitivity to changes in the timing of spring. *Philos. Trans. R. Soc. Lond. B Biol. Sci.* **2016**, *371*. [CrossRef]
42. Westerling, A.L.; Hidalgo, H.G.; Cayan, D.R.; Swetnam, T.W. Warming and earlier spring increase western U.S. forest wildfire activity. *Science* **2006**, *313*, 940–943. [CrossRef]
43. Abatzoglou, J.T.; Williams, A.P. Impact of anthropogenic climate change on wildfire across western US forests. *Proc. Natl. Acad. Sci. USA* **2016**, *113*, 11770–11775. [CrossRef]
44. Goss, M.; Swain, D.L.; Abatzoglou, J.T.; Sarhadi, A.; Kolden, C.A.; Williams, A.P.; Diffenbaugh, N.S. Climate change is increasing the likelihood of extreme autumn wildfire conditions across California. *Environ. Res. Lett.* **2020**, *15*, 094016. [CrossRef]
45. Allen, C.D.; Breshears, D.D.; McDowell, N.G. On underestimation of global vulnerability to tree mortality and forest die-off from hotter drought in the Anthropocene. *Ecosphere* **2015**, *6*, 121–155. [CrossRef]
46. Park Williams, A.; Cook, B.I.; Smerdon, J.E.; Bishop, D.A.; Seager, R.; Mankin, J.S. The 2016 Southeastern U.S. Drought: An Extreme Departure From Centennial Wetting and Cooling. *J. Geophys. Res. Atmos.* **2017**, *122*, 10888–10905. [CrossRef]
47. Nagy, R.C.; Fusco, E.; Bradley, B.; Abatzoglou, J.T.; Balch, J. Human-Related Ignitions Increase the Number of Large Wildfires across U.S. Ecoregions. *Fire* **2018**, *1*, 4. [CrossRef]
48. Balch, J.; Schoennagel, T.; Williams, A.; Abatzoglou, J.; Cattau, M.; Mietkiewicz, N.; St. Denis, L. Switching on the Big Burn of 2017. *Fire* **2018**, *1*, 17. [CrossRef]

49. Abatzoglou, J.T.; Balch, J.K.; Bradley, B.A.; Kolden, C.A. Human-related ignitions concurrent with high winds promote large wildfires across the USA. *Int. J. Wildland Fire* **2018**, *27*, 377–386. [CrossRef]
50. Williams, A.P.; Abatzoglou, J.T.; Gershunov, A.; Guzman-Morales, J.; Bishop, D.A.; Balch, J.K.; Lettenmaier, D.P. Observed impacts of anthropogenic climate change on wildfire in California. *Earth's Future* **2019**, *7*. [CrossRef]
51. Ganteaume, A.; Barbero, R. Contrasting large fire activity in the French Mediterranean. *Nat. Hazards Earth Syst. Sci.* **2019**, *19*, 1055–1066. [CrossRef]
52. Barbero, R.; Abatzoglou, J.; Larkin, N.; Kolden, C.; Stocks, B. Climate change presents increased potential for very large fires in the contiguous United States. *Int. J. Wildland Fire* **2015**, *10*, 1071. [CrossRef]
53. Krawchuk, M.A.; Moritz, M.A. Constraints on global fire activity vary across a resource gradient. *Ecology* **2011**, *92*, 121–132. [CrossRef]
54. Moritz, M.A.; Batllori, E.; Bradstock, R.A.; Gill, A.M.; Handmer, J.; Hessburg, P.F.; Leonard, J.; McCaffrey, S.; Odion, D.C.; Schoennagel, T.; et al. Learning to coexist with wildfire. *Nature* **2014**, *515*, 58–66. [CrossRef]
55. Holden, Z.A.; Swanson, A.; Luce, C.H.; Jolly, W.M.; Maneta, M.; Oyler, J.W.; Warren, D.A.; Parsons, R.; Affleck, D. Decreasing fire season precipitation increased recent western US forest wildfire activity. *Proc. Natl. Acad. Sci. USA* **2018**. [CrossRef]
56. Uhl, J.; Leyk, S.; McShane, C.; Braswell, A.; Connor, D.; Balk, D. Fine-grained, spatio-temporal datasets measuring 200 years of land development in the United States. *Earth Syst. Sci. Data* **2020**. [CrossRef]
57. Donovan, G.H.; Brown, T.C. Be careful what you wish for: The legacy of Smokey Bear. *Front. Ecol. Environ.* **2007**, *5*, 73–79. [CrossRef]
58. Brenkert-Smith, H.; Dickinson, K.L.; Champ, P.A.; Flores, N. Social Amplification of Wildfire Risk: The Role of Social Interactions and Information Sources. *Risk Anal.* **2013**, *33*, 800–817. [CrossRef] [PubMed]
59. Anderson, S.E.; Bart, R.R.; Kennedy, M.C.; MacDonald, A.J.; Moritz, M.A.; Plantinga, A.J.; Tague, C.L.; Wibbenmeyer, M. The dangers of disaster-driven responses to climate change. *Nat. Clim. Chang.* **2018**. [CrossRef]
60. Cohen, J.D. Preventing disaster: Home ignitability in the wildland-urban interface. *J. For.* **2000**, *98*, 15–21.
61. Cohen, J.D. Relating flame radiation to home ignition using modeling and experimental crown fires. *Can. J. For. Res.* **2004**, *34*, 1616–1626. [CrossRef]
62. Syphard, A.D.; Brennan, T.J.; Keeley, J.E. The importance of building construction materials relative to other factors affecting structure survival during wildfire. *Int. J. Disaster Risk Reduct.* **2017**, *21*, 140–147. [CrossRef]
63. Hamilton, M.; Fischer, A.P.; Guikema, S.D.; Keppel-Aleks, G. Behavioral adaptation to climate change in wildfire-prone forests. *Wiley Interdiscip. Rev. Clim. Chang.* **2018**, *9*, e553. [CrossRef]
64. Moritz, M.A.; Knowles, S.G. Coexisting with wildfire. *Am. Sci.* **2016**, *104*, 220–227. [CrossRef]
65. Kolden, C.A. We're Not Doing Enough Prescribed Fire in the Western United States to Mitigate Wildfire Risk. *Fire* **2019**, *2*, 30. [CrossRef]
66. Tedim, F.; Leone, V.; Amraoui, M.; Bouillon, C.; Coughlan, R.M.; Delogu, M.G.; Fernandes, M.P.; Ferreira, C.; McCaffrey, S.; McGee, K.T.; et al. Defining Extreme Wildfire Events: Difficulties, Challenges, and Impacts. *Fire* **2018**, *1*, 9. [CrossRef]
67. Joseph, M.B.; Rossi, M.W.; Mietkiewicz, N.P.; Mahood, A.L.; Cattau, M.E.; St. Denis, L.A.; Nagy, R.C.; Iglesias, V.; Abatzoglou, J.T.; Balch, J.K. Understanding and predicting extreme wildfires in the contiguous United States. *Ecol. Appl.* **2019**, *29*, e01898. [PubMed]
68. Koo, E.; Pagni, P.J.; Weise, D.R.; Woycheese, J.P. Firebrands and spotting ignition in large-scale fires. *Int. J. Wildland Fire* **2010**, *19*, 818–843. [CrossRef]

© 2020 by the authors. Licensee MDPI, Basel, Switzerland. This article is an open access article distributed under the terms and conditions of the Creative Commons Attribution (CC BY) license (http://creativecommons.org/licenses/by/4.0/).

Article

Parcel-Level Risk Affects Wildfire Outcomes: Insights from Pre-Fire Rapid Assessment Data for Homes Destroyed in 2020 East Troublesome Fire

James R. Meldrum [1,*], Christopher M. Barth [2], Julia B. Goolsby [1], Schelly K. Olson [3], Adam C. Gosey [3], James (Brad) White [3], Hannah Brenkert-Smith [4], Patricia A. Champ [5] and Jamie Gomez [6]

1 U.S. Geological Survey, Fort Collins Science Center, Fort Collins, CO 80526, USA; julia.goolsby@colorado.edu
2 U.S. Bureau of Land Management—Montana/Dakotas, Fire and Aviation Management, Billings, MT 59105, USA; cbarth@blm.gov
3 Grand County Wildfire Council and Grand Fire Protection District No. 1, Granby, CO 80446, USA; solson@grandfire.org (S.K.O.); agosey@grandfire.org (A.C.G.); bwhite@grandfire.org (J.W.)
4 Institute of Behavioral Science, University of Colorado-Boulder, Boulder, CO 80309, USA; hannahb@colorado.edu
5 U.S. Department of Agriculture, Forest Service, Rocky Mountain Research Station, Fort Collins, CO 80526, USA; patricia.a.champ@usda.gov
6 West Region Wildfire Council, Montrose, CO 81401, USA; jamie.gomez@cowildfire.org
* Correspondence: jmeldrum@usgs.gov

Abstract: Parcel-level risk (PLR) describes how wildfire risk varies from home to home based on characteristics that relate to likely fire behavior, the susceptibility of homes to fire, and the ability of firefighters to safely access properties. Here, we describe the WiRē Rapid Assessment (RA), a parcel-level rapid wildfire risk assessment tool designed to evaluate PLR with a small set of measures for all homes in a community. We investigate the relationship between 2019 WiRē RA data collected in the Columbine Lake community in Grand County, Colorado, and whether assessed homes were destroyed in the 2020 East Troublesome Fire. We find that the overall parcel-level risk scores, as well as many individual attributes, relate to the chance that a home was destroyed. We also find strong evidence of risk spillovers across neighboring properties. The results demonstrate that even coarsely measured RA data capture meaningful differences in wildfire risk across a community. The findings also demonstrate the importance of accounting for multiple aspects of PLR, including both hazards and susceptibility, when assessing the risk of wildfire to homes and communities. Finally, the results underscore that relatively small actions by residents before a fire can influence wildfire outcomes.

Keywords: parcel-level risk; risk assessment; post-fire analysis; risk mitigation; rapid assessment; natural hazards

1. Introduction

In October 2020, Grand County, Colorado, was severely affected by the 193,812-acre East Troublesome Fire. Fueled by drought, beetle-killed trees, and red flag weather conditions (i.e., high winds, dry fuels, and low relative humidity), the fire grew rapidly, including a spread of 87,093 acres in the 24 h starting on the afternoon of October 21. The fire resulted in two deaths and destroyed 366 homes. This paper investigates whether parcel-level rapid wildfire risk assessment data collected before the fire can help to explain why these homes were destroyed while others in the fire's path were not. In so doing, it more broadly investigates the relevance of parcel-level characteristics in determining wildfire risk to homes and people.

The East Troublesome Fire was one of many devastating wildfires in 2020. The general risk that wildfire presents to society is well-recognized, e.g., [1–3], as are the roles of development patterns [4–6] and climate change [7,8] in contributing to increasing risk.

Reflecting broad concern, many organizations with interests in reducing wildfire risk conduct some version of wildfire risk assessment. The U.S. federal government supports two databases with national coverage that seek to describe wildfire risk at either the census tract (Federal Emergency Management Agency's National Risk Index, https://hazards.fema.gov/nri/, accessed on 11 February 2022) or municipality scale (United States Department of Agriculture's Wildfire Risk to Communities, https://wildfirerisk.org/, accessed on 11 February 2022). State-wide assessments include the Oregon Wildfire Risk Explorer (https://oregonexplorer.info/topics/wildfire-risk?ptopic=62, accessed on 11 February 2022), and the Colorado Forest Atlas (https://coloradoforestatlas.org/, accessed on 11 February 2022). These assessments tend to follow the risk framework laid out by Finney [9], Thompson and Calkin [10], and Scott et al. [11] and packaged for a public audience at https://wildfirerisk.org. This framework, which is generally consistent with other established risk reduction frameworks [12,13], recognizes wildfire *risk* as the intersection of wildfire *hazard* and *vulnerability* to fire. Wildfire *hazard* encompasses the *likelihood* of a fire at a given location and the *intensity* of that fire if it occurs. *Vulnerability* encompasses both the *exposure* and *susceptibility* of people or assets to expected fire behavior. For example, people are exposed to hazards if they live in homes that are at risk, while their susceptibility depends in part on whether they would be able to safely evacuate during a wildfire. As another example, two homes might be similarly exposed to wildfire hazards but differ widely in susceptibility, and thus overall risk, if one is a wooden cabin and the other is a concrete bunker.

Here, we refine this general wildfire risk framework to focus on parcel-level wildfire risk to homes and residents. Specifically, we define *parcel-level risk* (PLR) as the combination of the local wildfire *hazard* posed to a residential parcel and the *vulnerabilities* of people and property to that hazard (gold boxes in Figure 1). PLR emphasizes intra-community heterogeneity in risk and is embedded within broader-scale contexts that also might influence risk to homes and residents, such as general social vulnerability or determinants of landscape-level hazards such as proximity to wildland vegetation (light green with dashed lines in Figure 1).

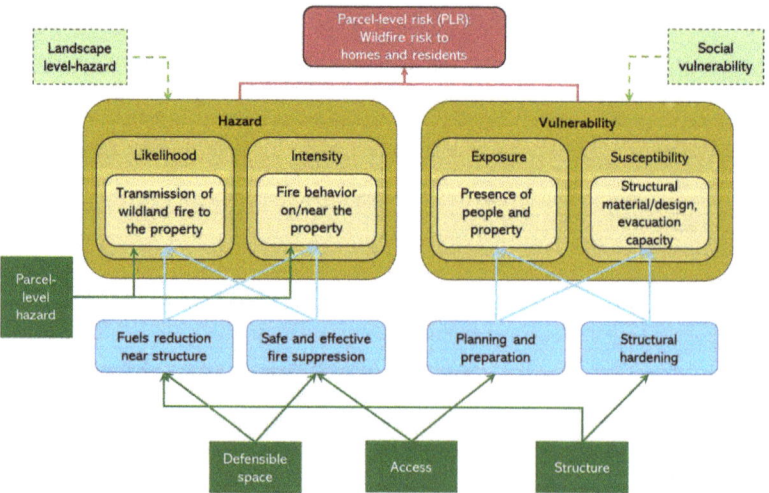

Figure 1. Conceptual model of parcel-level risk (PLR) that shows wildfire risk to homes and residents parsed into hazard and vulnerability (gold); actions that can reduce risk by addressing its components (light blue); and categories useful for a parcel-level wildfire risk assessment such as the WiRē RA (green). PLR is embedded within broader-scale contexts (light green with dashed lines) that influence risk to homes and residents but are typically not measured at scales sufficiently refined for capturing relevant parcel-level variation.

As described in the subsections below, four categories of action (blue in Figure 1) can address these aspects of risk at the parcel level. To summarize, fuel reduction near the structure and support for safe and effective fire suppression can reduce both the likelihood and intensity of wildfire on and near the property; planning and preparation before a wildfire can reduce the exposure of people and property as well as the susceptibility of exposed people and property; and structural hardening can reduce the vulnerability of homes and the people living in them. Thus, in the PLR conceptual model, risk varies at the level of individual parcels as a function not only of conditions associated with the possibility of wildfire and conditional fire behavior at that parcel, but also of conditions related to structural vulnerability and the potential for safe and effective response by fire suppression resources.

As such, PLR provides a framework for describing wildfire risk to homes and residents, built on a foundation of fire science that combines theoretical modeling, laboratory experiments, and post-fire investigations. Although embedded within the general framework of risk described above, PLR is motivated by the recognition that large scale wildfire risk assessments tend to have limited coverage of conditions that vary at the scale of individual properties and treat entire classes of assets, such as residential property, as responding uniformly to a hazard. In contrast, PLR offers an explicit focus on the heterogeneity of risk among the residential parcels that comprise a geographic community, consistent with fire science establishing differences in wildfire vulnerability [14] and recent studies emphasizing the influence of factors at multiple scales on the effectiveness of fire risk reduction strategies [15]. By design, PLR focuses on property conditions that can be influenced by residents or property owners.

The concept of PLR can be implemented by the *parcel-level wildfire risk assessment* of a relatively small set of attributes on residential properties. Specifically, four categories of observable attributes (green in Figure 1)—parcel-level hazard, defensible space, access, and structure—succinctly describe the status of the parcel as it relates to PLR. In practice, parcel-level wildfire risk assessments tend to take one of two forms, each filling a separate dimension of programmatic need: in-depth evaluations conducted on-site, or rapid assessments conducted from the road or sidewalk. An in-depth property evaluation is intended to provide detailed guidance (e.g., remove four tagged trees) for mitigating risk on a property and typically entails on-site engagement between a wildfire practitioner and a property owner. A rapid assessment, when conducted as a census of all properties within a community, can serve numerous purposes, including: identifying risk reduction priorities; generating material for outreach and education; tracking changes in vegetation and mitigation over time; informing suppression strategies; exploring gaps between perceived and assessed risk; and conducting post-fire analyses such as this paper. To these ends, some organizations with interests in wildfire risk mitigation, such as fire departments, state agencies, and non-governmental organizations, conduct parcel-level rapid wildfire risk assessments [16]. Although past research supports the relevance of detailed property characteristics assessed in an in-depth assessment [17], such as whether windows are double-paned and whether roofing includes a non-combustible fiberglass underlayment [18], such attributes are considered beyond the scope of parcel-level rapid wildfire risk assessment. Instead, rapid assessments typically measure relatively few attributes within each category, prioritizing those that can be evaluated rapidly and at a distance (i.e., from the road or sidewalk), and with only a small set of possible responses for each attribute.

Using a quick, systematic approach facilitates the assessment of PLR for all residential properties within a given community. However, one might question whether a small set of coarsely measured attributes, often measured imperfectly due to privacy and resource constraints, can be useful for representing a property's risk. This paper addresses that question. Specifically, we test whether parcel-level rapid wildfire risk assessment data, collected before the fire, explain the destruction of structures from a wildfire. To do so, we make the most of an unfortunate opportunity arising from the East Troublesome Fire. A year before that fire, member organizations of Grand County Wildfire Council (GCWC) conducted a parcel-level rapid wildfire risk assessment for a select set of communities in Grand County, Colorado. They employed the WiRē RA, a parcel-level rapid wildfire risk assessment approach that has been developed into a standardized tool by the authors and partners. The WiRē RA was developed over more than 15 years of collaboration by a team of researchers and practitioners, referred to as the Wildfire Research (WiRē) Team, as part of a systematic data collection and integration approach (the WiRē Approach) intended to inform local wildfire risk education efforts and allow for the monitoring of community adaptation over time [19]. For all residential homes within the selected communities, the WiRē RA recorded data on parcel-level attributes pertaining to defensible space, structural hardening, property access and identification, and parcel-level hazard. One of the selected communities, Columbine Lake, lay in the path of the East Troublesome Fire's rapid spread. On the night of 21 October 2020, the fire destroyed 30 structures. (Note that thirty structures in the Columbine Lake community were reported destroyed in the East Troublesome Fire. Of these, 26 of these were included in the rapid assessment data from 2019. Of these 26 assessed but destroyed structures, three have at least one attribute marked as unobservable in the rapid assessment data, resulting in 23 completely assessed structures destroyed by the fire. None of these numbers include the many structures that sustained damage during the fire (e.g., burns, smoke damage) but were not considered complete losses) in that community).

Few studies investigate the relationship between pre-fire parcel-level attributes and the likelihood of a structure being destroyed in a fire [17]. However, such analysis can be valuable. While some parcel-level attributes are impossible to determine after the fact if structures are destroyed in a fire [20], post-fire analysis is uniquely positioned to capture the dynamics of an event, which include the complexity of fire behavior as well as decisions and constraints regarding fire suppression and response. Post-fire analyses using pre-fire parcel attributes are rare for at least three reasons. First, the probability of a wildfire spreading into any community in any given year is quite low, suggesting a small pool of potential study locations. Second, the data must be recent, because many fire-relevant attributes (e.g., vegetation around the home) change over time. Third, sample sizes (i.e., the number of homes exposed to the fire) tend to be low, a constraint we seek to overcome by estimating the effect of overall risk scores, or individual assessed attributes, on whether a structure was destroyed in a series of individual models, each of which accounts for spatial correlations in both dependent and independent variables.

Our results demonstrate that assessing PLR using a parcel-level rapid wildfire risk assessment helps to explain the outcomes for individual residential properties within the Columbine Lake community. These results underscore the importance of evaluating both wildfire hazard and vulnerability when considering the risk of wildfire to people, homes, and communities. They also demonstrate the utility of a parcel-level rapid wildfire risk assessment, such as the WiRē RA, in representing PLR and its heterogeneity across parcels. Furthermore, the results suggest risk interdependencies across parcels but also that relatively small actions taken by residents can affect home survivability during a wildfire event.

2. Parcel-Level Wildfire Risk Assessment

This section summarizes wildland fire science relating to the four categories of observable attributes and describes their implementation in our parcel-level rapid wildfire risk assessment, the WiRē RA. This section also notes which of these attributes can be directly influenced by the residents or owners of individual properties.

2.1. Parcel-Level Hazard

In the general model of wildfire risk, likelihood refers to the probability of wildfire at a location, and intensity refers to fire behavior conditional on wildfire. Three main factors generally influence wildfire behavior: weather, fuel type, and topography [9]. In the PLR conceptual model, likelihood refers to the probability of a wildfire in the vicinity being transmitted to the property, and intensity refers to fire behavior on and near the property, if that were to occur. So defined, both likelihood and intensity are influenced by variation in fuels and topography at the parcel level. For example, post-fire simulations have found different fuel types to predict home destruction [21]. Grasses typically support fast-moving, low-intensity fires, whereas denser fuels support higher-intensity flames; both can pose high hazards in specific conditions [22]. Post-fire simulation has found slope, relative to the direction of a spreading fire, to predict home destruction [21]. A steep topography can increase the rate of wildfire spread or cause erratic behavior [23–25], and rough terrain can impede firefighter access [26]. Thus, a steep topography places nearby homes at risk [27], especially during rapid, wind-driven fires when suppression is particularly challenging [20,28].

Furthermore, nearby structures can themselves become ignition sources. Post-fire studies have found that the spatial arrangement of buildings, including building density [26] and distance to nearby buildings [29,30], predicted housing damage and that close home proximity allowed wildfire to spread quickly between homes [31]. Fuels and structure proximity can also interact; one post-fire study found that low-density, rural developments were at higher risk from wildfire than high-density developments, with possible explanations including the associated increased exposure to flammable vegetation, limited road access, and complex terrain [27].

Thus, local geographic characteristics affect PLR by influencing both the likelihood and intensity of wildfire on a property. The WiRē RA operationalizes parcel-level hazard with field-identified attributes measuring the distance to hazardous topography (e.g., valleys, cliffs, chimneys), the slope of the ground near the structure, and the general type and density of fuels on and around the property (i.e., attributes 1 through 3 on Table 1). Although not assessed during the initial Grand County project, distance to the closest home has since been integrated into the WiRē RA, and it was calculated retroactively for Grand County properties using spatial data (i.e., attribute 4 on Table 1). Furthermore, although weather conditions can vary systematically at the parcel level, they are not separately assessed during the WiRē RA.

Table 1. Description of attributes collected in the Wildfire Research (WiRē) Rapid Assessment and descriptive statistics for structures in the Columbine Lake community, Grand Lake, CO, with complete assessments, by whether structures were destroyed in the East Troublesome Fire or not. Shading depicts grouping of attributes into categories (i.e., 1–4: parcel-level hazard; 5–6: defensible space; 7–9: access; 10–12: structure).

Attribute Name and Description	Attribute Levels	Points	Not Destroyed ($n = 329$)	Destroyed ($n = 23$)	Moran Test p-Value
1: Distance to hazardous topography	More than 150 feet	0	87.8%	91.3%	<0.001
Distance from residence to ridge, steep	Between 50 and 150 feet	25	8.2%	8.7%	
drainage, or narrow canyon	Less than 50 feet	50	4.0%	0.0%	
2: Slope	Gentle—Less than 20%	0	81.8%	100.0%	<0.001
Overall slope of the property near	Moderate—Between 20% and 45%	10	17.3%	0.0%	
the residence	Steep—Greater than 45%	20	0.9%	0.0%	
3: Adjacent fuels	Light—Grasses	10	0.6%	0.0%	<0.001
Dominant vegetation on the property and	Medium—Light brush and/or isolated trees	20	75.7%	87.0%	
those properties immediately surrounding it	Dense—Dense brush and/or dense trees	40	23.7%	13.0%	
4: Distance to nearest home	More than 100 feet	0	1.8%	0.0%	<0.001
Closest distance to a	Between 30 and 100 feet	50	59.3%	47.8%	
neighboring residence	Between 10 and 30 feet	100	35.6%	39.1%	
	Less than 10 feet	200	3.3%	13.0%	

Table 1. Cont.

Attribute Name and Description	Attribute Levels	Points	Not Destroyed (n = 329)	Destroyed (n = 23)	Moran Test p-Value
5: Defensible space (vegetation) *Distance to overgrown, dense, or unmaintained vegetation*	More than 150 feet	0	10.3%	8.7%	<0.001
	Between 31 and 150 feet	50	43.5%	52.2%	
	Between 10 and 30 feet	75	33.4%	13.0%	
	Less than 10 feet	100	12.8%	26.1%	
6: Defensible space (other combustibles) *Distance to other combustible items (e.g., lumber, firewood, propane tank, hay bales)*	More than 30 feet	0	30.4%	13.0%	0.694
	Between 10 and 30 feet	40	45.3%	43.5%	
	Less than 10 feet	80	24.3%	43.5%	
7: Ingress/egress *Roads available in case one is blocked*	Two or more roads in/out	0	62.9%	60.9%	<0.001
	One road in/out	10	37.1%	39.1%	
8: Driveway clearance *Width of the driveway at the narrowest point*	More than 26 feet wide	0	9.4%	4.4%	0.881
	Between 20 and 26 feet wide	5	31.6%	34.8%	
	Less than 20 feet wide	10	59.0%	60.9%	
9: Address visibility *Visibility of house number at the end of the driveway*	House number is visible and reflective	0	8.5%	4.4%	<0.001
	House number is visible but not reflective	5	36.8%	34.8%	
	House number is not visible	10	54.7%	60.9%	
10: Roof material *Most vulnerable roofing material*	Tile, metal, or asphalt shingles	0	99.7%	95.7%	0.436
	Wood (shake shingles)	300	0.3%	4.4%	
11: Siding material *Most vulnerable siding material*	Noncombustible (e.g., stucco, brick, stone)	0	2.7%	0.0%	0.223
	Log or heavy timbers	35	13.4%	21.7%	
	Wood or vinyl siding	70	83.9%	78.3%	
12: Attachments *Combustible items attached to structure*	No balcony, deck, porch, or fence	0	5.8%	8.7%	0.444
	Combustible balcony, deck, porch, or fence	100	94.2%	91.3%	

2.2. Defensible Space

Wildfire likelihood and intensity are also affected by the presence and quality of defensible space around the structure. The phrase 'defensible space' originates from the reference to this zone as supporting safe and effective fire suppression activities around a structure. Indeed, defensible space has been found to reduce entrapment risk for firefighters, ostensibly increasing their effectiveness [32]. Given that fuels in the defensible space zone can pass flame to a structure through direct point of contact, radiative heat, or firebrands (airborne embers) [14], defensible space also provides passive protection in terms of reducing fuels available near the structure.

While firefighters' experience can confirm the value of a zone that supports safe fire suppression activities (i.e., the active mechanism), laboratory experiments support the value of reducing fuels that could transmit fire to the home (i.e., the passive mechanism). In experiments, structures burned upon direct contact with flames [33], but mock home structures survived radiant heat from an active crown fire that was at least 10 m away [34]. Empirical post-fire case studies also indicate the role of defensible space in reducing fire transmission from nearby vegetation. Analysis of vegetative conditions derived from pre-fire aerial imagery in southern Australia demonstrated a significant impact of 40 m defensible space on home survival, because it distanced the home from the higher radiative heat and ember density near burning vegetation; increased distance from upwind shrubs and trees up to 100 m away had a lesser but still significant effect on home survival [35]. Two other post-fire studies found that pre-fire vegetation conditions near the home, measured with aerial and satellite imagery, helped to determine structure damage during fires in eastern Australia [36] and northern California [30]. Aerial imagery after fires in San Diego County, CA, suggested that up to 30 m of defensible space increased home survival, with little added benefit beyond that [22]. The same study suggested that only 30 to 40 percent of vegetation must be removed to achieve defensible space protection, provided the remaining vegetation is properly spaced and does not overhang the structure [22,27]. However, effectiveness can depend on specific fire conditions, as some studies found vegetation measures, including defensible space, to be relevant but less important predictors of home destruction than other factors such as topography in two wildfires in San Diego, CA and Boulder, CO [26].

Although these empirical studies tend to focus on the passive value of fuels reduction when interpreting their findings, the passive effectiveness of fuel reduction is confounded by active effectiveness in terms of how it affects decisions about suppression made dynamically during an event. Indeed, defensive actions have also strongly predicted structure damages [36]. Thus, results regarding defensible space in one incident may have limited transferability to other contexts, particularly if fire behavior and response differ.

To represent these dual mechanisms of risk reduction, the PLR conceptual model (Figure 1) links defensible space to fuels reduction near the structure and support for safe and effective fire suppression. The WiRē RA operationalizes defensible space in two attributes. The first, vegetative defensible space, measures the closest distance from the residence to overgrown, dense, or unmaintained vegetation (i.e., attribute 5 on Table 1). The second, other combustible materials, measures the closest distance from the residence to other combustible materials, such as wood piles, wicker furniture, propane tanks (i.e., attribute 6 on Table 1). The defensible space category is differentiated from parcel-level hazard not only because of the relationship with supporting safe and effective fire suppression near the structure but also because, in contrast to the parcel-level hazard category, the two attributes measuring defensible space reflect the outcomes of actions taken (or not) by the residents of a property. Thus, attributes in the defensible space category offer opportunities for residents to intervene in the wildfire risk to their homes or themselves.

2.3. Access

Access considerations make up the third category of the PLR conceptual model (Figure 1). Access considerations relate to two interventions in wildfire risk to a home and its residents: the ability of suppression resources to safely access the home during a wildfire, which can reduce the hazard, and the ability of residents to evacuate safely, which can reduce their exposure to the hazard.

Ingress/egress refers to available routes for connecting a property to a reasonably distant, safe location. Many fire-prone communities, including in Colorado, have just one path of ingress/egress [37], making firefighter access and resident evacuation difficult [38]. A study of various southern California wildfires found distance to major roads to be an important predictor of home destruction in low-density communities, perhaps due to firefighter response times [27]. Similarly, safe ingress/egress achieved via preparatory fuel clearing allowed firefighters to access burning properties in a southern Californian fire [32].

Driveway clearance, including width, length, and the presence of a turnaround, affects the ability for fire engines to enter a property—and rapidly exit if necessary. Ideally, the driveway or cleared space around it are wide enough for two vehicles to pass each other and for an emergency vehicle to turn around [39,40]. Driveway width requirements were found to increase firefighter safety in southern California, allowing personnel to return to burning houses behind the fire front [32].

Address visibility can also support safe and effective fire response. At night or in smoky conditions, visible addressing, GPS address data [41], and paper maps [32] can all help firefighters to orient themselves and reach the correct destination. During a fire, personnel may need to call in additional resources to specific locations; visible addressing allows firefighters to quickly locate homes and respond to an active fire. This can be particularly important during large fires when personnel come from other locations and are not familiar with the area. Accordingly, community wildfire preparedness plans or codes often include standards pertaining to address visibility, such as reflectivity, font size, posting location, and combustibility. Despite this perceived importance, we know of no study that assesses the efficacy of address visibility in influencing the risk to homes during a wildfire.

The WiRē RA operationalizes access considerations into three attributes: ingress/egress, driveway clearance, and address visibility (i.e., attributes 7 through 9 in Table 1). In some cases, ingress/egress is determined at the community level; however, individual homes within communities with otherwise good ingress/egress options might have only one feasible route to the property due to, for example, cul-de-sacs or dead-end roads. Unless enforced by local codes and zoning, driveway clearance and address visibility will vary at the parcel level and are often only observable by visual observation of the property. These latter two attributes are also within the control of residents and/or property owners.

2.4. Structure

Structural characteristics make up the fourth category of the PLR conceptual model (Figure 1). Depending on materials, design, and condition, roofing, siding, and items attached to a home (e.g., decks, fences), all can be vulnerable to firebrands and can ignite the rest of a home [18]. However, some studies have found siding to be less important to wildfire mitigation than other home characteristics, because most siding materials can resist ignition, given adequate distance from vegetation or other combustibles [18,27]. Decks and fences can pass flames to the home or to other combustibles [18,42], even with "non-combustible" materials; debris accumulated in cracks or underneath a deck are more common sources of deck ignition than the flammability of the deck itself [43]. Post-fire analysis of two fast-moving southern California wildfires found attached wooden fencing to significantly predict home destruction, regardless of whether homes were in a high- or low-exposure environment [20]. Although analysis was constrained by a lack of pre-fire data, that study also suggested that risk is reduced by clearing debris from gutters, eaves, and roofs and by using fire-resistant or non-combustible materials wherever possible [20].

The WiRē RA operationalizes structural considerations with three attributes: roofing material, most vulnerable siding material, and the presence and combustibility of attachments (i.e., attribute 10 through 12 on Table 1). These attributes primarily relate to structural hardening and associated reductions in vulnerability to wildfire, but the interface between fuels and attachments also affects the localized hazard. As with access considerations, these attributes typically vary at the parcel level. In some cases, existing databases (e.g., county assessor records) include relevant information, but these attributes are often only discernable by visual observation of the property. Structural attributes can be changed by residents and/or property owners through renovation or replacing materials, although initially building with less combustible materials tends to be significantly less costly [44].

2.5. Overall Risk

Notwithstanding its emphasis on understanding and communicating the individual attributes, the WiRē RA enables calculating an overall parcel-level wildfire risk score. Each possible response for the thirteen attributes is assigned a point value based on subjective expert judgment following the fire science described above, with the weighted sum providing an overall wildfire risk score (see Tables 1 and 2). For example, out of 1000 possible points, a wood shake-shingle roof is worth 300, whereas insufficient driveway clearance is worth 10, reflecting the general understanding that a shake-shingle roof influences risk much more than driveway clearance does. This weighting system can be adjusted to specific programmatic needs and local contexts, and indeed, the authors have worked with numerous programs to adjust attributes and their weights accordingly. The "Assessor Reference Guide" (ARG) presented in Figure S1 in the Supplementary Materials provides rationale and additional considerations for attributes and response categories as collaboratively determined for the WiRē RA project in Grand County Colorado.

Table 2. Description of summary scores calculated from the Wildfire Research (WiRē) Rapid Assessment and statistics (mean, Moran Test *p*-value) for structures in the Columbine Lake community, Grand Lake, CO with complete assessments, by whether structures were destroyed in the East Troublesome Fire or not. A Benjamini–Hochberg procedure [45] suggests that only *p*-values of $p < 0.001$ on this table (bolded) are significant after adjusting for multiple comparisons with an assumed 10% false discovery rate.

Attribute Name and Description	Attribute Levels	Not Destroyed ($n = 329$)	Destroyed ($n = 23$)	Moran Test *p*-Value
13: Category score: Parcel-level hazard	Sum of points for attributes 1 through 4	102.5	113.9	0.166
14: Category score: Defensible space	Sum of points for attributes 5 and 6	97.1	114.1	0.064
15: Category score: Access	Sum of points for attributes 7 through 9	18.9	19.6	0.672
16: Category score: Structure	Sum of points for attributes 10 through 12	158.5	166.7	0.644
17: Overall risk score	Sum of points for attributes 1 through 12	277.1	414.3	**<0.001**

Similarly, weighted sums can be calculated for each of the four categories contributing to PLR shown on Figure 1 (i.e., parcel-level hazard, defensible space, access, and hazard) by summing the point values for included attributes. Attributes and point values have been developed iteratively through more than 15 years of collaborative partnerships between the WiRē Team and wildfire practitioners throughout the western United States. Although the WiRē RA aggregates individual attributes into overall risk using a simple weighted sum, the literature suggests that some attributes may interact. For example, flammable siding has been found to pose less risk with proper defensible space [18], and homes on steep slopes have been found to benefit from greater defensible space distances than homes on shallower slopes [22]. While the linear sum can be modified to represent such interactions, we believe that doing so in any rigorous way, including generalized across different fuel and fire behavior contexts, would require a substantially more nuanced understanding of the interactions than currently exists.

The WiRē RA is focused on identifying parcel-level heterogeneity within a community. Thus, it typically does not explicitly include information from broader-scale assessments, such as wildfire behavior modeling that relates to hazards at the landscape scale or social vulnerability assessments that relate to vulnerability at the community scale. However, we acknowledge that comprehensive assessment of the wildfire risk to homes and residents entails consideration of the broader-scale contexts in which the community, and thus the PLR, is embedded (see Figure 1, light green boxes with dashed borders). Depending on purposes, the assessment of risk might also benefit from additional considerations not addressed here, such as the potential for parcel-level variation in typical weather conditions or in weather conditions expected during extreme events.

Finally, we note that the concept of PLR expands upon the related concept of the home ignition zone (HIZ), which emphasizes that reducing the susceptibility of homes to wildfire, by appropriately managing the materials, design, and maintenance of the home in relation to the immediate surroundings, is key to reducing the risk of home loss [33,46]. While the HIZ concept focuses on susceptibility, PLR further encompasses localized wildfire hazard and exposure to provide a more comprehensive assessment of risk at the parcel level. This accommodates the inclusion of not only structural conditions and defensible space but also other localized influences on the wildfire's likelihood and intensity (i.e., parcel-level hazard) and conditions that interface with suppression and preparation (i.e., access). PLR also recognizes the role of defensible space in supporting safe and effective fire suppression, which relates more to the localized wildfire hazard than to the susceptibility of a home. Most importantly, these modifications recognize that the interventions possible for one actor, at one scale, can interact with the interventions possible for other actors and at other scales; for example, a resident's decision to widen their driveway can improve a fire department's ability to safely protect that resident's home.

3. Materials and Methods

3.1. Study Context

Columbine Lake is a small community approximately two miles away from downtown Grand Lake, CO, a gateway town on the western side of the Rocky Mountain National Park (see Figure 2). In October of 2020, this community was one of many in Grand County, CO, affected by the East Troublesome Fire. The fire was reported in the Arapaho National Forest on 14 October and spread to over 10,000 acres in the first three days. The fire then grew rapidly, covering 187,964 acres by the end of 23 October. The fire destroyed an estimated 366 residences and 214 outbuildings and commercial structures [47] and damaged many more. Despite the destruction of 26 out of 461 assessed structures in the community of Columbine Lake, the community received somewhat of a glancing blow from the fire, in comparison to other communities in the area that received the full brunt of the extreme fire conditions and consequently faced near total destruction of their structures. Note that the number of destroyed structures reported throughout this paper pertains to complete losses among structures with complete rapid assessment data from 2019; this number does not include numerous structures that sustained damage (e.g., burns, smoke damage) but were not considered complete losses, nor does it include structures destroyed by the fire but not covered by the 2019 rapid assessment. This total does include 3 structures with only partially complete assessments, as described in more detail below.

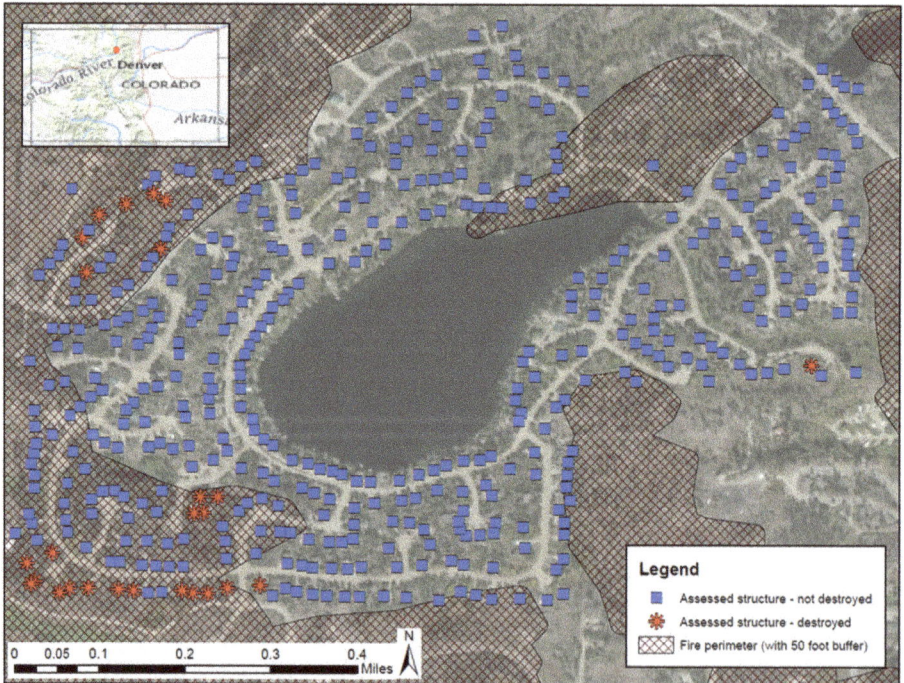

Figure 2. Map of East Troublesome Fire perimeter [48] and structures in the community of Columbine Lake, Grand Lake, CO that were assessed by Grand County Wildfire Council in 2019 and either destroyed (red) or not (blue) in the East Troublesome Fire. Inset shows the location within Colorado. Basemap image is the intellectual property of Esri and is used herein under license. Copyright © 2022 Esri and its licensors. All rights reserved.

3.2. Data

Previously, GCWC had selected Columbine Lake as one of six communities to be the site of a detailed data collection effort in collaboration with the Wildfire Research (WiRē) Center. (The Wildfire Research (WiRē) Center (https://wildfireresearchcenter.org, accessed on 11 February 2022) is a non-profit organization that works with wildfire practitioners to seek locally tailored pathways to create fire-adapted communities through the WiRē Approach described above and by ref. [19].) As part of this effort, GCWC used the WiRē RA tool to assess 461 homes in the community of Columbine Lake in the summer of 2019. Data collection was conducted by individuals from each of the five fire protection districts who had been trained to conduct the WiRē RA. The assessment was conducted from the roadside and supplemented using online imagery where necessary. As is common in roadside assessments, numerous attributes could not be clearly observed for all parcels and were marked as unobservable in the WiRē RA dataset. While no more than three parcels were marked as unobservable for 10 out of the 12 attributes, two attributes ("other combustibles" and "attachments") were frequently unobservable (see Table S1 for details). Omitting parcels with unobservable attributes truncates the dataset by 24 percent, down to a total of 352 fully assessed structures, 23 of which were destroyed in the fire. To avoid introducing unnecessary measurement error bias, we used this set of 352 structures with complete assessment data for our main analysis. However, because the large proportion of incomplete assessments also introduced gaps in the spatially lagged measures, Table S2 replicates our primary results for the full dataset with unobservable attributes coded as having the riskiest rating for that attribute. Because proportionally fewer destroyed structures had incomplete data versus structures not destroyed (12% vs. 24%), this coding was conservative with respect to identifying meaningful attributes for explaining structures destroyed in the East Troublesome Fire. The results shown in Table S2 are generally consistent with the main results presented below.

GCWC assessed twelve attributes in 2019: the eleven represented by attributes 1–3 and 5–12 in Table 1, plus an additional attribute describing the length of the driveway that was omitted from the analysis here due to a lack of variation. However, the WiRē Team continues to revise the assessment tool as necessary based on new understandings. Based on revisions to the standard WiRē RA implemented in March 2020 (i.e., before the East Troublesome Fire), we supplemented the WiRē RA data collected by GCWC with an estimate of the distance between structures (attribute 4 in Table 1). We calculated the distance to the nearest home using point data for the locations of assessed structures. To adjust for point versus polygon data, 40 feet were subtracted from all calculated values (based on an assumed 20 feet from centroid to exterior wall for each of two homes) before converting to categorical ratings for inclusion in the overall risk score. We then calculated the overall risk score using the relative point values agreed upon by wildfire mitigation experts on the WiRē Team on March 2020, shown in Table 1.

Table 1 provides descriptive statistics for data pertaining to structures with complete assessment data destroyed and not destroyed within the Columbine Lake community. After the East Troublesome Fire, GCWC identified which structures within the WiRē RA dataset for Columbine Lake had been destroyed, as shown on the map in Figure 2. For reference, we overlaid spatial data on the fire perimeter [48], along with a 50-foot buffer to accommodate imprecision in spatial data, upon a point layer representing the locations of all assessed structures. Although 21 out of 23 structures with complete assessments were identified as within the burn perimeter, we note that all structures in the community were within a quarter mile or less from this active burn perimeter and that the entire community was exposed to wind-driven ember showers and active suppression activity during the events of 21 October 2020. Further, we note that although the fire perimeter is influenced by fire behavior at broader, landscape-level scales, the location of the perimeter within the community is not exogenous, but rather is at least partially simultaneously determined by structure destruction, due to parcel-level conditions, suppression activity, and because structures themselves can act as a source of fuel for the fire. Accordingly, we considered the entire community as being exposed to the fire and estimate our main models using the full dataset. To test the robustness of this decision, we replicated our main models with a subset of the data, constrained to the 116 structures within the burn perimeter; the results (shown in Table S3) are similar in direction to those from the preferred models but generally greater in magnitude while being considerably less precise.

3.3. Empirical Analysis

We focused our primary analysis on all structures within the community with complete WiRē RA data, modeling the influence of assessed variables and summary scores on whether a structure was lost in the East Troublesome Fire. Although, ideally, we would model the likelihood of a structure being destroyed as a function of all assessed attributes jointly, the small number of observations limited the degrees of freedom to do so. Instead, we estimated a series of separate models, each with one assessed attribute or summary score as the main independent variable.

Past research [49], practitioner experience, and the spatial patterns apparent in Figure 2 all suggested the possibility of spatial clustering of not only fire outcomes but also many of the assessed attributes. We tested and found strong evidence for spatial dependence in whether structures were destroyed (Moran test χ^2 = 187.96, $p < 0.001$) and in data for seven of the 12 assessed attributes (last column of Table 1). Thus, we estimated relationships using a series of spatial Durbin models [50], given by:

$$y = \alpha + \beta X + WX\gamma + \rho Wy + \epsilon$$
$$\epsilon \sim N(0, \sigma^2 I_n) \quad (1)$$

In all models, the dependent variable was defined as $y = 1$ if the assessed structure was destroyed and $y = 0$ otherwise (although the binary outcome variable suggests that a logit or probit specification would be more appropriate, the theory and implementation of a spatial Durbin logit/probit model is not well established. Regarding the main limitation of the linear probability model approach, Woodridge [51], (p. 455) notes, "... If the main purpose is to estimate the partial effect of [the independent variable] on the response probability, averaged across the distribution of [the independent variable], then the fact that some predicted values are outside the unit interval may not be very important." See Table S4 for a comparison of an ordinary versus logistic regression for all attributes in the dataset, without controlling for spatial effects). Each model included: a constant for the intercept (α); a slope (β) multiplied by a sole independent variable (X) (corresponding to the numerical score for one of the twelve assessed attributes shown in Table 1 or the five summary scores shown in Table 2); a spatial weights matrix (W), calculated as the inverse-distances among the centroids of all assessed structures; a spatial lag (γ) for the weighted independent variable WX; a spatial lag (ρ) for the weighted dependent variable (Wy); and an error term (ϵ) consisting of a normally distributed mean-zero disturbance with constant variance (σ^2) multiplied by the n-dimensional identity matrix (I_n), where n is equal to the number of observations.

We estimated models in Stata/SE 16.0 (any use of trade, firm, or product names is for descriptive purposes only and does not imply endorsement by the U.S. Government) using the *spregress* command, employing the generalized spatial two-stage least-squares estimator and treating errors as heteroskedastic. Because of the spatial relationships, parameter estimates for the spatial Durbin model do not have straightforward interpretations and only make intuitive sense in terms of combined effects. Accordingly, we reported and focused on the estimated impacts, as defined by LeSage and Pace [52], which summarize the average marginal effects of the independent variable on the reduced form mean and are calculated with the *estat impacts* command in Stata. The reported impacts included direct impacts, which report the average of the own-marginal effects of the independent variable; indirect impacts, which report the average of the spillover effects of the independent variable, and total impacts, which are the sum of the other two and thus of primary interest.

That is, direct impacts reflect the average of the effects of each property's attributes directly on whether that property's own structure was destroyed, ignoring any spillover effects across properties. Indirect impacts can be interpreted as reflecting the average of either the spillover from one property's attribute upon the outcomes for all its neighbors or the spillover from all of a property's neighbors' attributes upon the outcome for that property. As the sum of direct and indirect impacts, total impacts thus reflect the average overall effect of a property's attributes on whether that property's structure was destroyed, accounting for the interactions amongst neighbors.

4. Results: Risk Assessment Data Help Explain Destroyed Structures

In this section, we investigate the relationships between WiRē RA data and whether homes in the Columbine Lake community were destroyed by the East Troublesome Fire using the modeling approach described above. Tables 3 and 4 show the results for seventeen separate spatial Durbin models for structures in the Columbine Lake community with complete assessments.

Table 3 reports estimated parameters for each model. In all models, the coefficient on the spatial lag on the dependent variable (Wy) is large, positive, and strongly significant, consistent with the spatial clustering of structures that were lost and with the possible effect of structures acting as a source of fuel and increasing the hazard to their neighbors during the event. The strong significance of the spatial lag on the independent variable (WX) in nearly all models provides further support for estimating spatial effects with the spatial Durbin model.

Table 3. Coefficients estimated for 17 separate spatial Durbin models, one for each independent variable listed, of whether a structure was destroyed (y) as a function of the independent variable from the Wildfire Research (WiRē) risk assessment (RA) for structures with a complete assessment (n = 352; $\bar{y}$ = 0.07). Shading depicts grouping of attributes into categories (i.e., 1–4: parcel-level hazard; 5–6: defensible space; 7–9: access; 10–12: structure). A Benjamini–Hochberg procedure [45] suggests that all p-values of $p = 0.077$ or less on this table (bolded) are significant after adjusting for multiple comparisons with an assumed 10% false discovery rate.

$y = 1$ If Structure Destroyed; $y = 0$ Otherwise $n = 352$	Constant (α)			Independent Variable (X)			Spatial Lag on Independent Variable (WX)			Spatial Lag on Dependent Variable (Wy)		
	coef.	std.err.	$p > |z|$	coef.	std.err.	$p > |z|$	coef.	std.err.	$p > |z|$	coef.	std.err.	$p > |z|$
1: Distance to hazardous topography	−0.149	0.053	**0.005**	−0.0015	0.0008	**0.077**	0.0178	0.0110	0.104	2.473	0.470	**<0.001**
2: Slope	0.000	0.031	0.998	−0.0024	0.0017	0.155	−0.0201	0.0167	0.229	1.654	0.337	**<0.001**
3: Adjacent fuels	0.506	0.112	**<0.001**	−0.0005	0.0011	0.635	−0.0261	0.0054	**<0.001**	3.036	0.448	**<0.001**
4: Distance to nearest home	0.253	0.065	**<0.001**	0.0004	0.0004	0.273	−0.0062	0.0013	**<0.001**	3.670	0.584	**<0.001**
5: Defensible space (vegetation)	0.348	0.124	**0.005**	0.0007	0.0005	0.138	−0.0083	0.0023	**<0.001**	2.392	0.410	**<0.001**
6: Defensible space (other combustibles)	0.304	0.085	**<0.001**	0.0006	0.0004	0.101	−0.0129	0.0028	**<0.001**	3.550	0.575	**<0.001**
7: Ingress/egress	0.142	0.062	**0.021**	0.0086	0.0026	**0.001**	−0.0580	0.0177	**0.001**	1.545	0.334	**<0.001**
8: Driveway clearance	0.403	0.095	**<0.001**	0.0014	0.0028	0.607	−0.0749	0.0150	**<0.001**	3.125	0.471	**<0.001**
9: Address visibility	0.303	0.093	**0.001**	0.0069	0.0033	**0.038**	−0.0647	0.0147	**<0.001**	2.738	0.465	**<0.001**
10: Roof material	−0.033	0.013	**0.011**	0.0013	0.0011	0.239	−0.0142	0.0068	**0.036**	1.942	0.443	**<0.001**
11: Siding material	0.439	0.116	**<0.001**	−0.0001	0.0008	0.922	−0.0091	0.0018	**<0.001**	3.036	0.491	**<0.001**
12: Attachments	0.419	0.105	**<0.001**	−0.0003	0.0006	0.584	−0.0058	0.0012	**<0.001**	3.209	0.484	**<0.001**
13: Category score: Parcel-level hazard	0.337	0.082	**<0.001**	0.0002	0.0003	0.572	−0.0051	0.0011	**<0.001**	3.547	0.546	**<0.001**
14: Category score: Defensible space	0.337	0.112	**0.003**	0.0006	0.0003	**0.071**	−0.0054	0.0013	**<0.001**	2.969	0.466	**<0.001**
15: Category score: Access	0.356	0.105	**0.001**	0.0034	0.0015	**0.019**	−0.0285	0.0060	**<0.001**	2.611	0.411	**<0.001**
16: Category score: Structure	0.335	0.126	**0.008**	0.0005	0.0006	0.425	−0.0035	0.0007	**<0.001**	3.169	0.479	**<0.001**
17: Overall risk score (100 points)	0.284	0.120	**0.018**	0.0366	0.0216	0.090	−0.1525	0.0311	**<0.001**	3.260	0.502	**<0.001**

Table 4. Impacts estimated from 17 separate spatial Durbin models, one for each independent variable listed, of whether a structure was destroyed (y) as a function of the independent variable from the Wildfire Research (WiRē) risk assessment (RA) for structures with a complete assessment ($n = 352$; $\bar{y} = 0.07$). Shading depicts grouping of attributes into categories (i.e., 1–4: parcel-level hazard; 5–6: defensible space; 7–9: access; 10–12: structure). A Benjamini–Hochberg procedure [45] suggests that all p-values of $p = 0.021$ or less on this table (bolded) are significant after adjusting for multiple comparisons with an assumed 10% false discovery rate.

y = 1 If Structure Destroyed; y = 0 Otherwise $n = 352$	Total Impact			Direct Impact			Indirect Impact		
	dy/dx	std.err.	$p > |z|$	dy/dx	std.err.	$p > |z|$	dy/dx	std.err.	$p > |z|$
1: Distance to hazardous topography	−0.0115	0.0060	0.053	−0.0003	0.0161	0.987	−0.0113	0.0170	0.508
2: Slope	0.0286	0.0307	0.352	−0.0027	0.0018	0.131	0.0313	0.0294	0.287
3: Adjacent fuels	0.0132	0.0023	**<0.001**	0.0003	0.0052	0.959	0.0129	0.0042	**0.002**
4: Distance to nearest home	0.0022	0.0003	**<0.001**	0.0004	0.0006	0.496	0.0018	0.0004	**<0.001**
5: Defensible space (vegetation)	0.0057	0.0018	**0.002**	0.0005	0.0008	0.524	0.0052	0.0018	**0.004**
6: Defensible space (other combustibles)	0.0049	0.0011	**<0.001**	0.0009	0.0021	0.690	0.0041	0.0015	**0.005**
7: Ingress/egress	0.0813	0.0587	0.166	0.0083	0.0024	**<0.001**	0.0730	0.0592	0.218
8: Driveway clearance	0.0347	0.0060	**<0.001**	0.0020	0.0053	0.706	0.0327	0.0063	**<0.001**
9: Address visibility	0.0340	0.0075	**<0.001**	0.0067	0.0041	0.104	0.0273	0.0087	**0.002**
10: Roof material	0.0286	0.2200	0.897	0.0023	0.0153	0.879	0.0262	0.2047	0.898
11: Siding material	0.0046	0.0010	**<0.001**	0.0002	0.0021	0.926	0.0044	0.0019	**0.021**
12: Attachments	0.0027	0.0005	**<0.001**	−0.0003	0.0007	0.649	0.0031	0.0009	**<0.001**
13: Category score: Parcel-level hazard	0.0020	0.0005	**<0.001**	0.0003	0.0010	0.769	0.0017	0.0006	**0.007**
14: Category score: Defensible space	0.0025	0.0007	**<0.001**	0.0011	0.0140	0.936	0.0014	0.0135	0.919
15: Category score: Access	0.0160	0.0040	**<0.001**	0.0038	0.0025	0.128	0.0122	0.0041	**0.003**
16: Category score: Structure	0.0014	0.0004	**<0.001**	0.0005	0.0006	0.419	0.0009	0.0009	0.305
17: Overall risk score (100 points)	0.0512	0.0120	**<0.001**	0.0363	0.0222	0.101	0.0149	0.0321	0.643

That said, as described above, coefficients on the independent variable (X) and its spatial lag (WX) cannot be separately interpreted in an intuitive way. Thus, we focus on the average impacts shown in Table 4 as our main results. Each row reports the average total impact, the average direct impact, and the average indirect impact, with the average estimated impact, standard error, and p-value shown for each measure. Impacts pertain to marginal effects, meaning that they describe the average change in the probability of the structure being destroyed for a 1-point change in the independent variable, calculated at the reduced form mean. As shown in Table 1, attribute ratings are scored anywhere from 5 to 300 points, depending on the attribute, such that higher points always pertain to higher expected risk.

Table 4 shows that eight of the twelve individual attributes measured by the WiRē RA are estimated as having positive and significant total impacts on the likelihood of a structure being destroyed. Strikingly, none of these variables are estimated as having significant direct impacts upon a given structure, whereas all of the assessed attributes with significant total estimates are estimated as having significant indirect impacts. Indirect impacts can be interpreted in two ways: the perspective *from* an observation, which relates to how a change in a single parcel influences the risk to all other structures, or the perspective *to* an observation, which relates to how changes in all other parcels influence a single parcel [52]. From either perspective, the results imply strong spillovers in risk across properties, in which the hazardous conditions on individual parcels play a large role in determining the risk to their neighbors as well as to themselves through the interactions of fire risk between structures.

Given the small sample size, the linear probability model specification, and the role of expert-assigned weights in determining parameter magnitudes, we focus primarily on the direction and significance of the estimated average total impacts for interpretation. Among the four variables for the parcel-level hazard (1–4), adjacent fuels and the distance to nearest home are estimated as positive and significant. Both defensible space attributes, pertaining to vegetation (5) or other combustibles near the home (6), have positive total effects on risk. Driveway clearance (8) and address visibility (9), two out of three attributes related to access, are positive and significant, as are both siding (11) and attachments (12) for the structure category. In addition, all five summary measures (13–17), including the overall risk score (variable 17), are also found to have a positive and significant total impact. The mean estimated impact of 0.0512 from the (scaled) overall risk score variable suggests an approximate 5% increase in the likelihood of a structure being lost for every 100-point increase in risk, with the result coming from the combined direct and indirect impacts. For further intuition, Figure 3 depicts the distributions of overall wildfire risk scores for structures destroyed versus not destroyed within the entire community (a) as well as within the fire perimeter (b). The figure suggests a meaningful relationship in both cases, with destroyed structures on average having higher overall risk scores.

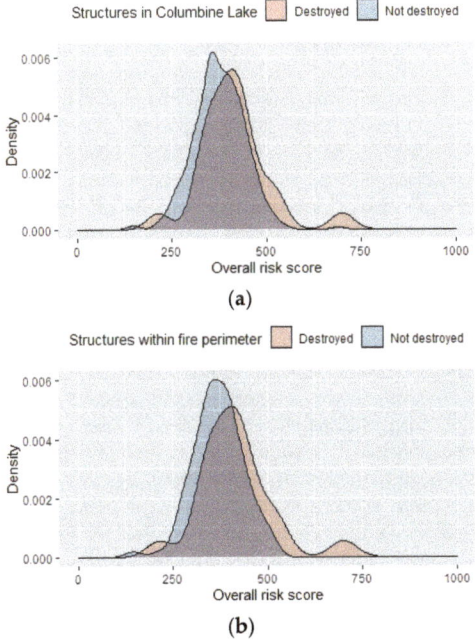

Figure 3. Comparison of the distribution of overall risk score for structures with complete assessments that were destroyed (orange) versus not destroyed (blue) in the East Troublesome Fire, for the entire community of Columbine Lake, Grand Lake, CO (**a**) and within the fire perimeter (**b**).

Finally, Tables S5 and S6 replicate the main results of Table 4 for all available indicators or for binary indicators equal to 0 for attributes rated at the lowest risk level and equal to 1 otherwise, respectively. These results do not rely on the relative weights assigned for calculating the summary risk scores. Although nuanced differences exist, these replications demonstrate the general robustness of the implied linear response to the expert-judgment score values assigned to attribute levels. In particular, the results for the binary measures for the measured attributes, shown in Table S6, are quite strong and generally analogous to the main results of Table 4.

5. Discussion

The results suggest that the WiRē RA provides strong explanatory power for whether structures in the Columbine Lake community were destroyed in the East Troublesome Fire. The overall wildfire risk score from the WiRē RA strongly relates to whether a structure was destroyed. Because the summary risk score from the WiRē RA is explicitly constructed to represent overall PLR, this suggests that PLR meaningfully describes variation in wildfire risk across homes within a community. However, it is also notable that the distributions for overall risk scores overlap substantially for structures destroyed versus not destroyed, demonstrating that although the WiRē RA data provide insights for whether structures were destroyed by the fire or not, the results are far from deterministic. Some structures with relatively very low risk scores were destroyed in the fire, while other structures with relatively high risk scores were spared, and this is true both within the fire perimeter and across the entire community.

Given the additional attribute- or category-specific results discussed above, and the fact that results pertaining to the overall risk score rely on the subjective expert judgment of relative attribute weights, we maintain that a single summary measure of overall risk does not tell the entire story, but rather, it can mask important—and actionable—information. All four category-specific summaries have significant total impacts on the likelihood that a structure was destroyed in the East Troublesome Fire, suggesting that each of these four categories represents a relevant component of PLR. This suggests numerous points of entry for reducing PLR, including common recommendations related to defensible space and structural hardening as well as access considerations. Most of the individual attributes measured in the WiRē RA relate to whether structures were destroyed or not, offering nuanced insights. For example, one of the most common recommendations for reducing wildfire risk to a home is the maintenance of a defensible space around the home; however, details of recommendations vary. Recent guidance, e.g., [53], often emphasizes the importance of the nearest zone, within approximately 5 feet of the structure. Past research from other contexts found no measurable differences in risk reduction beyond about 100 feet [22] or 130 feet [35] from the home. Here, although our main results generally find increasing risk as the distance to dense or unmaintained vegetation decreases, the results estimated with separate indicator variables suggest similar increases in risk for any measured amount of defensible space less than the maximum observation of 150 feet. In other words, our results suggest that effective defensible space possibly extends to at least 150 feet from the structure. Our results also suggest the importance of removing other combustible items, including but not limited to flammable furniture, woodpiles, and propane tanks, from the vicinity of the home. Indeed, our binary-indicator model suggests a similar increase in the likelihood of a home being destroyed due to having less than 30 feet of distance between any such items and the structure as compared to having less than 150 feet of distance from maintained vegetation.

We also find strong evidence of the interdependence of risk among structures, in which the decisions made by one's neighbors can affect the risk to one's own property [49,54]. Evidence includes the significant impact of the proximity to the nearest home attribute, which is consistent with other research that found the distance between structures to be an important component of wildfire risk to homes, e.g., [29,31]. Evidence also comes from the spatial correlations in whether structures were destroyed, demonstrated robustly across all estimated models, which suggests that burned structures tend to be clustered. Further evidence for risk interdependencies comes from the significantly estimated average indirect impacts for many attributes, which signify risk spillovers in terms of the attributes' influence on properties. This all underscores the importance of considering homes and other structures as not only *recipients* of fire, but also as *drivers* of wildfire behavior. These results also suggest important limitations for the applicability of landscape-level hazard models to a populated built environment, due in part to the practice of masking "urban" or "developed" pixels as "unburnable" terrain in the fuel models that underly such modeling [55,56].

In addition to missing heterogeneity in parcel-level hazard or the local fuel characteristics represented by the defensible space category, landscape-level models that treat all residential property the same neglect the heterogeneity in vulnerability that is driven by differences in access or structure characteristics. Our results are consistent with the expectations set by the literature reviewed above in demonstrating the value of structural hardening for reducing risk, such as having less vulnerable siding materials and avoiding combustible attachments such as decks, fences, or porches. We also find that two of the access-related attributes, driveway clearance and address visibility, have significant impacts on the likelihood that a structure was destroyed.

Because it cannot be assessed with post-fire data, the role of address visibility is worth considering further. Our models suggest that whether an address was visibly posted or not is significantly related to the likelihood that a structure was destroyed in the fire. As discussed above, it is quite plausible that address visibility supports a safe and effective response to a given property, and the strength of this result suggests value from further investigation. However, operational memory suggests that address visibility was not particularly important in the conditions of this fire, and thus this attribute might serve as a proxy for variation in risk reduction effort not otherwise reflected in the coarse measurements of the rapid assessment. For example, residents who increase their address visibility might also be more likely to remove debris from porches and gutters. Either way, this result suggests that small actions taken by residents can make a real difference in their wildfire risk. Further, these results reflect the potential importance of the complex dynamics of a wildfire event, such as the interactions between parcel-level characteristics and suppression decisions, in determining the realized wildfire risk.

Finally, decisions about wildfire risk mitigation actions are influenced by many factors that often interact in complex ways, including but not limited to risk perceptions, costs, and perceived effectiveness, e.g., [44,57–60]. Because rapid assessments represent parcel-level heterogeneity in risk within a community, rapid assessments can inform the development and implementation of programs that encourage or support these decisions. For example, an organization might decide to use rapid assessment data to prioritize properties for more time-intensive detailed site assessments or for more costly cost-share incentive programs. Of course, comprehensive parcel-scale assessments may face challenges for any given community in terms of practicality, cost-effectiveness, and coverage. Practitioners who are interested in utilizing such approaches must consider the tradeoffs involved in implementing such assessments and whether the benefits provided, in terms of the potential uses, warrant the financial and time costs to their organizations.

6. Conclusions

The WiRē RA conducted by GCWC in the summer of 2019 captured elements of the risk of a home being destroyed during the East Troublesome Fire of October 2020. Despite small sample sizes and the coarse measurements underlying the WiRē RA, the overall risk score constructed from the weighted sum of assessed attributes meaningfully depicts heterogeneity in the risk that the East Troublesome Fire presented to homes in the Columbine Lake community. Furthermore, seven out of the twelve attributes assessed in the original WiRē RA, including attributes from each of the four categories (i.e., parcel-level hazard, defensible space, access, and structure), help to explain which homes were destroyed. The results also support incorporating proximity to the nearest home into parcel-level wildfire risk assessment, as that attribute is also strongly related to the likelihood of whether a home was destroyed.

Thus, the parcel-level conditions described by the PLR conceptual model are important for understanding and communicating about wildfire risk to homes. These results are consistent with other recent works, e.g., [14,33,46], in arguing for the importance of conditions within the home ignition zone for understanding wildfire risk to homes. That said, this paper reports on an empirical analysis of a unique event; we do not claim that the attributes found to be meaningful here are the most important for reducing risks, nor that these results should supersede those found from other contexts, whether in terms of geographic, social, or fire conditions. In terms of generalization, the details of which specific attributes mattered in this case study are less important than the basic fact that some property-level attributes were found important. Since parcel-level characteristics as measured by the WiRē RA tool matter here, there is reason to believe they might matter in the next fire. These findings suggest that having this type of data for all parcels in a community could help community programs to identify where to start, or how best to prioritize resources, for reducing wildfire risk to homes and their residents.

Our results underscore that parcel-level variation is an important consideration for understanding risk to homes for homeowners, communities, fire managers, and policy makers. Indeed, our results suggest that parcel-level WiRē RA data offer insights not available from landscape-level hazard modeling outputs. In other words, landscape-level hazard information cannot replace detailed parcel-level information in terms of understanding wildfire risk to a home. This suggests the importance of working toward the greater integration of parcel-level variation in risk-related characteristics (i.e., about PLR) into existing landscape-level wildfire risk assessments. Given the robustness of the results on the distance between structures here and in the past literature, a promising first step could be more consistently including such a measure—assessed at the level of individual parcels, rather than simply as an average housing density measure—into broader-scale assessments.

The results also demonstrate the usefulness of rapidly collected, coarse data in representing a property's wildfire risk. Measurement for the WiRē RA is imperfect, based on professional judgment with at most four levels for a given attribute, and conducted from the roadside to accommodate privacy concerns and resource constraints. The overall risk score is based on subjective expert judgment pertaining to the relative weights of different attributes. It is reasonable to ask whether such imperfect data are useful for planning or prioritization; our results suggest that they can be. As our results are robust to the use of binary versus expanded indicator variables for each attribute, future research might explore the potential for efficiency gains through further simplifying current approaches to parcel-level rapid wildfire risk assessment, such as employing binary rather than three- or four-leveled indicators.

That said, the lack of relationship found here of course does not imply that a variable might not be critically important in a different context. For example, numerous studies support the importance of noncombustible roofing materials for reducing risk despite the lack of observed importance in our case study—but only two assessed structures in our dataset had wooden roofs; one of them burned down and the other did not. Our estimation approach cannot overcome the limited support for many of the possible measurement levels in the rapid assessment. Most of our estimated average effect sizes are relatively low, reflecting the fact that numerous other considerations not represented by the data here also contribute to whether a home may be destroyed. As parcel-level rapid assessment data become more widely collected, there will likely be more opportunities to investigate the effectiveness of this type of data in representing wildfire risk to homes; we hope such unfortunate intersections of past data collection and hazardous events can be seized upon to reduce the social costs of wildfires.

Overall, this study finds that resident or homeowner actions matter for reducing wildfire risk to the home. While the resident of a property might have limited capacity to change conditions underlying a hazardous burn probability or conditional flame length rating, that same resident likely can influence many of the attributes measured by the WiRē RA. One of the many influential attributes in our model—address visibility—is also an easy, low-cost attribute to change. Furthermore, action at the parcel and community scales constitutes a critical piece of the story for assessing the full wildfire risk to homes, even though landscape-level wildfire hazard assessments might not be able to capture these efforts. As such, assessed risk levels should not be taken as given at the scale of an individual property; rather, residents can reduce their own risk, and that of their neighbors, by focusing on aspects of PLR over which they have some control. We hope such information empowers wildfire risk practitioners, and the residents they serve, to take meaningful actions to reduce the wildfire risk to homes and communities.

Supplementary Materials: The following supporting information can be downloaded at: https://www.mdpi.com/article/10.3390/fire5010024/s1, Table S1: Number of structures for which attributes were marked as "unobservable" in the Wildfire Research Risk Assessment (WiRē RA); Table S2: Replication of Table 4 using full dataset with unobservable attributes for structures coded as having the riskiest rating for that attribute instead of being coded as missing and dropped from subsequent analysis; Table S3: Replication of Table 4 constrained to structures within the burn perimeter; Table S4: Comparison of ordinary versus logistic regression for all assessed attributes (jointly modeled), without considering spatial effects; Table S5: Replication of Table 4 with full indicators for the levels of each attribute of a structure, rather than using the numerical score as an implied linear measure; Table S6: Replication of Table 4 with binary indicators for each attribute of a structure, coded to 0 for the lowest-risk level and 1 for all others, rather than using the numerical score as an implied linear measure; Figure S1: Assessor Reference Guide (ARG) developed for Grand County WiRē RA.

Author Contributions: Conceptualization, analysis, visualization: J.R.M.; methodology: J.R.M., C.M.B., S.K.O., A.C.G., H.B.-S., P.A.C., J.G.; writing—original draft preparation: J.R.M., J.B.G.; writing—review and editing: all. All authors have read and agreed to the published version of the manuscript. The findings and conclusions in this article are those of the authors; they do not necessarily represent the views of the USDA and should not be construed to represent USDA agency determination or policy.

Funding: This research was funded, in part, by USDA Forest Service, State and Private Forestry. Rapid assessment data and data on destroyed structures were funded by Grand County Wildfire Council with the support of the Fire Districts of Grand County.

Data Availability Statement: Rapid assessment data and data on destroyed structures are owned by Grand County Wildfire Council (https://bewildfireready.org/, accessed on 11 February 2022) with the support of the Fire Districts of Grand County and are available upon request; other data are available from NIFC [47] as described in the text.

Conflicts of Interest: The authors declare no conflict of interest. The funders had no role in the design of the study; in the collection, analyses, or interpretation of data; in the writing of the manuscript, or in the decision to publish the results. The Wildfire Research (WiRē) Center (https://wildfireresearchcenter.org, accessed on 11 February 2022) is a nonprofit organization that works with wildfire practitioners and collaborated with Grand County Wildfire Council for data collection; the author Brenkert-Smith is a member of the Wildfire Research (WiRē) Center Board, and the authors Meldrum, Champ, and Barth are on the Wildfire Research (WiRē) Advisory Committee.

References

1. Gill, A.M.; Stephens, S.L. Scientific and social challenges for the management of fire-prone wildland–urban interfaces. *Environ. Res. Lett.* **2009**, *4*, 034014. [CrossRef]
2. Schoennagel, T.; Balch, J.K.; Brenkert-Smith, H.; Dennison, P.E.; Harvey, B.J.; Krawchuk, M.A.; Mietkiewicz, N.; Morgan, P.; Moritz, M.A.; Rasker, R.; et al. Adapt to more wildfire in western North American forests as climate changes. *Proc. Natl. Acad. Sci. USA* **2017**, *114*, 4582–4590. [CrossRef] [PubMed]
3. Moritz, M.A.; Batllori, E.; Bradstock, R.A.; Gill, A.M.; Handmer, J.; Hessburg, P.F.; Leonard, J.; McCaffrey, S.; Odion, D.C.; Schoennagel, T.; et al. Learning to coexist with wildfire. *Nature* **2014**, *515*, 58–66. [CrossRef]
4. Radeloff, V.C.; Helmers, D.P.; Kramer, H.A.; Mockrin, M.H.; Alexandre, P.M.; Bar-Massada, A.; Butsic, V.; Hawbaker, T.J.; Martinuzzi, S.; Syphard, A.D.; et al. Rapid growth of the US wildland-urban interface raises wildfire risk. *Proc. Natl. Acad. Sci. USA* **2018**, *115*, 3314–3319. [CrossRef] [PubMed]
5. Caggiano, M.D.; Hawbaker, T.J.; Gannon, B.M.; Hoffman, C.M. Building Loss in WUI Disasters: Evaluating the Core Components of the Wildland–Urban Interface Definition. *Fire* **2020**, *3*, 73. [CrossRef]
6. Syphard, A.D.; Bar Massada, A.; Butsic, V.; Keeley, J.E. Land use planning and wildfire: Development policies influence future probability of housing loss. *PLoS ONE* **2013**, *8*, e71708. [CrossRef]
7. Westerling, A.L.; Hidalgo, H.G.; Cayan, D.R.; Swetnam, T.W. Warming and earlier spring increase western U.S. forest wildfire activity. *Science* **2006**, *313*, 940–943. [CrossRef]
8. Abatzoglou, J.T.; Williams, A.P. Impact of anthropogenic climate change on wildfire across western US forests. *Proc. Natl. Acad. Sci. USA* **2016**, *113*, 11770–11775. [CrossRef]
9. Finney, M.A. The challenge of quantitative risk analysis for wildland fire. *For. Ecol. Manag.* **2005**, *211*, 97–108. [CrossRef]
10. Thompson, M.P.; Calkin, D.E. Uncertainty and risk in wildland fire management: A review. *J. Environ. Manag.* **2011**, *92*, 1895–1909. [CrossRef]

11. Scott, J.H.; Thompson, M.P.; Calkin, D.E. *A Wildfire Risk Assessment Framework for Land and Resource Management*; U.S. Department of Agriculture, Forest Service, Rocky Mountain Research Station: Fort Collins, CO, USA, 2013; p. 83.
12. *Sendai Framework for Disaster Risk Reduction 2015–2030*; United Nations Office for Disaster Risk Reduction: Sendai, Japan, 2015; p. 32.
13. Ludwig, K.A.; Ramsey, D.W.; Wood, N.J.; Pennaz, A.B.; Godt, J.W.; Plant, N.G.; Luco, N.; Koenig, T.A.; Hudnut, K.W.; Davis, D.K.; et al. *Science for a Risky World—A U.S. Geological Survey Plan for Risk Research and Applications*; U.S. Geological Survey: Reston, VA, USA, 2018; p. 57.
14. Caton, S.E.; Hakes, R.S.P.; Gorham, D.J.; Zhou, A.; Gollner, M.J. Review of Pathways for Building Fire Spread in the Wildland Urban Interface Part I: Exposure Conditions. *Fire Technol.* **2016**, *53*, 429–473. [CrossRef]
15. Syphard, A.D.; Rustigian-Romsos, H.; Keeley, J.E. Multiple-Scale Relationships between Vegetation, the Wildland–Urban Interface, and Structure Loss to Wildfire in California. *Fire* **2021**, *4*, 12. [CrossRef]
16. FAC-LN. Promoting Fire Adapted Communities through Property Assessments: Data & Tools. *Fire Adapted Community Learning Network: A Quick Guide for Community Leaders, Number 2.1*. 2015, p. 2. Available online: https://fireadaptednetwork.org/wp-content/uploads/2015/12/FACQuickGuide2.1.pdf (accessed on 11 February 2022).
17. Hakes, R.S.P.; Caton, S.E.; Gorham, D.J.; Gollner, M.J. A Review of Pathways for Building Fire Spread in the Wildland Urban Interface Part II: Response of Components and Systems and Mitigation Strategies in the United States. *Fire Technol.* **2016**, *53*, 475–515. [CrossRef]
18. Quarles, S.L.; Valachovic, Y.; Nakamura, G.M.; Nader, G.A.; De Lasaux, M.J. *Home Survival in Wildfire-Prone Areas: Building Materials and Design Considerations*; Publication 8393; University of California Agriculture and Natural Resources: St. Davis, CA, USA, 2010; 22p. [CrossRef]
19. Champ, P.; Barth, C.; Brenkert-Smith, H.; Falk, L.; Gomez, J.; Meldrum, J. Putting people first: Using social science to reduce risk. *Wildfire Magazine*. International Association of Wildland Fire, Missoula, MT, USA. 2021, pp. 30–34. Available online: https://www.iawfonline.org/article/putting-people-first-using-social-science-to-reduce-risk/ (accessed on 11 February 2022).
20. Maranghides, A.; McNamara, D.; Mell, W.; Trook, J.; Toman, B. *A Case Study of a Community Affected by the Witch and Guejito Fires: Report# 2: Evaluating the Effects of Hazard Mitigation Actions on Structure Ignitions*; National Institute of Standards and Technology, US Department of Commerce and US Forest Service: Gaithersburg, MD, USA, 2013.
21. Duff, T.J.; Penman, T.D. Determining the likelihood of asset destruction during wildfires: Modelling house destruction with fire simulator outputs and local-scale landscape properties. *Saf. Sci.* **2021**, *139*, 105196. [CrossRef]
22. Syphard, A.D.; Brennan, T.J.; Keeley, J.E. The role of defensible space for residential structure protection during wildfires. *Int. J. Wildland Fire* **2014**, *23*, 1165–1175. [CrossRef]
23. Dupuy, J. Slope and Fuel Load Effects on Fire Behavior: Laboratory Experiments in Pine Needles Fuel Beds. *Int. J. Wildland Fire* **1995**, *5*, 153–164. [CrossRef]
24. Rodrigues, A.; Ribeiro, C.; Raposo, J.; Viegas, D.X.; André, J. Effect of Canyons on a Fire Propagating Laterally Over Slopes. *Front. Mech. Eng.* **2019**, *5*, 41. [CrossRef]
25. Viegas, D.X.; Simeoni, A. Eruptive Behaviour of Forest Fires. *Fire Technol.* **2011**, *47*, 303–320. [CrossRef]
26. Alexandre, P.M.; Stewart, S.I.; Keuler, N.S.; Clayton, M.K.; Mockrin, M.H.; Bar-Massada, A.; Syphard, A.D.; Radeloff, V.C. Factors related to building loss due to wildfires in the conterminous United States. *Ecol. Appl.* **2016**, *26*, 2323–2338. [CrossRef]
27. Syphard, A.D.; Brennan, T.J.; Keeley, J.E. The importance of building construction materials relative to other factors affecting structure survival during wildfire. *Int. J. Disaster Risk Reduct.* **2017**, *21*, 140–147. [CrossRef]
28. Graham, R.T. *Hayman Fire Case Study*; US Department of Agriculture, Forest Service, Rocky Mountain Research Station: Fort Collins, CO, USA, 2003.
29. Penman, S.H.; Price, O.F.; Penman, T.D.; Bradstock, R.A. The role of defensible space on the likelihood of house impact from wildfires in forested landscapes of south eastern Australia. *Int. J. Wildland Fire* **2019**, *28*, 4–14. [CrossRef]
30. Knapp, E.E.; Valachovic, Y.S.; Quarles, S.L.; Johnson, N.G. Housing arrangement and vegetation factors associated with single-family home survival in the 2018 Camp Fire, California. *Fire Ecol.* **2021**, *17*, 25. [CrossRef]
31. Cohen, J.D.; Stratton, R.D. *Home Destruction Examination: Grass Valley Fire, Lake Arrowhead, California*; Technol Paper R5-TP-026b; US Department of Agriculture, Forest Service, Pacific Southwest Region (Region 5): Vallejo, CA, USA, 2008; 26p.
32. Kolden, C.A.; Henson, C. A socio-ecological approach to mitigating wildfire vulnerability in the wildland urban interface: A case study from the 2017 Thomas fire. *Fire* **2019**, *2*, 9. [CrossRef]
33. Cohen, J.D. Preventing disaster: Home ignitability in the wildland-urban interface. *J. For.* **2000**, *98*, 15–21.
34. Alexander, M.E.; Stocks, B.J.; Wotton, B.M.; Flannigan, M.D.; Todd, J.B. The international crown fire modelling experiment: An overview and progress report. In Proceedings of the Second Symposium on Fire and Forest Meteorology, Phoenix, AZ, USA, 11–16 January 1998; pp. 20–23.
35. Gibbons, P.; Van Bommel, L.; Gill, A.M.; Cary, G.J.; Driscoll, D.A.; Bradstock, R.A.; Knight, E.; Moritz, M.A.; Stephens, S.L.; Lindenmayer, D.B. Land management practices associated with house loss in wildfires. *PLoS ONE* **2012**, *7*, e29212. [CrossRef] [PubMed]
36. Price, O.F.; Whittaker, J.; Gibbons, P.; Bradstock, R. Comprehensive Examination of the Determinants of Damage to Houses in Two Wildfires in Eastern Australia in 2013. *Fire* **2021**, *4*, 44. [CrossRef]
37. Cova, T.J.; Theobald, D.M.; Norman, J.B.; Siebeneck, L.K. Mapping wildfire evacuation vulnerability in the western US: The limits of infrastructure. *GeoJournal* **2013**, *78*, 273–285. [CrossRef]

38. Cova, T.J. Public safety in the urban–wildland interface: Should fire-prone communities have a maximum occupancy? *Nat. Hazards Rev.* **2005**, *6*, 99–108. [CrossRef]
39. McGee, T.K.; McFarlane, B.L.; Varghese, J. An examination of the influence of hazard experience on wildfire risk perceptions and adoption of mitigation measures. *Soc. Nat. Resour.* **2009**, *22*, 308–323. [CrossRef]
40. Nelson, K.C.; Monroe, M.C.; Johnson, J.F. The Look of the Land: Homeowner Landscape Management and Wildfire Preparedness in Minnesota and Florida. *Soc. Nat. Resour.* **2005**, *18*, 321–336. [CrossRef]
41. Li, D.; Cova, T.J.; Dennison, P.E.; Wan, N.; Nguyen, Q.C.; Siebeneck, L.K. Why do we need a national address point database to improve wildfire public safety in the U.S.? *Int. J. Disaster Risk Reduct.* **2019**, *39*, 101237. [CrossRef]
42. Westhaver, A. *Why Some Homes Survived: Learning from the Fort Mcmurray Wildland/Urban Interface Fire Disaster*; Institute for Catastrophic Loss Reduction: Toronto, ON, Canada, 2017.
43. Quarles, S.L.; Standoher-Alfano, C.D. *Wildfire Research: Ignition Potential of Decks Subjected to an Ember Exposure*; Insurance Institute for Business & Home Safety: Richburg, SC, USA, 2018; p. 39.
44. Quarles, S.; Pohl, K. Costs of WUI Codes and Standards for New Construction. In *Encyclopedia of Wildfires and Wildland-Urban Interface (WUI) Fires*; Manzello, S.L., Ed.; Springer International Publishing: Cham, Switzerland, 2020; pp. 1–11.
45. Benjamini, Y.; Hochberg, Y. Controlling the False Discovery Rate: A Practical and Powerful Approach to Multiple Testing. *J. R. Stat. Soc. Ser. B* **1995**, *57*, 289–300. [CrossRef]
46. Calkin, D.E.; Cohen, J.D.; Finney, M.A.; Thompson, M.P. How risk management can prevent future wildfire disasters in the wildland-urban interface. *Proc. Natl. Acad. Sci. USA* **2014**, *111*, 746–751. [CrossRef] [PubMed]
47. InciWeb. East Troublesome Fire Information. Available online: https://inciweb.nwcg.gov/incient/7242 (accessed on 12 October 2020).
48. NIFC. Interagency Fire Perimeter History—All Years. National Interagency Fire Center (NIFC). 2021. Available online: https://data-nifc.opendata.arcgis.com/datasets/nifc::interagency-fire-perimeter-history-all-years/about (accessed on 11 February 2022).
49. Warziniack, T.; Champ, P.; Meldrum, J.; Brenkert-Smith, H.; Barth, C.M.; Falk, L.C. Responding to Risky Neighbors: Testing for Spatial Spillover Effects for Defensible Space in a Fire-Prone WUI Community. *Environ. Resour. Econ.* **2018**, *73*, 1023–1047. [CrossRef]
50. Anselin, L. *Spatial Econometrics: Methods and Models*; Kluwer Academic Publishers: Boston, MA, USA, 1988; p. 284.
51. Wooldridge, J.M. *Econometric Analysis of Cross Section and Panel Data*; MIT Press: Cambridge, MA, USA, 2002.
52. LeSage, J.; Pace, R.K. *Introduction to Spatial Econometrics*; CRC Press: New York, NY, USA, 2009.
53. CSFS. *The Home Ignition Zone: A Guide to Preparing Your Home for Wildfire and Creating Defensible Space*; Colorado State University: Fort Collins, CO, USA, 2021; p. 15.
54. Butry, D.; Donovan, G.H. Protect thy neighbor: Investigating the spatial externalities of community wildfire hazard mitigation. *For. Sci.* **2008**, *54*, 417–428.
55. Scott, J.H.; Burgan, R.E. *Standard Fire Behavior Fuel Models: A Comprehensive Set for Use with Rothermel's Surface Fire Spread Model*; U.S. Department of Agriculture, Forest Service, Rocky Mountain Research Station: Fort Collins, CO, USA, 2005; p. 72.
56. LANDFIRE. LANDFIRE 1.4.0 Scott and Burgan Fire Behavior Fuel Models (FBMF40) and Existing Vegetation Type (EVT) layers. 2017. Available online: https://landfire.gov/ (accessed on 11 February 2022).
57. Sisante, A.M.; Taylor, M.H.; Rollins, K.S. Understanding homeowners' decisions to mitigate wildfire risk and create defensible space. *Int. J. Wildland Fire* **2019**, *28*, 901. [CrossRef]
58. Meldrum, J.R.; Brenkert-Smith, H.; Champ, P.; Gomez, J.; Falk, L.; Barth, C. Interactions between resident risk perceptions and wildfire risk mitigation: Evidence from simultaneous equations modeling. *Fire* **2019**, *2*, 46. [CrossRef]
59. Paveglio, T.B.; Stasiewicz, A.M.; Edgeley, C.M. Understanding support for regulatory approaches to wildfire management and performance of property mitigations on private lands. *Land Use Policy* **2021**, *100*, 104893. [CrossRef]
60. Ghasemi, B.; Kyle, G.T.; Absher, J.D. An examination of the social-psychological drivers of homeowner wildfire mitigation. *J. Environ. Psychol.* **2020**, *70*, 101442. [CrossRef]

 *fire*

Article

Mitigating Source Water Risks with Improved Wildfire Containment

Benjamin M. Gannon [1,2,*], Yu Wei [1] and Matthew P. Thompson [3]

[1] Department of Forest and Rangeland Stewardship, Colorado State University, Fort Collins, CO 80523, USA; yu.wei@colostate.edu
[2] Colorado Forest Restoration Institute, Colorado State University, Fort Collins, CO 80523, USA
[3] Rocky Mountain Research Station, USDA Forest Service, Fort Collins, CO 80526, USA; matthew.p.thompson@usda.gov
* Correspondence: benjamin.gannon@colostate.edu

Received: 21 July 2020; Accepted: 20 August 2020; Published: 21 August 2020

Abstract: In many fire-prone watersheds, wildfire threatens surface drinking water sources with eroded contaminants. We evaluated the potential to mitigate the risk of degraded water quality by limiting fire sizes and contaminant loads with a containment network of manager-developed Potential fire Operational Delineations (PODs) using wildfire risk transmission methods to partition the effects of stochastically simulated wildfires to within and out of POD burning. We assessed water impacts with two metrics—total sediment load and frequency of exceeding turbidity limits for treatment—using a linked fire-erosion-sediment transport model. We found that improved fire containment could reduce wildfire risk to the water source by 13.0 to 55.3% depending on impact measure and post-fire rainfall. Containment based on PODs had greater potential in our study system to reduce total sediment load than it did to avoid degraded water quality. After containment, most turbidity exceedances originated from less than 20% of the PODs, suggesting strategic investments to further compartmentalize these areas could improve the effectiveness of the containment network. Similarly, risk transmission varied across the POD boundaries, indicating that efforts to increase containment probability with fuels reduction would have a disproportionate effect if prioritized along high transmission boundaries.

Keywords: water supply; erosion; wildfire containment; Potential fire Operational Delineations; Monte Carlo simulation; transmission risk

1. Introduction

Improved wildfire containment is an attractive strategy to mitigate the risk of degrading water quality beyond limits for treatment because of the potential to limit fire sizes and impacts to tolerable levels without the need to completely exclude fire from the landscape. Recent efforts to make containment planning more proactive, focus on zoning the landscape into fire management units called Potential fire Operational Delineations (PODs) using existing high probability control features such as roads, rivers, and fuel transitions [1,2]. Beyond the inherent value of engaging managers in the process to identify and critique potential control features, the resulting POD areas become relevant spatial units for pre-fire analysis of endogenous and transmitted wildfire risk to inform response strategies that are appropriate for the predicted direction and magnitude of fire effects to water supplies and other natural resources and human assets [2]. While there has been substantial progress engaging managers in the bottom up approach to develop and employ PODs and their associated response strategies [2–8], less attention has been paid to evaluating the risk mitigation effectiveness of containing wildfire within these units and what functional improvements should be made to the size and spatial arrangement of the containers to maximize their protection benefit for water supplies and other values, such as wildlife, that depend on the scale of fire activity.

Wildfire is often harmful to water quality because reductions in surface cover and infiltration cause increases in surface runoff and erosion that can mobilize and transport contaminants into surface drinking water sources [9–12]. While the specific contaminants and concentrations of concern may vary by watershed and water system [12–14], water quality degradation generally becomes problematic when large quantities of sediment are mobilized by intense rainfall causing contaminant concentrations to exceed thresholds for effective water treatment (e.g., [15]). Post-fire sediment loads are influenced by fire size and burn severity, topography, soil properties, and rainfall intensity [10,16,17]. Previous efforts to account for fire effects on watersheds and water supplies account for some of these factors [18,19], but the use of relative fire effects measures makes it difficult to evaluate whether a given fire will degrade water quality. This shortcoming has been addressed in recent years with increasing use of spatially explicit erosion and sediment transport models to make quantitative predictions of sediment yield from modeled wildfires (e.g., [20–24]). Sediment yield models have been widely used to examine the risk mitigation effectiveness of area-wide fuel treatments meant to reduce burn severity [23–26] but they have not yet been used to evaluate the performance of fire containment strategies to reduce area burned.

Some water systems have discrete features, such as terminal reservoirs, that could be targeted for protection within a single POD, but many municipal watersheds in the western USA are hundreds to thousands of square kilometers in size and therefore require some level of internal compartmentalization to protect water supplies. In theory, the size and spatial arrangement of PODs could be designed to mitigate the risk of water quality degradation by both containing fires with potential for large growth and subsequent contaminant loads near their ignition sources and ensuring that within-POD burning does not result in adverse consequence. Managers consider both values at risk and presence of control features when delineating PODs, which often results in smaller PODs near developed areas and larger PODs in the backcountry [2,4]. However, it is not clear that the size and configuration of manager-delineated PODs will reduce risk of wildfire-related water quality degradation. Several attempts have been made to automate the processes of identifying suitable control features and aggregating them into PODs [27,28] using roads, streams, watershed boundaries, and spatial models of suppression difficulty and potential for control [29–31], but data-driven approaches have yet to inform the desired size and spatial configuration of PODs to mitigate a particular risk.

Recognizing the importance of fire size, location, and burn severity for watershed response, several previous studies have employed Monte Carlo wildfire simulation to characterize watershed exposure and water supply risk [19,32–34]. Their results suggest that most risk to water supplies is associated with a small subset of total fire activity. Moreover, the source locations of damaging wildfires tend to cluster in certain parts of the landscape, which implies containment benefits will depend strongly on location. Simulated fire ignition locations and fire extents can be intersected with relevant management units to partition fire impacts from burning within the unit of origin and transmission to the surrounding landscape [2,35,36]. Analyzing risk transmission across a network of PODs could help to identify locations with high source risk that would benefit from investment in activities to improve containment probability, such as roadside fuels reduction. Areas with fuels conducive to fast fire spread tend to transmit the most fire [36], which will result in high water supply risk when adjacent areas have high erosion potential and/or short transport paths to water supplies. Analysis of water supply risk from self-burning could also identify high risk PODs that would benefit from further compartmentalization.

The goal of this study is to provide a proof of concept model to evaluate the effectiveness of a containment network at mitigating risk of source water quality degradation. The general approach should also be relevant for assessing risk to other resources that depend on disturbance size. We utilized Monte Carlo wildfire simulation, erosion, and sediment transport modeling to quantify the potential water supply impacts from a set of simulated wildfires with and without containment. We analyzed risk and risk mitigation with two measures of water supply impact—total sediment load and frequency of exceeding turbidity limits for treatment—to highlight how considering the scale-dependent effects of wildfire changes the perceived mitigation value of fire containment. Risk transmission was analyzed to identify possible improvements to the containment network with measures of transmitted risk highlighting those PODs and POD boundaries that could benefit from activities to improve containment probability and measures of self-burning indicating areas in need of further compartmentalization.

2. Materials and Methods

2.1. Evaluation Framework

The evaluation framework was designed to contrast the water quality impacts of uncontained wildfires and wildfires contained within the POD of origin in terms of total sediment load and average post-storm suspended sediment (Figure 1). Total sediment load is similar to the commonly used net value change measures in risk assessment [37,38] insomuch as more is interpreted as bad and any marginal reduction decreases risk. However, using change in total sediment load as a measure of risk has the potential to falsely assign mitigation benefit to containment when either the load from the uncontained wildfire is already below a meaningful threshold of water quality degradation or containment reduces erosion but the resulting load is still above the treatment threshold. Average post-storm suspended sediment concentration is used here to estimate whether fires will degrade water quality beyond limits for water treatment and whether degradation outcomes change with containment. This measure of risk better approximates the threshold-dependent nature of water quality degradation owing to the size of the receiving waterbody and the water system sensitivity to contaminants.

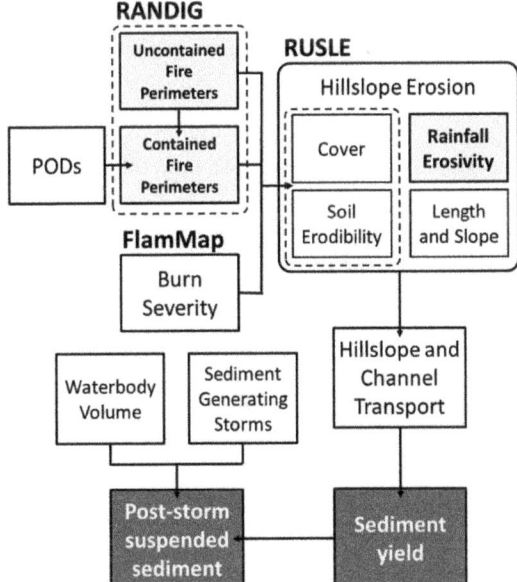

Figure 1. The evaluation framework focuses on total sediment yield and average post-storm suspended sediment as measures of water quality degradation risk. Variable inputs are in light grey. Stochastically simulated wildfire perimeters were combined with estimates of burn severity to model post-fire erosion and sediment transport to the water supply both with and without containment. Sediment yield was converted to average post-storm suspended sediment concentration using the receiving waterbody volume and the annual frequency of sediment generating storms.

Our evaluation framework focuses on the key uncertainties in wildfire-water quality degradation risk related to the extent of the watershed burned and post-fire rainfall (Figure 1). As further described in following sections, many plausible wildfire perimeters were simulated with the Monte Carlo ignition and spread model RANDIG [35,39], which were then clipped to their POD of origin to approximate a strategy of improved containment. Post-fire erosion was then simulated for each perimeter using crown fire activity predicted with FlamMap 5.0 [40] as a proxy for burn severity to modify the cover and soil variables in the Revised Universal Soil Loss Equation (RUSLE) [41]. We accounted for uncertainty in post-fire rainfall by modeling erosion for three rainfall scenarios ranging from common to extreme. We estimated annual sediment loads to the water supply based on the predicted proportion of sediment transported off hillslopes and through channels using Sediment Delivery Ratio (SDR) models [42,43]. Post-storm suspended sediment concentrations were estimated by assuming average storm sediment loads are diluted in the mean daily flow volume of the river during the May to October thunderstorm season, which is associated with most post-fire erosion and water quality degradation in the study region [15,16]. All analyses were completed with R version 3.5.3 [44] except where noted otherwise.

2.2. Study Area

The study area encompasses 3021 km^2 of the Front Range Mountains in Colorado, USA (Figure 2). The Front Range has a history of large and severe fires that have caused extreme erosion, reservoir sedimentation, and water quality degradation [15,45–48]. The names of the focal municipal water supply and other geographic features within the study area are withheld for security reasons. The extent of the study area was defined to include the contributing area to a municipal pipeline diversion (1254 km^2) and a network of PODs developed by the local National Forest and their partnering state and local fire management agencies (an additional 1767 km^2). PODs that intersected a 5 km buffer around the watershed were included to analyze fires that spread into the watershed from nearby areas. Elevation ranges from 1559 to 4135 m above sea level across the study area. The climate is continental with warm dry summers and cold winters. Most erosion in this region results from intense rainstorms during the summer and early fall [16,49]. The study area is primarily forest (71.7%, most of which is dominated by conifers) and the remainder is a mix of shrubland (9.0%), sparsely vegetated alpine (8.9%), and grassland (8.7%) [50]. Land ownership is split between the USDA Forest Service (55.3%), private (18.8%), National Park Service (18.1%), local government (6.4%), and state (1.4%).

2.3. Potential Fire Operational Delineations

PODs were developed by fire and resource management specialists from the local National Forest and external fire management partners from other federal, state, and local agencies. The PODs range in size from 502 to 23,672 ha with a mean of 4316 ha and a median of 3516 ha. The PODs tend to be smallest near human settlements due to both the increased presence of control features and greater need for fire containment around communities. PODs larger than 10,000 ha are clustered in the higher elevation, western portion of the of the study area where much of the land is publicly owned and the transportation network is sparse. PODs also tend to be large along the major river canyon that runs west to east across the study area (Figure 2) due to limited presence of high probability control features other than the river and highway in the canyon bottom.

The rugged topography, rocky soils, and dense forests of the study area are major constraints on firefighter and equipment accessibility and operability. Accordingly, managers preferentially chose roads as the control features to bound PODs; of the 1386 km of POD edge, 985 km are roads (71.0%), 167 km are trails (12.0%), 150 km are ridges (10.8%), 46 km are streams (3.3%), and the remaining 40 km are fuel transitions, lakes/reservoirs, or lacking defined control features (2.9%). Many of the trails and ridges selected as control features are in barren or sparsely vegetated areas of the alpine, so roads make up an even larger proportion of POD edges in the fuel types where wildfire transmission is a concern. Numerous observational studies have documented that roads benefit fire control by serving as hard fire breaks that either stop fires passively or in combination with suppression firing or holding activities [30,51–53]. The frequent use of roads in this POD network suggests containment probability should be high along most boundaries under low to moderate fire weather and many boundaries have potential for containment under more extreme conditions with well-coordinated suppression tactics.

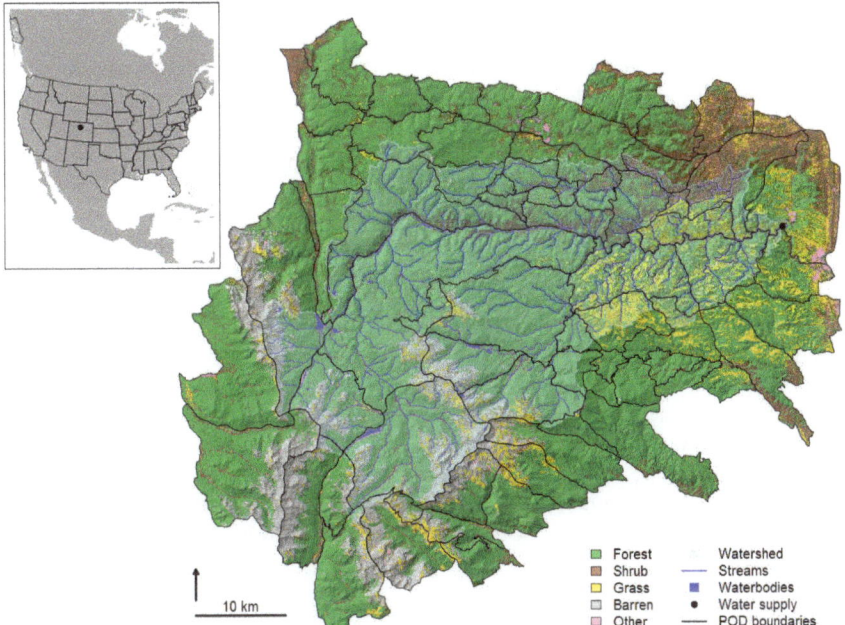

Figure 2. Map of the study area featuring the focal watershed and PODs that intersect a five km buffer around the watershed. Landcover is from LANDFIRE [50]. Barren is sparsely vegetated alpine. The inset maps the location of the study area in the USA.

2.4. Fire Occurrence

We used the Monte Carlo fire simulation program RANDIG, which is a command-line version of the FlamMap minimum travel time module [39], to model a plausible set of 5000 large fire growth events across the study area. The inputs to RANDIG include raster surfaces of fuels, topography, and ignition density, and a set of fire scenarios describing the fuel moisture, wind speed, wind direction, spot probability, and burn duration for the simulations and their probabilities of occurrence. The intent of our model parameterization is to approximate the distribution of potential area burned during the initial growth period of large fires owing to variation in wind direction and wind speed. We focused on the early growth period of fires to align with the desire to contain most fires before they leave the POD of origin. Modeling fire growth over longer periods would increase fire size and thus the avoided area burned and water quality impacts but would also introduce greater uncertainty about final fire extent as more potential containment features are encountered and weather conditions are likely to moderate.

Raster fuels and topography data representing landscape conditions circa 2014 were acquired from LANDFIRE [50] including canopy cover, canopy bulk density, canopy base height, canopy height, surface fire behavior fuel model [54], elevation, slope, and aspect. Fuels were adjusted in lodgepole pine (*Pinus contorta* var. *latifolia*) forests by lowering the canopy base height by 20% and changing the fire behavior fuel model to high load conifer litter (TL5 from [54]) to better match recent observations of extreme fire behavior in these forests [55]. The other spatial input is a raster surface of ignition density, which influences the relative probability of fire ignition across the modeling domain. Spatial point locations of historical fires from Short [56] were generalized into a raster surface of ignition density using a kernel density function with a search distance of 10 km in ArcGIS 10.3 [57].

Fuel moisture, wind speed, and wind direction for the fire scenarios (Table 1) were informed by data from a Remote Automated Weather Station [58] located in the northern half of the study area at 2500 m above sea level. Fuel moisture and wind speed percentiles were calculated with FireFamilyPlus 4.1 [59] and wind speed was converted from a 10-min to 1-min average based on Crosby and Chandler [60]. Most large fires in this region occur in early summer during drought years or in the fall when fuel moisture is extremely low. These conditions were approximated using the historical 3rd percentile fire season (1 April–31 October) fuel moistures, which are 2, 3, 6, 30, and 60 percent for the 1-h, 10-h, 100-h, herbaceous, and woody fuels, respectively. Fuel moisture was held constant across all wind scenarios because it exhibits little meaningful variation below the 10th percentile. Wind scenarios were designed to approximate the joint probability distribution of wind speeds and directions that are problematic for fire growth. The 50th, 90th, and 97th percentiles of 1-min average wind speeds are 19.3, 33.8, and 43.5 kph (at 6 m). We generalized these into three levels of wind speed (16.1, 32.2, and 48.3 kph) and their associated spotting probabilities (0.02, 0.05, and 0.10) that we assigned relative probabilities of occurrence of 0.90, 0.07, and 0.03. Previous large fires in this landscape are associated with strong westerly winds and our analysis of the historical record found that 74.1% of all winds greater than or equal to 16.1 kph were from the northwest, west, or southwest, which have relative probabilities of occurrence equal to 0.29, 0.48, and 0.23. We combined the three levels of wind speed and spotting probabilities with the three variations of wind direction into a total of nine fire scenarios (Table 1). Burn duration was set to four hours for all scenarios, which was determined by incrementally adjusting burn duration in 30 min time steps until the largest simulated fire was within ±5% of 20,000 ha, which we judge as a reasonable upper bound for fire size during a single burn period in this landscape based on other fires in the region [46].

Table 1. Fire scenarios used to simulate fires in RANDIG. Burn duration was set to 240 min and fuel moisture was held constant at the 3rd percentile of the historical record.

Scenario	Wind Speed (kph at 6 m)	Direction (deg)	Spot Probability	Scenario Probability
1	16.1	225	0.02	0.259
2	16.1	270	0.02	0.431
3	16.1	315	0.02	0.210
4	32.2	225	0.05	0.020
5	32.2	270	0.05	0.034
6	32.2	315	0.05	0.016
7	48.3	225	0.1	0.009
8	48.3	270	0.1	0.014
9	48.3	315	0.1	0.007

2.5. Fire Behavior and Severity

Crown fire activity [61] was modeled as a proxy for burn severity with FlamMap 5.0 [40] by mapping surface fire, passive crown fire, and active crown fire to low, moderate, and high severity, respectively. Crown fire activity is commonly used to estimate burn severity for watershed modeling [24,33,62] because it captures the trend of increasing fire intensity along the gradient of surface to active crown fire behavior. Fuel moisture was set to the same 3rd percentile fuel moisture described in the fire occurrence section. The same topography and modified fuels rasters were also used as the landscape inputs to FlamMap. To simplify the analysis, we modeled burn severity for the middle wind speed scenario (32.2 kph at 6 m) and used the wind blowing uphill option to represent a consistent worst-case scenario for all aspects.

2.6. Post-Fire Watershed Response

Post-fire erosion and sediment transport to the water diversion was predicted with a system of coupled hillslope erosion, hillslope sediment transport, and channel sediment transport models

(Figure 1) that has been calibrated to make reasonable predictions of post-fire sediment yields within the study region [24]. The NHDPlus raster and watershed network products [63] were used to represent the topological connections between upland sediment sources and the water diversion point via sub-catchment drainage paths to the flowline network and the series of intervening flowlines between each catchment and the diversion. First, gross hillslope erosion was modeled for each fire with a raster Geographic Information System implementation [64] of RUSLE [41]. Sediment transport to streams was predicted using an empirical model of post-fire hillslope sediment delivery ratio from the western USA [42] to estimate the proportion of sediment generated in each pixel that makes it to the flowline network. Third, the total sediment from each catchment was routed down the flowline network to the diversion point using a simple model of channel sediment delivery ratio [43] adapted for the channel types in the study area.

2.6.1. Hillslope Erosion

RUSLE predicts gross erosion (Mg ha^{-1} year^{-1}) as the product of factors for rainfall erosivity (R), soil erodibility (K), length and slope (LS), cover (C), and support practices (P) [41]. Rainfall erosivity is calculated as the product of storm maximum rainfall intensity and kinetic energy per unit area [41]. First year post-fire erosion was modeled at three levels of May to October rainfall erosivity—403, 887, 5168 MJ mm ha^{-1} h^{-1}—representing the 2, 10, and 100-year recurrence interval rainfall erosivity (hereafter "rainfall erosivity") for the regional climate [65,66]. The May through October period was selected because most post-fire erosion in this region occurs in response to high intensity summer rainfall [16]. LS was calculated from a 30 m resolution digital elevation model [63] following the methods of Winchell et al. [67] with a maximum limit on flow accumulation of 0.9-ha imposed to approximate the original hillslope length guidance in Renard et al. [41]. Baseline K came from the Soil Survey Geographic Database where available and the State Soil Geographic Database to fill missing data [68]. Post-fire erosion was simulated by modifying the K and C factors based on wildfire extent and burn severity [24,69]. No support practices were considered to model the unmitigated erosion hazard. Baseline erosion is not a major concern for water quality, so we focused our assessment on the post-fire increase in erosion. First-year post-fire increase in erosion (A) was calculated with Equation (1) for each level of rainfall erosivity.

$$A = R \times LS \times [(K_b \times C_b) - (K \times C)], \tag{1}$$

The subscript b indicates the burned condition for K and C factors. We limited hillslope erosion predictions to 100 Mg ha^{-1} year^{-1} based on the maximum observed values reported in the study region [49].

2.6.2. Hillslope Sediment Transport

An empirical model of post-wildfire hillslope sediment delivery ratio (hSDR) from the western USA [42] was used to estimate the proportion of sediment generated in each pixel that makes it to the stream network. The NHDPlus flowlines were first extended to include all pixels with a contributing area greater than 10.8 ha [70] to better approximate the extent of the post-fire channel network. Post-fire hSDR was then estimated with the annual length ratio model from Wagenbrenner and Robichaud [42]. We applied this model to predict hSDR as a function of the flow path length from each pixel to the nearest stream channel as the "catchment length" and the flow path length across the pixel as the "plot length" (Equation (2)). Flow path length to the nearest channel was calculated from a 30 m digital elevation model [63] in ArcGIS 10.3 [57]. We doubled the predicted hSDR to account for under-sampling of suspended sediment in the model training data and to roughly calibrate our net sediment yield predictions to the small catchment yields from the Hayman Fire in Colorado [42]. This increased the maximum hSDR from 0.27 to 0.54 for areas near streams and it increased the minimum hSDR from 0.05 to 0.10 for locations furthest from streams. We later compare our modeled gross and net hillslope sediment yields to relevant field observations in the discussion to demonstrate that this assumption is reasonable. Channel pixels were assigned hSDR of 1.

$$\log(hSDR) = -0.56 - 0.0094 \times (\text{flow path length to channel/flow path length across pixel}), \quad (2)$$

The first-year mass of sediment (Mg) delivered from a catchment to the stream network (TS) was calculated as the sumproduct of the post-fire hillslope erosion (A), the pixel area, and hSDR for all burned pixels (N) in the catchment (Equation (3)).

$$TS = SUM(A_i \times 0.09 \text{ ha/pixel} \times hSDR_i)|i = 1 \text{ to } i = N, \quad (3)$$

2.6.3. Channel Sediment Transport

Sediment was routed through the NHDPlus flowline network to the diversion by adapting the channel sediment delivery ratio (cSDR) model of Frickel et al. [43] to the channel types in the study watershed [24]. In montane streams of this region, sediment retention is generally highest in low order channels because of high roughness and limited transport capacity and very low in the high order channels with high transport capacity [45]. Observations of post-fire sediment transport in a similar watershed in Wyoming suggest transport of fine sediments in suspension should be very efficient in high order channels even during base flow conditions [71]. These trends are approximated in our model by assigning cSDRs of 0.75, 0.80, 0.85 and 0.95 per 10 km of stream length to 1st, 2nd, 3rd, and 4th or higher-order streams, respectively. Sediment retention in lakes and reservoirs was accounted for by assigning as a cSDR of 0.05 to the terminal flowline in each waterbody. The annual mass of fire-related sediment (Mg) delivered to the water diversion (TD) was calculated as the sum of sediment delivered to streams for all upstream catchments multiplied by the product of cSDRs for the intervening flowlines (Equation (4)).

$$TD = SUM(TS_j \times [PRODUCT(cSDR_k)|k = 1 \text{ to } k = P])|j = 1 \text{ to } j = O, \quad (4)$$

The subscript j is the index for the O upstream catchments and the subscript k is the index for the P intervening flowlines between catchment j and the pipeline diversion.

2.7. Water Supply Impacts

The first metric of water supply impact is the total wildfire related sediment delivered to the diversion (Mg). The second metric is the per-fire average post-storm suspended sediment concentration (SSC). Wilson et al. [66] found that a threshold rainfall intensity of 7 mm h^{-1} best predicts when post-fire hillslope erosion will occur in this region. This intensity is exceeded on average four times per year in the study watershed. We make the simplifying assumption that the first-year post-fire sediment load from the coupled erosion and sediment transport model is divided equally among four storms. We estimate that 35% of the hillslope erosion predicted by RUSLE is part of the fine-grained inorganic and organic components that contribute to suspended sediment based on observations of soil particle sizes generated from post-fire hillslope erosion and transported in suspension after summer thunderstorms in the region [71,72]. Post-fire water quality is usually degraded for short periods (hours to days) following rainstorms in this region [48,73], so we calculate post-storm suspended sediment concentrations using the average storm load of fine sediment and the daily flow volume past the diversion point, which averages 1.48×10^9 L per day for the May to October period (gage-adjusted estimates from [63]). Suspended sediment concentration is rarely monitored directly, so limits for treatment are more commonly expressed in turbidity. For this analysis, we use the high end of 100 Nephelometric Turbidity Units (NTU) reported in the literature [15,74] to be conservative in our judgement of exceeding limits for treatment. A conversion equation (Equation (5)) developed from post-fire monitoring of the Fourmile Canyon Fire was used to predict turbidity (NTU) from SSC (mg L^{-1}) [15].

$$NTU = (SSC - 2.84)/1.166, \qquad (5)$$

2.8. Containment Effectiveness Evaluation and Prioritization

To quantify the effectiveness of containment, we focused on the difference between the total water impact measures with and without containment including watershed area burned, sediment delivered to the diversion, and number of turbidity threshold exceedances. The difference between impact measures for the uncontained and contained scenarios is the avoided transmitted risk [36]. Total sediment load is a continuous value whereas turbidity exceedance is a binary outcome. Water quality degradation was only considered transmitted when the outcome changed from below 100 NTU for within POD burning to above 100 NTU for the entire fire footprint. To prioritize improvements along the potential control lines that bound PODs, we calculated risk transmission across the POD edges based on their proportional engagement with the fires that originate in their respective PODs; that is, the outcomes associated with fire spreading to the surrounding landscape were divided among the lines based on their intersected length. It is anticipated that the primary mitigation action would be fuels reduction along the control lines, so transmission risk was normalized by length to compare the relative benefit of hardening control lines.

3. Results

3.1. Fire Occurrence

Historical fire ignitions from the FOD [56] were concentrated in the lower and middle portions of the focal watershed and along the southern boundary of the study area (Figure 3a) reflecting both variation in fire season length and human use of the landscape. The 5000 wildfires simulated with RANDIG ranged in size from 0.09 to 20,868 ha with a mean of 1961 ha and a median of 1469 ha. We selected the 3040 fires that burned at least part of the focal watershed for further analysis. Their size distribution did not vary substantially from that of the full simulation set. The excluded fires either did not grow large enough to intercept the focal watershed, or the predominant wind direction caused them to spread away from it. The middle and lower portions of the watershed are predicted to burn most frequently due to both the greater ignition density and the presence of fuel types that promote faster spread (Figure 3b). The high elevations in the western half of the study area are predicted to burn infrequently due to low ignition density and sparse fuels. The southeast corner of the study area near the water diversion has low burn probability because the fuels have not yet recovered from a recent wildfire.

3.2. Fire Behavior and Severity

Crown fire activity is predicted to vary across the watershed due to differences in fuels and topography (Figure 4a). A notable portion of the alpine and some recently burned areas are mapped as non-burnable cover types (13.7%). Surface, passive crown, and active crown fire are predicted on 25.9%, 39.3%, and 21.1% of the watershed area respectively, which we use as proxies for low, moderate, and high burn severity. This translates to predictions of low severity effects in grass and shrub fuel types and moderate or high severity effects in most forests. High severity effects are most common in forests with high horizontal and vertical continuity on steep slopes. Our prediction that approximately 60% of the watershed should burn at moderate or high severity is in line with the observed severity of recent large wildfires in Colorado [75].

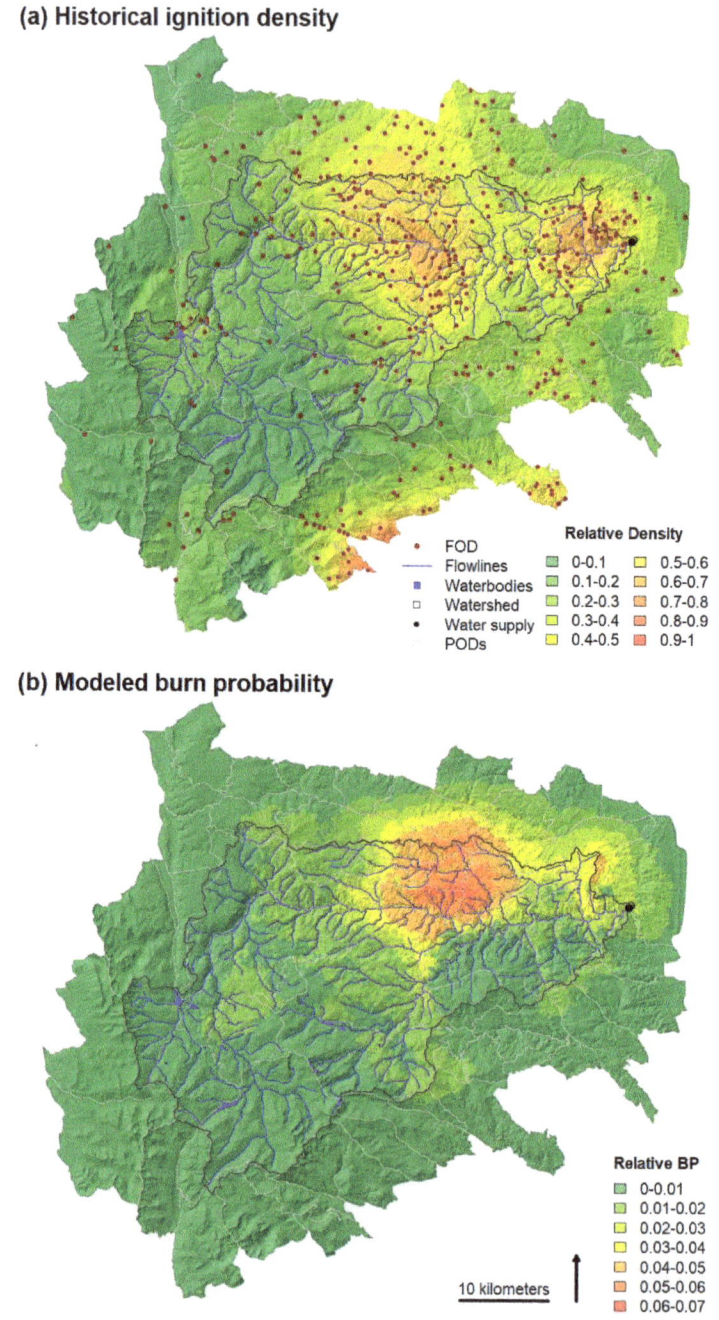

Figure 3. (a) Fire Occurrence Database (FOD) records of historical ignitions and interpolated surface of relative ignition density used in the RANDIG simulations. (b) Burn probability from the simulated fires that intercept the study watershed.

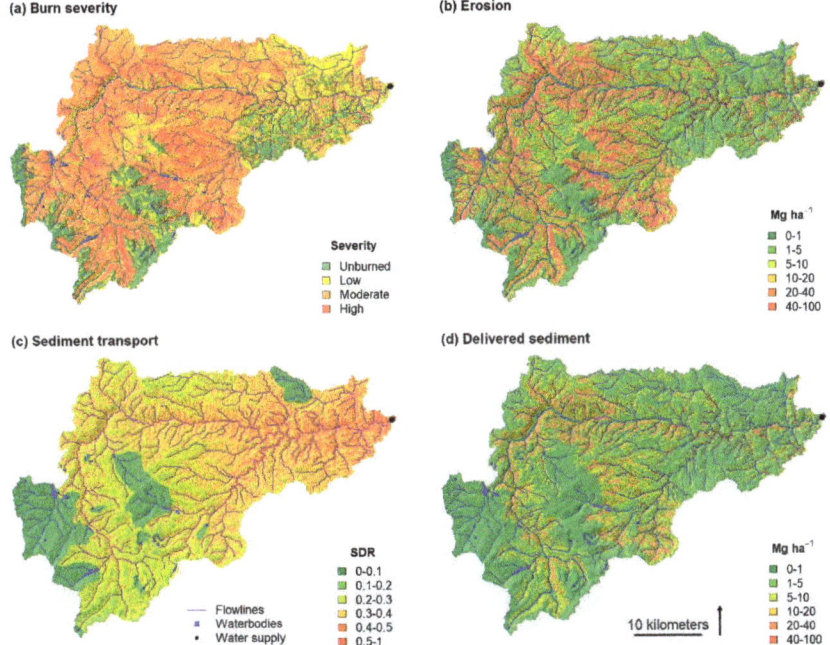

Figure 4. (**a**) Predicted burn severity using crown fire activity categories of surface, passive crown, and active crown fire as proxies for low, moderate, and high severity fire. (**b**) Predicted post-fire erosion with 2-year rainfall erosivity. (**c**) Combined Sediment Delivery Ratio (SDR) accounting for both hillslope and channel transport. (**d**) Predicted sediment delivery to the water supply diversion with 2-year rainfall erosivity.

3.3. Watershed Response

Like burn severity, the magnitudes of post-fire erosion and sediment transport vary widely across the watershed owing to variation in topography, soils, and proximity to the diversion. Figure 4 illustrates this for the 2-year rainfall erosivity. The greatest sediment hazard is associated with steep terrain near the major channels that is predicted to burn at moderate or high severity. Post-fire erosion and sediment transport potential is generally low in the flatter terrain in the northeast quadrant of the watershed, the high mountains above major waterbodies, and the recently burned areas. The spatial distribution of sediment hazard is similar for 10-year and 100-year rainfall erosivity, but the absolute magnitude increases considerably. Table 2 summarizes the distribution of predicted erosion, sediment delivery to streams, and sediment delivery to the diversion for the 3040 simulated wildfires that burned in the watershed. The predicted mean post-fire gross erosion for the simulated wildfires is 12.3, 20.4, and 46.4 Mg ha^{-1} for the 2, 10, and 100-year rainfall erosivity, respectively. Much of this sediment should be retained in the watershed, especially where waterbodies interrupt sediment transport (Figure 4c), so delivery to the diversion averages only 4.2, 7.0, and 15.9 Mg ha^{-1} for the 2, 10, and 100-year rainfall erosivity, respectively.

Table 2. Summary statistics of first-year post fire erosion, sediment delivery to streams, and sediment delivery to the water supply diversion (div.) in Mg ha^{-1} by rainfall erosivity for the simulated wildfires that burned into the watershed. These are total sediment yields including the coarse and fine fractions.

	2-Year Rainfall Erosivity			10-Year Rainfall Erosivity			100-Year Rainfall Erosivity		
Statistic	Erosion	To Streams	To div.	Erosion	To Streams	To div.	Erosion	To Streams	To div.
Lower decile	2.0	1.0	0.4	4.3	2.1	0.9	18.5	9.1	4.3
Lower quartile	5.0	2.6	1.6	9.8	5.0	3.2	32.3	16.5	11.0
Median	9.0	4.7	3.3	16.5	8.6	6.2	45.2	23.4	16.8
Mean	12.3	6.2	4.2	20.4	10.3	7.0	46.4	23.4	15.9
Upper quartile	16.8	8.6	6.0	28.1	14.3	9.9	60.8	30.7	21.5
Upper decile	27.7	13.7	8.7	42.9	20.9	13.6	75.3	36.8	24.7

3.4. Avoided Watershed Area Burned

For improved containment at POD boundaries to avoid water supply impacts, the target fires must leave the POD of origin under unmanaged conditions. Of the 3040 simulated wildfires that burned at least part of the focal watershed, 2351 of them (77.3%) burned at least some area outside the origin POD. Fires occasionally burned more than ten PODs, but of the fires that burned more than one POD, most burned between two and five PODs (77.9%). This suggests that most fire transmission during the initial burn period is between a POD and its adjacent neighbors, but some rare events may burn across multiple POD boundaries.

Containing all fires within their POD of origin would reduce the average watershed area burned from 1361 to 562 ha per fire, a 58.7% reduction (Table 3). The distributions of watershed area burned for the contained and uncontained scenarios are shown in Figure 5a. Containing large fires has the greatest potential to avoid watershed area burned; the 1396 fires that burned more than 1000 ha account for 93.8% of the avoided area burned. Containment in the POD of origin would eliminate fires that burn more than 10,000 ha in the watershed, which numbered 26 (0.9%) in the uncontained scenario. The percentage of fires burning greater than 5000 ha would be reduced from 4.0 to 0.2. Watershed area burned by fires that originate from PODs that are wholly or mostly outside the watershed should be reduced to negligible levels under the containment scenario, but these PODs account for only a small fraction of area burned when fires are allowed to grow freely (Figure 6a). Most fires start in the central and eastern portion of the watershed (Figure 3) and the predominant west winds means that PODs in the lower 2/3rds of the watershed are the source of fires that burn the greatest area (Figure 6a). All else equal, larger PODs are larger sources of fire because they have more ignitions. Containment reduced watershed area burned from fires that ignited in 61 of the 70 PODs, but some of the largest PODs still have substantial watershed area burned with containment (Figure 6a) because fires have room to grow large before encountering a potential control feature.

Table 3. Summary of water supply impacts across all fires by containment scenario and rainfall erosivity. A turbidity threshold of 100 NTU was used to compute the number of exceedances.

Watershed Area Burned (Mean ha per Fire)				
	Self-burning	Total	Avoided	Avoided (%)
	562	1361	799	58.7
Sediment to Diversion (Mean Mg per Fire)				
Rainfall Erosivity	Self-Burning	Total	Avoided	Avoided (%)
2-year	3031	6115	3085	50.4
10-year	4904	10,188	5284	51.9
100-year	10,411	23,273	12,863	55.3
Turbidity Exceedances (Count of Fires)				
Rainfall Erosivity	Self-Burning	Total	Avoided	Avoided (%)
2-year	1110	1668	558	33.5
10-year	1503	1910	407	21.3
100-year	1922	2210	288	13.0

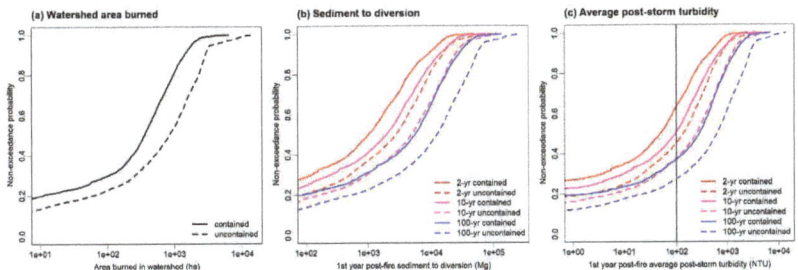

Figure 5. Summary of containment effects on distribution of fire-level indicators of water supply risk by rainfall erosivity including: (**a**) watershed area burned, (**b**) first-year post-fire sediment to the diversion, and (**c**) first-year post-fire average post-storm turbidity (vertical black line marks the 100 NTU threshold for treatment).

3.5. Avoided Sediment

Containment reduced the total sediment load to the pipeline diversion by 50.4–55.3% depending on rainfall erosivity from an average of 6.1–23.2 thousand Mg per fire to an average of 3.1–10.4 thousand Mg per fire (Table 3). The distributions of sediment delivered to the diversion for the contained and uncontained scenarios are shown in Figure 5b. Sediment loads vary across several orders of magnitude due to differences in fire size, erosion and sediment transport potential, and post-fire rainfall. The effect of containment on sediment load is roughly equivalent to reducing rainfall erosivity one level (Figure 5b). The spatial distribution of sediment source risk is similar to that of watershed area burned (Figure 6b). PODs that are partially or wholly outside the watershed are a minimal risk to water supplies after containment, but fire activity in the larger PODs situated in the middle of the watershed is still expected to produce large sediment loads.

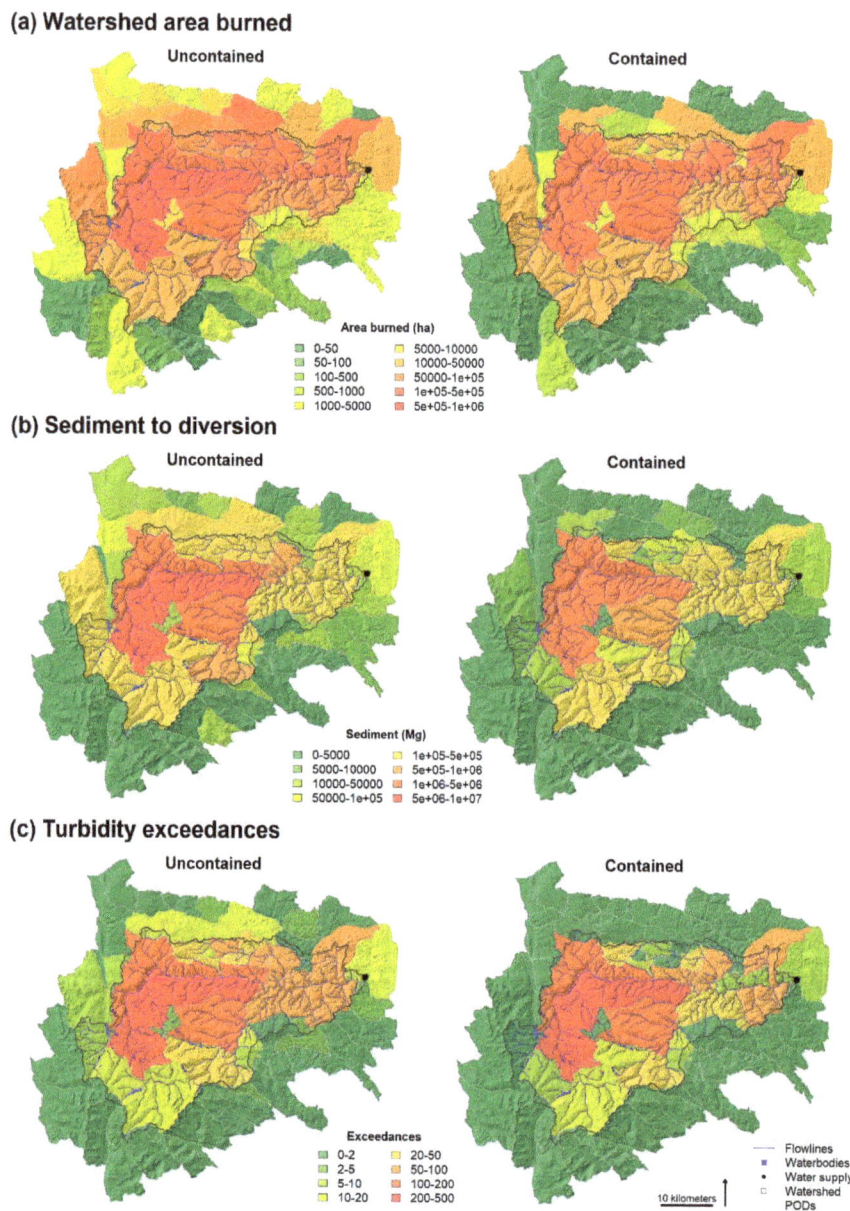

Figure 6. Spatial summary of containment effects on distribution of POD-level indicators of water supply risk for the 2-year rainfall erosivity including: (**a**) watershed area burned, (**b**) first-year post-fire sediment to the diversion, and (**c**) frequency of turbidity exceedances for fires that originate within each POD.

3.6. Avoided Water Quality Degradation

Containment effects on water quality degradation were less substantial than for watershed area burned and total sediment to the diversion (Table 3; Figure 5c); turbidity exceedances were reduced by 33.5, 21.3, and 13.0 percent for the 2, 10, and 100-year rainfall erosivity, respectively. With containment, 36.5, 49.4, and 63.2 percent of fires are predicted to exceed the 100 NTU threshold for the 2, 10, and 100-year rainfall erosivity, respectively. Most fires that caused turbidity to exceed limits for treatment originated in the large PODs in the middle of the watershed (Figure 6c). The three PODs with the most turbidity exceedances are all larger than 10,000 ha. Containment only reduced the number of turbidity exceedances from these PODs from 640 to 568 (an 11.3% reduction) for the 2-year rainfall erosivity, and containment offered almost no mitigation benefit (1.0% fewer exceedances) for these PODs under the most extreme rainfall scenario. In contrast, containment reduced turbidity exceedances by more than 50% in 33 of the 70 PODs under median rainfall conditions. These PODs range in size from 502 to 14,153 ha with a mean of 3548 ha. Many of these PODs are mostly or wholly outside the watershed, but some are smaller PODs inside the watershed.

3.7. Prioritizing POD Network Improvements

The limited effect of containment on turbidity exceedances highlights the need to break up the three large PODs with high source risk in the middle of the watershed (Figure 6c). These three PODs are also the top priorities for further compartmentalization based on watershed area burned and total sediment load from self-burning. With containment, an additional eight PODs were the source of 20 or more turbidity exceedances under median rainfall conditions. Cumulatively, these top 11 PODs account for 91.4% of the fires that degraded water quality in the contained scenario, so efforts to further reduce fire sizes in these PODs should have high benefit.

Prioritizing improvements along the potential control lines that bound PODs can be informed with measures of risk transmission (Figure 7). Total sediment to the diversion was transmitted at the highest rates along POD edges in the middle portion of the watershed (Figure 7a) where there is high potential for fires to spread into erosion prone terrain near the diversion (Figures 3b and 4). In contrast, transmitted water quality degradation was more concentrated along the edges associated with the smaller PODs in the north central portion of the watershed (Figure 7b). Transmission risk was also high for several control lines in the eastern half of the watershed that are nearly perpendicular to the dominant wind direction. Mitigation priorities differed depending on which metric of transmission risk was used (Figures 7 and 8). The two metrics both identify a similar order of priorities (Spearman's $\rho = 0.89$) but they have moderate disagreement about the magnitudes of potential risk mitigation (Pearson's $R = 0.71$), especially for the highest-ranking edges (Figure 8). Most notably, few of the POD edges associated with the three large PODs that are the source of most turbidity exceedances (Figure 6c) are high priorities for mitigation because containment at these locations infrequently changes the water quality outcome despite the potential to avoid large quantities of sediment.

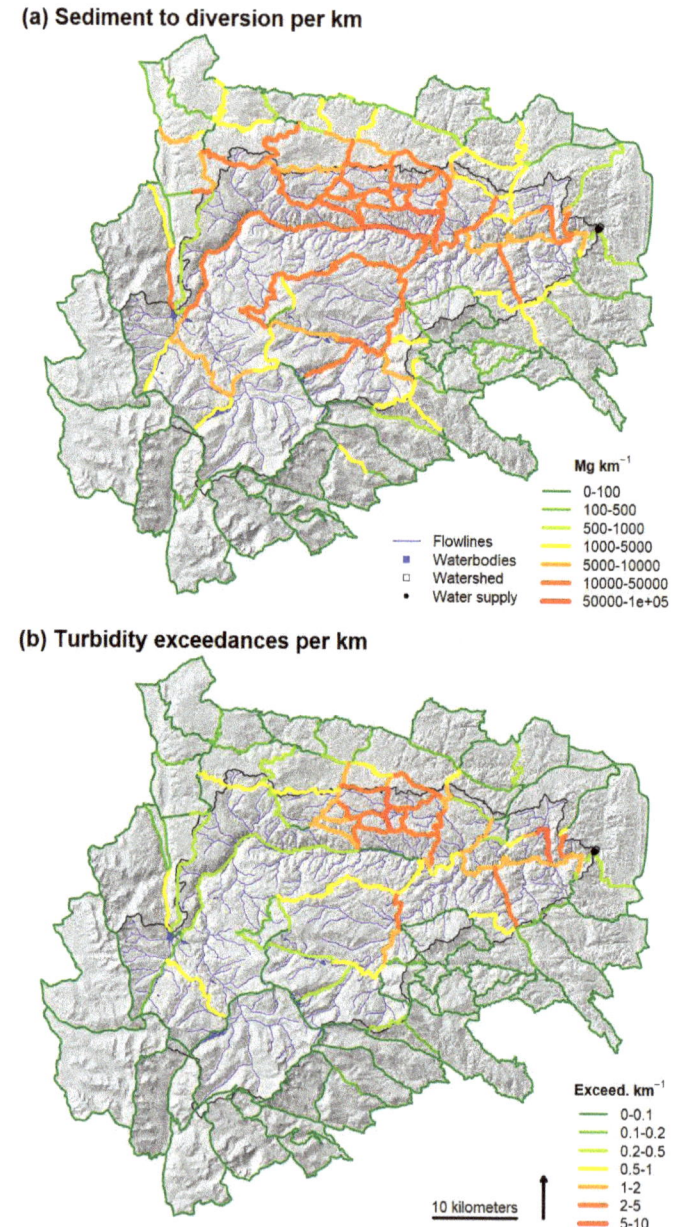

Figure 7. Total transmitted risk for all fires from (**a**) sediment to diversion and (**b**) turbidity exceedances normalized to edge length in kilometers for 2-year rainfall erosivity.

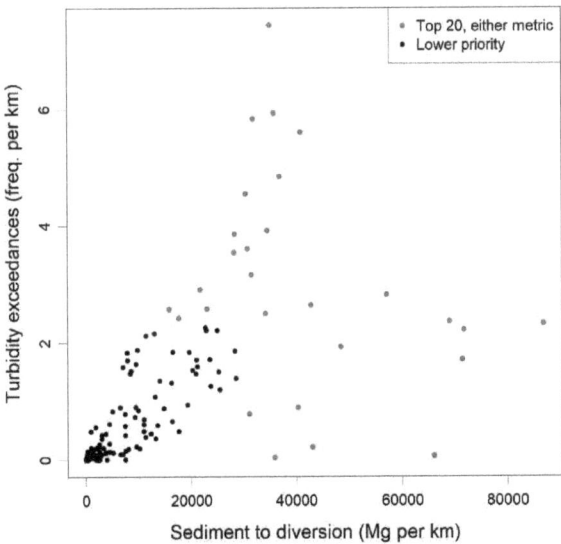

Figure 8. Edge transmission risk comparison for 2-year rainfall erosivity. Edges ranked in the top 20 using either metric are colored red.

4. Discussion

This proof of concept analysis demonstrates the potential for improved early containment of large fires to lower watershed area burned by 58.7% and to reduce risk to source water between 13.0% and 55.3% depending on impact metric considered. Proportional reductions in total sediment load to the diversion ranged between 50.4% and 55.3%, but the potential to avoid exceeding turbidity limits for treatment was notably lower—varying between 33.5% and 13.0% reduction for the 2- and 100-year rainfall erosivity, respectively (Table 3). The contrasting response of our water impact metrics to increasing rainfall erosivity (Table 3) reveals that avoiding large quantities of sediment may not translate to avoiding degraded water quality if the residual sediment load is still large. The sources of water supply risk and potential mitigation benefits of fire containment varied widely across the POD network (Figure 6) suggesting the potential to further improve mitigation effectiveness with targeted divisions to reduce the size of PODs with high risk from self-burning and fuels reduction to improve containment probability along high transmission boundaries (Figure 7).

Our analysis built on previous studies of wildfire-water supply risk and wildfire risk transmission to estimate the avoided water supply impacts from improved fire containment within pre-identified PODs. Omi [18] approached this issue from the related perspective of fuel break construction and maintenance in California using estimates of avoided area burned and a relative damage index to value fuel break benefits. Monte Carlo wildfire simulation and watershed effects analyses capture similar information on exposure and impacts with the added benefit of associating fire outcomes with their ignition locations and final extents [32,33]. A recent effort to zone the study landscape into PODs provided the operationally relevant fire containers used to estimate avoided water supply impacts using risk transmission methods [35,36] as suggested by Davis [76] to estimate the area saved from burning after encountering a control feature. The avoided area burned and sediment load measures we modeled are similar to the impact metrics used to value the benefit of containment in previous studies, but our evaluation of water quality degradation provided a unique opportunity to evaluate whether the size and spatial arrangement of the PODs are appropriate to mitigate a scale-dependent risk.

Our results suggest POD-based containment could meaningfully reduce risk of exceeding turbidity limits for treatment (Table 3), but the large percentage of unmitigated risk implies that the containment network could be more effective with smaller PODs.

Our estimates of avoided impacts are premised on the simplifying assumption that all fires are contained within their POD of origin, which is likely realistic for many of our modeled fires but optimistic for the most extreme fires in the region [35,46]. We chose not to address the probability of containment in this study because existing models focus on characteristics of the control features, surrounding fuels and topography, and fire behavior [30,77,78] but do not explicitly consider the effects of suppression [79]. Managers in this landscape primarily identified roads as control features because they aid firefighter access [29] and suppression firing [51]. It is also anticipated that proactively identifying control features and response strategies will lead to timely and well-coordinated tactics that increase the probability of containment. For example, extensive pre-season planning has been credited with improving the strategic use of suppression firing and aerial retardant drops to contain fire in PODs during extreme weather [80]. We did not account for suppression firing in this study, which can sometimes substantially increase area burned [81] and thus would dampen the contrast between our containment scenarios. However, managers ideally use backing fire to minimize adverse effects [80]. Improved modeling of suppression actions and effects would help to refine our estimates of risk mitigation.

The post-fire erosion and sediment transport modeling used here has several limitations that are important to acknowledge. First, the linked fire and erosion model system (Figure 1) is subject to multiple data, model, and model linkage uncertainties that have potential for prediction error as discussed extensively in previous publications [24,25]. Recent work has shown that water quality at the basin scale is sometimes minimally impacted despite modeled increases in hillslope erosion [82], emphasizing the need to test and refine erosion and sediment transport models with empirical observations at multiple scales [83]. Most of our predicted first-year post-fire hillslope erosion yields for the 2-year and 10-year rainfall erosivity scenarios (Table 2) are close to the study-wide means of 9.5–22.2 Mg ha^{-1} and the range of individual hillslope observations of 0.1–38.2 Mg ha^{-1} from previous fires in the region exposed to moderate rainfall [11,17,47,84]. Many of these studies had hillslope sediment fences fill and overtop, so the reported yields are usually interpreted as a lower bound estimate of the true erosion rate. For the 100-year rainfall erosivity, only the top decile of modeled fires exceed the 72 Mg ha^{-1} of rill and interrill erosion reported in the first year after the Buffalo Creek Fire in response to similarly extreme rainfall (converted from volume estimates of [45] using bulk density of 1.6 Mg m^{-3}). Despite doubling the efficiency of hillslope transport in this study, only the net sediment delivery to streams for the upper decile of fires with 10-year rainfall erosivity and the upper half of fires with 100-year rainfall erosivity (Table 2) approach the small catchment sediment yields of 22.0–38.6 Mg ha^{-1} observed in the first two years after the Hayman Fire [85,86]. This seems reasonable given the larger size of most catchments in this study. After our rough calibration, our combined hillslope and channel SDR values (Figure 4c) are close to SDR values estimated with similar travel time methods [87,88]. None of the simulated fires at any rainfall level (Table 2) are predicted to deliver sediment to the diversion at a rate close to the whole watershed sediment yield of 52.5 Mg ha^{-1} for the first year of the Buffalo Creek Fire [45], likely because we did not account for channel erosion.

Our water degradation analysis also layers on additional assumptions that the annual suspended sediment load is evenly divided among the annual average of four sediment-generating storms and the storm sediment load is evenly mixed in the average daily flow volume of the river during the thunderstorm season. Despite these approximations, the resulting turbidities—which averaged 309, 516, and 1181 NTU for the 2, 10, and 100-year rainfall erosivity, respectively—align well with common observations in the region of post-fire turbidities between 100 and 1000 NTU and occasional observations >1000 NTU [15,48,89]. The assumption that storm load is an equal division of annual load does not account for the substantial intra-annual variability in storm characteristics [15,83,90], seasonal trends in runoff and erosion [91], nor the interannual variability in the frequency of storms with sufficient intensity to cause erosion [34,66]. Similarly, unaccounted for variability in daily flow volume should influence the vulnerability of the water source. Given these simplifications, we have more confidence in our contrasts of containment benefits across scenarios than we do in our absolute estimates of degradation risk. Our analysis also focused exclusively on the acute periods of severe water quality degradation after rainstorms in the first year after fire, so it is unclear if containing fires to smaller sizes will avoid elevated carbon, nitrogen, phosphorus, manganese, and suspended solids concentrations that may persist for years after fires in Colorado [15,89], increasing treatment complexity and cost and raising concerns about the formation of disinfection byproducts [74,92]. Similar water quality responses and treatment challenges have been observed after wildfires in Canada, Australia, and Europe [12,13].

Despite uncertainties in the precise magnitude of risk reduction, improved containment appears promising compared to other mitigation strategies. We found that limiting fires to their POD of origin should reduce the total sediment load from wildfire between 50.4 and 55.3% (Table 3). Previous assessments of landscape-scale fuel treatments in the western USA predict long-term sediment reduction of 19% [25] and up to 34% reduction in sediment costs [24]. Salis et al. [93] project that treating 15% of a landscape in Sardinia, Italy would only reduce average sediment yield 4–12%, but their treatment scenarios were not prioritized to avoid erosion. Based on the narrowest contrast in these figures (34% for fuel treatment and 50.4% for containment), POD-based containment should compare favorably to landscape scale fuels reduction as long as the containment failure rate is less than 32%. Furthermore, compartmentalizing fire in small units of the landscape has the potential to avoid disrupting multi-source water systems by limiting fire impacts to a single source. The benefit of containing individual wildfires should vary widely (Figure 5), as fire encounters with control features and associated impacts beyond the POD of origin depend strongly on where the fire ignites.

We also demonstrated how risk transmission metrics could inform improvements to the POD network, which should be relevant to fire, land, and water managers engaged in spatial fire planning. The small number of PODs with high risk from self-burning are high priorities for further compartmentalization, which could require improving firefighter access and/or reducing fuels. Fine scale analyses of risk factors and containment opportunities would benefit these efforts. If further divisions are not feasible or practical (e.g., because of wilderness or wildlife habitat concerns), these PODs could be candidates for fuels reduction with prescribed or managed fire. It is also valuable for water managers to identify areas that are not conducive to proactive risk mitigation, so they can plan how to best respond to the anticipated effects of future fires. As previously discussed, we did not estimate the probability of containing wildfire at POD boundaries and how containment probability would change with fuels reduction, but managers are interested in identifying potential control lines in need of improvement to support safe and effective fire response. Measures of transmission risk across the POD edges (Figure 7) highlight where these efforts should be targeted to maximize their benefit. However, priorities differed depending on the water supply effects measure used (Figure 8); most notably, there is greater potential to avoid degradation by improving containment probability around the smaller PODs. Further analyses are needed to evaluate if fuel conditions around these POD edges necessitate treatment for firefighting effectiveness and safety.

The style of Monte Carlo exposure and effects analyses we present should also be useful for evaluating fire protection strategies for other high value resources and assets that depend on the scale of disturbance. For example, most ecological concerns relate to the area and spatial pattern of high severity effects on vegetation and the resulting consequences for wildlife and reforestation by dispersal-limited species (e.g., [94–96]). If intolerable levels of fire exposure or effects can be defined for ecological values, similar methods could be used to assess the protection value of POD-based containment. Wildfire impacts to homes and other values in the wildland-urban interface (WUI) are almost always negative, but consequences often become disastrous when the area and assets affected by fire overwhelm firefighting resources [97]. Managers intuitively design smaller PODs in the WUI [2], but it has not been tested whether these PODs are appropriately sized to avert WUI disasters—i.e., whether asset exposure for most fires is below the fire protection capacity. Similarly, wildfire impacts to transportation networks may cross thresholds of concern for evacuation when traffic exceeds the capacity of the available routes. Explicitly defining performance objectives for these and other fire protection concerns could help to tailor POD size and spatial arrangement in future fire planning efforts.

5. Conclusions

Improved wildfire containment has potential to meaningfully reduce wildfire risk to water supplies, but these effects are scale dependent. In our test cases, approximately 75% of fires intersected potential control features and, if these fires were contained within their POD of origin, watershed area burned would be reduced by 58.7%, total sediment load to the diversion would be reduced between 50.4 and 55.3%, and water quality degradation beyond limits for treatment would be reduced between 13.0 and 33.5%. Risk mitigation was higher for total sediment load than water quality degradation because containment did not always change water quality outcomes. Moreover, priorities to improve the network design by modifying the size of the PODs or improving containment probability along their edges differ depending on the effects measure used. This highlights the importance of properly defining water supply impacts for wildfire risk assessment and mitigation effectiveness studies. Similar analyses could be applied to other scale-dependent resources at risk of wildfire to inform containment network design.

Author Contributions: Conceptualization, B.M.G., Y.W. and M.P.T.; methodology, B.M.G.; software, B.M.G.; formal analysis, B.M.G.; investigation, B.M.G.; data curation, B.M.G.; writing—original draft preparation, B.M.G.; writing—review and editing, B.M.G., Y.W. and M.P.T.; visualization, B.M.G.; supervision, Y.W.; project administration, Y.W.; funding acquisition, Y.W. and M.P.T. All authors have read and agreed to the published version of the manuscript.

Funding: This research was funded by joint venture agreement 19-JV-11221636-170 between the USDA Forest Service Rocky Mountain Research Station and Colorado State University; cost share agreement number 17-CS-11021000-032 between the USDA Forest Service, Arapaho and Roosevelt National Forests and Pawnee National Grassland and the Colorado Forest Restoration Institute at Colorado State University; and agreement number 19-DG-11031600-062 between the USDA Forest Service, Southwestern Region and the Colorado Forest Restoration Institute at Colorado State University.

Acknowledgments: The authors thank Codie Wilson for sharing the rainfall data used in the analysis.

Conflicts of Interest: The authors declare no conflict of interest. The funders had no role in the design of the study; in the collection, analyses, or interpretation of data; in the writing of the manuscript, or in the decision to publish the results.

References

1. O'Connor, C.D.; Thompson, M.P.; Rodríquez y Silva, F. Getting ahead of the wildfire problem: Quantifying and mapping management challenges and opportunities. *Geosciences* **2016**, *6*, 35. [CrossRef]
2. Thompson, M.P.; Bowden, P.; Brough, A.; Scott, J.H.; Gilbertson-Day, J.; Taylor, A.; Anderson, J.; Haas, J.R. Application of wildfire risk assessment results to wildfire response planning in the Southern Sierra Nevada, California, USA. *Forests* **2016**, *7*, 64. [CrossRef]
3. Thompson, M.P.; MacGregor, D.G.; Dunn, C.J.; Calkin, D.E.; Phipps, J. Rethinking the wildland fire management system. *J. For.* **2018**, *116*, 382–390. [CrossRef]
4. Caggiano, M.D. *Collaboratively Engaging Stakeholders to Develop Potential Operational Delineations*; Report CFRI-1908; Colorado Forest Restoration Institute: Fort Collins, CO, USA, 2019.
5. Caggiano, M.D.; O'Connor, C.D.; Sack, R.B. *Potential Operational Delineations and Northern New Mexico's 2019 Fire Season*; Report CFRI-2002; Colorado Forest Restoration Institute: Fort Collins, CO, USA, 2019.
6. Greiner, M.; Kooistra, C.; Schultz, C. *Pre-Season Planning for Wildland Fire Response: An Assessment of the US Forest Service's Potential Operational Delineations (PODs)*; Practitioner Paper #05; Public Lands Policy Group at Colorado State University: Fort Collins, CO, USA, 2020.
7. Dunn, C.J.; O'Connor, C.D.; Abrams, J.; Thompson, M.P.; Calkin, D.E.; Johnston, J.D.; Stratton, R.; Gilbertson-Day, J. Wildfire risk science facilitates adaptation of fire-prone social-ecological systems to the new fire reality. *Environ. Res. Lett.* **2020**, *15*, 025001. [CrossRef]

8. Stratton, R.D. The path to strategic wildland fire management planning. *Wildfire Mag.* **2020**, *29*, 24–31.
9. DeBano, L.F.; Neary, D.G.; Ffolliott, P.F. Soil physical processes. In *Wildland Fire in Ecosystems: Effects of Fire on Soils and Water*; Neary, D.G., Ryan, K.C., Eds.; General Technical Report RMRS-GTR-42; USDA Forest Service, Rocky Mountain Research Station: Ogden, UT, USA, 2005; Volume 4, pp. 29–51.
10. Shakesby, R.A.; Doerr, S.H. Wildfire as a hydrological and geomorphological agent. *Earth Sci. Rev.* **2006**, *74*, 269–307. [CrossRef]
11. Larsen, I.J.; MacDonald, L.H.; Brown, E.; Rough, D.; Welsh, M.J.; Pietraszek, J.H.; Libohova, Z.; Benavides-Solorio, J.D.; Schaffrath, K. Causes of post-fire runoff and erosion: Water repellency, cover, or soil sealing? *Soil Sci. Soc. Am. J.* **2009**, *73*, 1393–1407. [CrossRef]
12. Smith, H.G.; Sheridan, G.J.; Lane, P.N.J.; Nyman, P.; Haydon, S. Wildfire effects on water quality in forest catchments: A review with implications for water supply. *J. Hydrol.* **2011**, *396*, 170–192. [CrossRef]
13. Emelko, M.B.; Silins, U.; Bladon, K.D.; Stone, M. Implications of land disturbance on drinking water treatability in a changing climate: Demonstrating the need for "source water supply and protection" strategies. *Water Res.* **2011**, *45*, 461–472. [CrossRef]
14. Abraham, J.; Dowling, K.; Florentine, S. Risk of post-fire metal mobilization into surface water resources: A review. *Sci. Total Environ.* **2017**, *599–600*, 1740–1755. [CrossRef]
15. Murphy, S.F.; Writer, J.H.; McCleskey, R.B.; Martin, D.A. The role of precipitation type, intensity, and spatial distribution in source water quality after wildfire. *Environ. Res. Lett.* **2015**, *10*, 084007. [CrossRef]
16. Benavides-Solorio, J.D.; MacDonald, L.H. Measurement and prediction of post-fire erosion at the hillslope scale, Colorado Front Range. *Int. J. Wildland Fire* **2005**, *14*, 457–474. [CrossRef]
17. Schmeer, S.R.; Kampf, S.K.; MacDonald, L.H.; Hewitt, J.; Wilson, C. Empirical models of annual post-fire erosion on mulched and unmulched hillslopes. *Catena* **2018**, *163*, 276–287. [CrossRef]
18. Omi, P.N. Planning future fuelbreak strategies using mathematical modeling techniques. *Environ. Manag.* **1979**, *3*, 73–80. [CrossRef]
19. Thompson, M.P.; Scott, J.; Langowski, P.G.; Gilbertson-Day, J.W.; Haas, J.R.; Bowne, E.M. Assessing watershed-wildfire risks on national forest system lands in the Rocky Mountain region of the United States. *Water* **2013**, *5*, 945–971. [CrossRef]
20. Cannon, S.H.; Gartner, J.E.; Rupert, M.G.; Michael, J.A.; Rea, A.H.; Parrett, C. Predicting the probability and volume of post-wildfire debris flows in the intermountain western United States. *Geol. Soc. Am. Bull.* **2010**, *122*, 127–144. [CrossRef]
21. Miller, M.E.; MacDonald, L.H.; Robichaud, P.R.; Elliot, W.J. Predicting post-fire hillslope erosion in forest lands of the western United States. *Int. J. Wildland Fire* **2011**, *20*, 982–999. [CrossRef]
22. Miller, M.E.; Elliot, W.J.; Billmire, M.; Robichaud, P.R.; Endsley, K.A. Rapid-response tools and datasets for post-fire remediation: Linking remote sensing and process-based hydrological models. *Int. J. Wildland Fire* **2016**, *25*, 1061–1073. [CrossRef]
23. Sidman, G.; Guertin, D.P.; Goodrich, D.C.; Thoma, D.; Falk, D.; Burns, I.S. A coupled modelling approach to assess the effect of fuel treatments on post-wildfire runoff and erosion. *Int. J. Wildland Fire* **2016**, *25*, 351–362. [CrossRef]
24. Gannon, B.M.; Wei, Y.; MacDonald, L.H.; Kampf, S.K.; Jones, K.W.; Cannon, J.B.; Wolk, B.H.; Cheng, A.S.; Addington, R.N.; Thompson, M.P. Prioritising fuels reduction for water supply protection. *Int. J. Wildland Fire* **2019**, *28*, 785–803. [CrossRef]
25. Elliot, W.J.; Miller, M.E.; Enstice, N. Targeting forest management through fire and erosion modelling. *Int. J. Wildland Fire* **2016**, *25*, 876–887. [CrossRef]
26. Jones, K.W.; Cannon, J.B.; Saavedra, F.A.; Kampf, S.K.; Addington, R.N.; Cheng, A.S.; MacDonald, L.H.; Wilson, C.; Wolk, B. Return on investment from fuel treatments to reduce severe wildfire and erosion in a watershed investment program in Colorado. *J. Environ. Manag.* **2017**, *198*, 66–77. [CrossRef] [PubMed]
27. Thompson, M.P.; Liu, Z.; Wei, Y.; Caggiano, M.D. Analyzing wildfire suppression difficulty in relation to protection demand. In *Environmental Risks*; Mihai, F.-C., Grozavu, A., Eds.; IntechOpen Limited: London, UK, 2018; pp. 45–64. [CrossRef]
28. Wei, Y.; Thompson, M.P.; Haas, J.R.; Dillon, G.K.; O'Connor, C.D. Spatial optimization of operationally relevant large fire confine and point protection strategies: Model development and test cases. *Can. J. For. Res.* **2018**, *48*, 480–493. [CrossRef]

29. Rodríguez y Silva, F.; Molina Martínez, J.R.; González-Cabán, A. A methodology for determining operational priorities for prevention and suppression of wildland fires. *Int. J. Wildland Fire* **2014**, *23*, 544–554. [CrossRef]
30. O'Connor, C.D.; Calkin, D.E.; Thompson, M.P. An empirical machine learning method for predicting potential fire control locations for pre-fire planning and operational fire management. *Int. J. Wildland Fire* **2017**, *26*, 587–597. [CrossRef]
31. Rodríguez y Silva, F.; O'Connor, K.; Thompson, M.P.; Molina Martínez, J.R.; Calkin, D.E. Modelling suppression difficulty: Current and future applications. *Int. J. Wildland Fire* **2020**, in press. [CrossRef]
32. Thompson, M.P.; Gilbertson-Day, J.W.; Scott, J.H. Integrating pixel- and polygon-based approaches to wildfire risk assessment: Applications to a high-value watershed on the Pike and San Isabel National Forests, Colorado, USA. *Environ. Model. Assess.* **2016**, *21*, 1–15. [CrossRef]
33. Haas, J.R.; Thompson, M.; Tillery, A.; Scott, J.H. Capturing spatiotemporal variation in wildfires for improving post-wildfire debris-flow hazard assessments. In *Natural Hazard Uncertainty Assessment: Modeling and Decision Support, Geophysical Monograph 223*; Riley, K., Webley, P., Thompson, M., Eds.; John Wiley & Sons: Hoboken, NJ, USA, 2017; pp. 301–317.
34. Gannon, B.M.; Wei, Y.; Thompson, M.P.; Scott, J.H.; Short, K.C. System analysis of wildfire-water supply risk in Colorado, U.S.A. with Monte Carlo wildfire and rainfall simulation. *Risk Anal.* **2020**. in review.
35. Haas, J.R.; Calkin, D.E.; Thompson, M.P. Wildfire risk transmission in the Colorado Front Range, USA. *Risk Anal.* **2015**, *35*, 226–240. [CrossRef]
36. Ager, A.A.; Palaiologou, P.; Evers, C.R.; Day, M.A.; Barros, A.M.G. Assessing transboundary wildfire exposure in the southwestern United States. *Risk Anal.* **2018**, *38*, 2105–2127. [CrossRef]
37. Finney, M.A. The challenge of quantitative risk analysis for wildland fire. *For. Ecol. Manag.* **2005**, *211*, 97–108. [CrossRef]
38. Scott, J.H.; Thompson, M.P.; Calkin, D.E. *A Wildfire Risk Assessment Framework for Land and Resource Management*; General Technical Report RMRS-GTR-315; USDA Forest Service, Rocky Mountain Research Station: Fort Collins, CO, USA, 2013.
39. Finney, M.A. An overview of FlamMap fire modeling capabilities. In Proceedings of the Fuels Management-How to Measure Success Conference, Portland, OR, USA, 28–30 March 2006; Andrews, P.L., Butler, B.W., Eds.; Proceedings RMRS-P-41. USDA Forest Service, Rocky Mountain Research Station: Fort Collins, CO, USA, 2006; pp. 213–220.
40. Finney, M.A.; Brittain, S.; Seli, R.C.; McHugh, C.W.; Gangi, L. *FlamMap: Fire Mapping and Analysis System, Version 5.0*; USDA Forest Service, Rocky Mountain Research Station: Fort Collins, CO, USA, 2015. Available online: http://www.firelab.org/document/flammap-software (accessed on 1 November 2019).
41. Renard, K.G.; Foster, G.R.; Weesies, G.A.; McCool, D.K.; Yoder, D.C. *Predicting Soil Erosion by Water: A Guide to Conservation Planning with the Revised Universal Soil Loss Equation (RUSLE)*; Handbook no. 703; USDA Agricultural Research Service Agricultural: Washington, DC, USA, 1997.
42. Wagenbrenner, J.W.; Robichaud, P.R. Post-fire bedload sediment delivery across spatial scales in the interior western United States. *Earth Surf. Process. Landf.* **2014**, *39*, 865–876. [CrossRef]
43. Frickel, D.G.; Shown, L.M.; Patton, P.C. *An Evaluation of Hillslope and Channel Erosion Related to Oil-Shale Development in the Piceance Basin, North-Western Colorado*; Colorado Department of Natural Resources, Colorado Water Resources Circular 30: Denver, CO, USA, 1975.
44. R Core Team. *R: A Language and Environment for Statistical Computing, Version 3.5.3*; R Foundation for Statistical Computing: Vienna, Austria, 2019. Available online: https://www.R-project.org/ (accessed on 1 May 2019).
45. Moody, J.A.; Martin, D.A. Initial hydrologic and geomorphic response following a wildfire in the Colorado Front Range. *Earth Surf. Process. Landf.* **2001**, *26*, 1049–1070. [CrossRef]
46. Graham, R.T. *Hayman Fire Case Study*; USDA Forest Service, Rocky Mountain Research Station, General Technical Report RMRS-GTR-114: Ogden, UT, USA, 2003.
47. Wagenbrenner, J.W.; MacDonald, L.H.; Rough, D. Effectiveness of three post-fire rehabilitation treatments in the Colorado Front Range. *Hydrol. Process.* **2006**, *20*, 2989–3006. [CrossRef]
48. Oropeza, J.; Heath, J. *Effects of the 2012 Hewlett and High Park Wildfires on Water Quality of the Poudre River and Seaman Reservoir*; City of Fort Collins Utilities Report: Fort Collins, CO, USA, 2013.
49. Moody, J.A.; Martin, D.A. Synthesis of sediment yields after wildland fire in different rainfall regimes in the western United States. *Int. J. Wildland Fire* **2009**, *18*, 96–115. [CrossRef]

50. LANDFIRE. *Fuel, Topography, Existing Vegetation Type, and Fuel Disturbance Layers, Version 1.4.0.*; USDOI Geological Survey: Washington, DC, USA, 2016. Available online: http://landfire.cr.usgs.gov/viewer/ (accessed on 23 August 2016).
51. Price, O.; Bradstock, R. The effect of fuel age on the spread of fire in sclerophyll forest in the Sydney region of Australia. *Int. J. Wildland Fire* **2010**, *19*, 35–45. [CrossRef]
52. Narayanaraj, G.; Wimberly, M.C. Influences of forest roads on the spatial pattern of wildfire boundaries. *Int. J. Wildland Fire* **2011**, *20*, 792–803. [CrossRef]
53. Yocum, L.L.; Jenness, J.; Fulé, P.Z.; Thode, A.E. Previous fires and roads limit wildfire growth in Arizona and New Mexico, U.S.A. *For. Ecol. Manag.* **2019**, *449*, 117440. [CrossRef]
54. Scott, J.H.; Burgan, R.E. *Standard Fire Behavior Fuel Models: A Comprehensive Set for Use with Rothermel's Surface Fire Spread Model*; General Technical Report RMRS-GTR-153; USDA Forest Service, Rocky Mountain Research Station: Fort Collins, CO, USA, 2005.
55. Moriarty, K.; Cheng, A.S.; Hoffman, C.M.; Cottrell, S.P.; Alexander, M.E. Firefighter observations of "surprising" fire behavior in mountain pine beetle-attacked lodgepole pine forests. *Fire* **2019**, *2*, 34. [CrossRef]
56. Short, K.C. *Spatial Wildfire Occurrence Data for the United States, 1992–2015*, 4th ed; USDA Forest Service Research Data Archive: Fort Collins, CO, USA, 2017. [CrossRef]
57. ESRI. *ArcGIS, Version 10.3*; Environmental Systems Research Institute: Redlands, CA, USA, 2015. Available online: https://www.esri.com/en-us/home (accessed on 1 July 2017).
58. NWCG. *Remote Automated Weather Station Data*; National Wildfire Coordinating Group: Washington, DC, USA, 2018. Available online: https://fam.nwcg.gov/fam-web/weatherfirecd/index.htm (accessed on 6 June 2018).
59. Bradshaw, L.; McCormick, E. *FireFamily Plus User's Guide, Version 2.0*; General Technical Report RMRS-GTR-67WWW; USDA Forest Service, Rocky Mountain Research Station: Ogden, UT, USA, 2000.
60. Crosby, J.S.; Chandler, C.C. Get the most from your windspeed observation. *Fire Control Notes* **1966**, *27*, 12–13.
61. Scott, J.H.; Reinhardt, E.D. *Assessing Crown Fire Potential by Linking Models of Surface and Crown Fire Behavior*; General Technical Research Paper RMRS-RP-29; USDA Forest Service, Rocky Mountain Research Station: Fort Collins, CO, USA, 2001.
62. Tillery, A.C.; Haas, J.R.; Miller, L.W.; Scott, J.H.; Thompson, M.P. *Potential Post-Wildfire Debris-Flow Hazards—A Pre-Wildfire Evaluation for the Sandia and Manzano Mountains and Surrounding Areas, Central New Mexico*; Scientific Investigations Report 2014-5161; US Geological Survey: Albuquerque, NM, USA, 2014.
63. USEPA; USGS. *National Hydrography Dataset Plus—NHDPlus, Version 2.1*; US Environmental Protection Agency and USDOI Geological Survey: Washington, DC, USA, 2012. Available online: http://www.horizon-systems.com/NHDPlus/index.php (accessed on 23 August 2016).
64. Theobald, D.M.; Merritt, D.M.; Norman, J.B. *Assessment of Threats to Riparian Ecosystems in the Western U.S.*; Report to the Western Environmental Threats Assessment Center by the USDA Stream Systems Technology Center and Colorado State University: Fort Collins, CO, USA, 2010.
65. Perica, S.; Martin, D.; Pavlovic, S.; Roy, I.; St. Laurent, M.; Trypaluk, C.; Unruh, D.; Yekta, M.; Bonnin, G. *NOAA Atlas 14, Volume 8 Precipitation-Frequency Atlas of the United States, Midwestern States, Version 2*; US National Oceanic and Atmospheric Administration: Silver Spring, MD, USA, 2013.
66. Wilson, C.; Kampf, S.K.; Wagenbrenner, J.W.; MacDonald, L.H. Rainfall thresholds for post-fire runoff and sediment delivery from plot to watershed scales. *For. Ecol. Manag.* **2018**, *430*, 346–356. [CrossRef]
67. Winchell, M.F.; Jackson, S.H.; Wadley, A.M.; Srinivasan, R. Extension and validation of a geographic information system-based method for calculating the Revised Universal Soil Loss Equation length-slope factor for erosion risk assessments in large watersheds. *J. Soil Water Conserv.* **2008**, *63*, 105–111. [CrossRef]
68. NRCS Soil Survey Staff. *Web Soil Survey*; USDA Natural Resources Conservation Service: Washington, DC, USA, 2016. Available online: https://websoilsurvey.nrcs.usda.gov/ (accessed on 23 August 2016).
69. Larsen, I.J.; MacDonald, L.H. Predicting post-fire sediment yields at the hillslope scale: Testing RUSLE and disturbed WEPP. *Water Resour. Res.* **2007**, *43*, W11412. [CrossRef]
70. Henkle, J.E.; Wohl, E.; Beckman, N. Locations of channel heads in the semiarid Colorado Front Range, USA. *Geomorphology* **2011**, *129*, 309–319. [CrossRef]
71. Ryan, S.E.; Dwire, K.A.; Dixon, M.K. Impacts of wildfire on runoff and sediment loads at Little Granite Creek, western Wyoming. *Geomorphology* **2011**, *129*, 113–130. [CrossRef]
72. Schmeer, S.R. Post-Fire Erosion Response and Recovery, High Park Fire, Colorado. Master's Thesis, Colorado State University, Fort Collins, CO, USA, 2014.

73. Sham, C.H.; Tuccillo, M.E.; Rooke, J. *Effects of Wildfire on Drinking Water Utilities and Best Practices for Wildfire Risk Reduction and Mitigation*; Report 4482; Water Research Foundation: Denver, CO, USA, 2013.
74. Writer, J.H.; Hohner, A.; Oropeza, J.; Schmidt, A.; Cawley, K.M.; Rosario-Ortiz, F.L. Water treatment implications after the High Park Wildfire, Colorado. *J. Am. Water Works Assn.* **2014**, *106*, 189–199. [CrossRef]
75. Sherriff, R.L.; Platt, R.V.; Veblen, T.T.; Schoennagel, T.L.; Gartner, M.H. Historical, observed, and modeled wildfire severity in montane forests of the Colorado Front Range. *PLoS ONE* **2014**, *9*, e106971. [CrossRef]
76. Davis, L.S. *The Economics of Wildfire Protection with Emphasis on Fuel Break Systems*; California Division of Forestry: Sacramento, CA, USA, 1965.
77. Wilson, A.A.G. Width of firebreak that is necessary to stop grass fires: Some field experiments. *Can. J. For. Res.* **1988**, *18*, 682–687. [CrossRef]
78. Mees, R.; Strauss, D.; Chase, R. Modeling wildland fire containment with uncertain flame length and fireline width. *Int. J. Wildland Fire* **1993**, *3*, 179–185. [CrossRef]
79. Agee, J.K.; Bahro, B.; Finney, M.A.; Omi, P.N.; Sapsis, D.B.; Skinner, C.N.; van Wagtendonk, J.W.; Weatherspoon, C.P. The use of shaded fuelbreaks in landscape fire management. *For. Ecol. Manag.* **2000**, *127*, 55–66. [CrossRef]
80. O'Connor, C.D.; Calkin, D.E. Engaging the fire before it starts: A case study from the 2017 Pinal Fire (Arizona). *Wildfire Mag.* **2019**, *28*, 14–18.
81. Ingalsbee, T. Ecological fire use for ecological fire management: Managing large wildfires by design. In Proceedings of the Large Wildland Fires Conference, Missoula, MT, USA, 19–23 May 2014; Keane, R.E., Matt, J., Parsons, R., Riley, K., Eds.; Proceedings RMRS-P-73. USDA Forest Service Rocky Mountain Research Station: Fort Collins, CO, USA, 2015; pp. 120–127.
82. Blake, D.; Nyman, P.; Nice, H.; D'Souza, F.M.L.; Kavazos, C.R.J.; Horwitz, P. Assessment of post-wildfire erosion risk and effects on water quality in south-western Australia. *Fire* **2020**, *29*, 240–257. [CrossRef]
83. Moody, J.A.; Shakesby, R.A.; Robichaud, P.R.; Cannon, S.H.; Martin, D.A. Current research issues related to post-wildfire runoff and erosion processes. *Earth Sci. Rev.* **2013**, *122*, 10–37. [CrossRef]
84. Robichaud, P.R.; Lewis, S.A.; Wagenbrenner, J.W.; Ashmun, L.E.; Brown, R.E. Post-fire mulching for runoff and erosion mitigation Part I: Effectiveness at reducing hillslope erosion rates. *Catena* **2013**, *105*, 75–92. [CrossRef]
85. Robichaud, P.R.; Wagenbrenner, J.W.; Brown, R.E.; Wohlgemuth, P.M.; Beyers, J.L. Evaluating the effectiveness of contour-felled log erosion barriers as a post-fire runoff and erosion mitigation treatment in the western United States. *Int. J. Wildland Fire* **2008**, *17*, 255–273. [CrossRef]
86. Robichaud, P.R.; Wagenbrenner, J.W.; Lewis, S.A.; Ashmun, L.E.; Brown, R.E.; Wohlgemuth, P.M. Post-fire mulching for runoff and erosion mitigation Part II: Effectiveness in reducing runoff and sediment yields from small catchments. *Catena* **2013**, *105*, 93–111. [CrossRef]
87. Ferro, V.; Porto, P. Sediment Delivery Distributed (SEDD) Model. *J. Hydrol. Eng.* **2000**, *5*, 411–422. [CrossRef]
88. Fernandez, C.; Wu, J.Q.; McCool, D.K.; Stöckle, C.O. Estimating water erosion and sediment yield with GIS, RUSLE, and SEDD. *J. Soil Water Conserv.* **2003**, *58*, 128–136.
89. Rhoades, C.C.; Entwistle, D.; Butler, D. The influence of wildfire extent and severity on streamwater chemistry, sediment and temperature following the Hayman Fire, Colorado. *Int. J. Wildland Fire* **2011**, *20*, 430–442. [CrossRef]
90. Kampf, S.K.; Brogan, D.J.; Schmeer, S.; MacDonald, L.H.; Nelson, P.A. How do geomorphic effects of rainfall vary with storm type and spatial scale in a post-fire landscape? *Geomorphology* **2016**, *273*, 39–51. [CrossRef]
91. Diodato, N.; Bellocchi, G. Reconstruction of seasonal net erosion in a Mediterranean landscape (Alento River Basin, Southern Italy) over the past five decades. *Water* **2019**, *11*, 2306. [CrossRef]
92. Hohner, A.K.; Cawley, K.; Oropeza, J.; Summers, R.S.; Rosario-Ortiz, F.L. Drinking water treatment response following a Colorado wildfire. *Water Res.* **2016**, *105*, 187–198. [CrossRef]
93. Salis, M.; Del Giudice, L.; Robichaud, P.R.; Ager, A.A.; Canu, A.; Duce, P.; Pellizzaro, G.; Ventura, A.; Alcasena-Urdiroz, F.; Spano, D.; et al. Coupling wildfire spread and erosion models to quantify post-fire erosion before and after fuel treatments. *Int. J. Wildland Fire* **2019**, *28*, 687–703. [CrossRef]
94. Chambers, M.E.; Fornwalt, P.J.; Malone, S.L.; Battaglia, M.A. Patterns of conifer regeneration following high severity wildfire in ponderosa pine-dominated forests of the Colorado Front Range. *For. Ecol. Manag.* **2016**, *378*, 57–67. [CrossRef]

95. Fornwalt, P.J.; Huckaby, L.S.; Alton, S.K.; Kaufmann, M.R.; Brown, P.M.; Cheng, A.S. Did the 2002 Hayman Fire, Colorado, USA, burn with uncharacteristic severity? *Fire Ecol.* **2016**, *12*, 117–132. [CrossRef]
96. Collins, B.M.; Stevens, J.T.; Miller, J.D.; Stephens, S.L.; Brown, P.M.; North, M.P. Alternative characterization of forest fire regimes: Incorporating spatial patterns. *Landsc. Ecol.* **2017**, *32*, 1543–1552. [CrossRef]
97. Calkin, D.E.; Cohen, J.D.; Finney, M.A.; Thompson, M.P. How risk management can prevent future wildfire disasters in the wildland-urban interface. *Proc. Natl. Acad. Sci. USA* **2014**, *111*, 746–751. [CrossRef]

 © 2020 by the authors. Licensee MDPI, Basel, Switzerland. This article is an open access article distributed under the terms and conditions of the Creative Commons Attribution (CC BY) license (http://creativecommons.org/licenses/by/4.0/).

Article

Assessing Potential Safety Zone Suitability Using a New Online Mapping Tool

Michael J. Campbell [1,*], Philip E. Dennison [1], Matthew P. Thompson [2] and Bret W. Butler [3]

[1] Department of Geography, University of Utah, Salt Lake City, UT 84112, USA; dennison@geog.utah.edu
[2] USDA Forest Service, Rocky Mountain Research Station, Fort Collins, CO 80526, USA; matthew.p.thompson@usda.gov
[3] USDA Forest Service, Rocky Mountain Research Station, Missoula, MT 59801, USA; bretwbutler@gmail.com
* Correspondence: mickey.campbell@geog.utah.edu

Citation: Campbell, M.J.; Dennison, P.E.; Thompson, M.P.; Butler, B.W. Assessing Potential Safety Zone Suitability Using a New Online Mapping Tool. *Fire* **2022**, *5*, 5. https://doi.org/10.3390/fire5010005

Academic Editor: James R. Meldrum

Received: 12 November 2021
Accepted: 5 January 2022
Published: 7 January 2022

Publisher's Note: MDPI stays neutral with regard to jurisdictional claims in published maps and institutional affiliations.

Copyright: © 2022 by the authors. Licensee MDPI, Basel, Switzerland. This article is an open access article distributed under the terms and conditions of the Creative Commons Attribution (CC BY) license (https://creativecommons.org/licenses/by/4.0/).

Abstract: Safety zones (SZs) are critical tools that can be used by wildland firefighters to avoid injury or fatality when engaging a fire. Effective SZs provide safe separation distance (SSD) from surrounding flames, ensuring that a fire's heat cannot cause burn injury to firefighters within the SZ. Evaluating SSD on the ground can be challenging, and underestimating SSD can be fatal. We introduce a new online tool for mapping SSD based on vegetation height, terrain slope, wind speed, and burning condition: the Safe Separation Distance Evaluator (SSDE). It allows users to draw a potential SZ polygon and estimate SSD and the extent to which that SZ polygon may be suitable, given the local landscape, weather, and fire conditions. We begin by describing the algorithm that underlies SSDE. Given the importance of vegetation height for assessing SSD, we then describe an analysis that compares LANDFIRE Existing Vegetation Height and a recent Global Ecosystem Dynamics Investigation (GEDI) and Landsat 8 Operational Land Imager (OLI) satellite image-driven forest height dataset to vegetation heights derived from airborne lidar data in three areas of the Western US. This analysis revealed that both LANDFIRE and GEDI/Landsat tended to underestimate vegetation heights, which translates into an underestimation of SSD. To rectify this underestimation, we performed a bias-correction procedure that adjusted vegetation heights to more closely resemble those of the lidar data. SSDE is a tool that can provide valuable safety information to wildland fire personnel who are charged with the critical responsibility of protecting the public and landscapes from increasingly intense and frequent fires in a changing climate. However, as it is based on data that possess inherent uncertainty, it is essential that all SZ polygons evaluated using SSDE are validated on the ground prior to use.

Keywords: firefighter safety; safe separation distance; safety zones; LCES; Google Earth Engine; lidar; LANDFIRE; Landsat; GEDI

1. Introduction

Wildland firefighters are tasked with a wide variety of fire management duties, many of which place them in close proximity to flames. One of the primary tasks is the removal of fuels and construction of containment lines in order to limit the potential damage to lives, property, and other critical resources [1–3]. Particularly when engaged in a direct attack, whereby firefighters may be working within a few meters or less of the flaming front, the potential risk for safety incidents is elevated [4]. Sudden or unexpected changes in fire behavior can have devastating effects to vulnerable fire personnel on the ground [5]. Events such as the Yarnell Hill fire in 2013, which claimed the lives of 19 firefighters and the South Canyon fire, which resulted in 14 firefighter fatalities, demonstrate the tragedy that can occur in the wildland fire profession [6–8]. Beyond these well-known, high-fatality events, there is an additional and significant background level of mortality that occurs among on-duty wildland firefighters [9]. The causes of death are varied, and include heart attacks, vehicular and aircraft accidents, falling trees, and smoke inhalation, to name a few. Between 1990 and 2016, there were 480 wildland firefighter fatalities, nearly one fifth (19%) of which were due to burnovers or entrapments. Burnovers are events in which flames overcome fire personnel, either on the ground or while in a vehicle, and entrapments are when firefighters become trapped by surrounding flames, unable to evacuate to safety [10]. As wildland fires increase in frequency, extent, and intensity, wildland firefighters may be put at heightened risk while working in the increasingly complex fire environment [11–15]. Passage of the Infrastructure Investment and Job Act in the US developing a requirement for federal agencies to "develop and adhere to recommendations for mitigation strategies for wildland firefighters to minimize exposure due to line-of-duty environmental hazards" further underscores the importance and continued policy relevance of wildland firefighter safety (see H.R. 3684 §40803(d)(5)(A)).

To avoid injury or fatality, firefighters employ a range of safety measures in anticipation of and throughout fire management operations. Safety is often considered the first priority of wildland firefighters, as evident in the 10 Standard Firefighting Orders and the 18 Watch Out Situations, colloquially referred to as the "10s and 18s" and known by all fire crews in the US [16,17]. One of the most important safety protocols is the interconnected and interdependent system of lookouts, communications, escape routes, and safety zones (LCES) [18]. First formalized by Paul Gleason in 1991, and later modified to include anchor points by Thorburn and Alexander in 2001 [19], proper implementation and continued reevaluation of LCES is essential to firefighter survival. Lookouts are members of a fire crew well-trained in the interpretation of fire behavior and weather conditions who are placed at a vantage point in the fire environment that ensures continued visual observation of the fire crew, the fire itself, and the surrounding landscape. Communications ensure that critical information is conveyed in a clear, timely, and orderly manner between various resources deployed on a fire, including ground crews, lookouts, vehicular and aerial assets, and incident command. Escape routes are pre-defined evacuation pathways that enable crews to retreat to a safety zone or other safe area. And finally, safety zones (SZs), which are the primary focus of this paper, are areas on the landscape that firefighters can retreat to in dangerous situations in order to avoid bodily harm. SZs are large, open areas with little or no flammable material contained within. They can take a variety of forms, including naturally sparsely or unvegetated areas and areas that have had fuel removed either mechanically, or through burning.

The effectiveness of SZs is defined by the extent to which they can provide safe separation distance (SSD) from surrounding or nearby flames [4,5,20–22]. SSD is defined as the distance one must maintain from fire to avoid burn injury, without deployment of a fire shelter [5,22]. The current operational guideline in the US for defining SSD emerged from the work of Butler and Cohen in 1998, who used radiative heat transfer modeling to determine that firefighters should maintain a distance of at least 4 times the height of the proximal flames [22]. Although this guideline has since been widely adopted [23], the research that underlies it is based solely on one heat transfer mechanism: radiation. Heat transfer by convection is also a major—sometimes dominant—force, particularly in the presence of steep slopes and high winds [24–26]. In the presence of such convective heat, particularly if a fire crew is upslope and/or downwind of flames, SSD will increase [5,21]. Thus, the four times flame height rule is likely insufficient in these conditions. Recent work by Butler et al. has sought to update this guideline with the inclusion of a "slope-wind factor", which adds a multiplicative term to the SSD equation to account for the effects of convective heat transfer [5,21,26–28]. In addition, given that SZs should be designated prior to, rather than during, the presence of flames, the four times flame height rule requires firefighters to predict how tall the flames might eventually be, which is a challenging endeavor. Accordingly, the newly proposed guidelines assume that, in a crown fire, flame height is approximately equal to twice the vegetation height [28]. As a result, the new SSD equation is defined as:

$$SSD = 8 \times VH \times \Delta, \qquad (1)$$

where VH is vegetation height and Δ is the slope-wind factor. Butler recently defined these slope-wind factors seen in Table 1, based not only on slope and wind speeds, but also on the burning conditions, as dictated by fuel conditions (e.g., moisture) and weather (e.g., relative humidity) [28].

Table 1. Slope-wind factors (Δ) from Butler, colored on a scale from blue (low Δ) to white (moderate Δ) to red (high Δ) [28].

		Slope					
		Flat (0–7.5%)	Low (7.6–22.5%)	Moderate (22.6–40%)	Steep (>40%)		
Wind Speed	Light (0–4.5 m/s)	0.8	1	1	2	Low	
		1	1	1.5	2	Moderate	
		1	1.5	1.5	3	Extreme	
	Moderate (4.6–8.9 m/s)	1.5	2	3	4	Low	Burning Condition
		2	2	4	6	Moderate	
		2	2.5	5	6	Extreme	
	High (>8.9 m/s)	2.5	3	4	6	Low	
		3	3	5	7	Moderate	
		3	4	5	10	Extreme	

Although guidelines for use on the ground are valuable, they still require the firefighters themselves to make the calculation of SSD on the ground while engaged in other fire management activities. This requires the ability to accurately estimate vegetation height and terrain slope and anticipate wind speed and fire intensity. Moreover, even if these difficult interpretations and predictions can be made, an even more challenging endeavor is to identify an area on the ground cleared of vegetation that provides the calculated SSD in all directions. In order to reduce the subjectivity of this process and the potential for interpretation error, Campbell et al. sought to develop a robust, geospatially driven approach for identifying and assessing the suitability of SZs in advance of a fire using high-resolution airborne lidar, which provides detailed terrain and vegetation height information [20]. Their method automatically identified open areas, determined the height of surrounding trees, and calculated a score for the relative suitability of potential safety zones within the open areas based on tree height and SSD. The potential for broad applicability of their approach, however, is limited by three main factors: (1) the lack of widely available airborne lidar data, (2) the temporal relevancy of and lack of scheduled updates to available lidar data, and (3) the fact that only existing clearings could be assessed. With respect to the first limitation, at present, high-quality, publicly accessible airborne lidar data is particularly sparse in the Western US, where fires are most frequent, large, and intense. Regarding the second limitation, vegetation structure is dynamic in fire-prone environments, where SZ assessment is most important. Thus, having outdated lidar data potentially limits accurate SSD calculation. Lastly, although existing open areas may be viable SZs, firefighters often use areas that have already burned ("black" areas) or create SZs through mechanical fuel removal.

To resolve the limitations of this previous work and to improve wildland firefighter safety, we are introducing a new, interactive, web-based, open-access mapping tool for estimating SSD and evaluating potential SZ effectiveness through geospatial analysis. Instead of relying on lidar, this tool uses LANDFIRE Existing Vegetation Height data, which is both nationally available in the contiguous US and is updated every few years. Additionally, instead of only assessing SSD-driven suitability on clearings that already exist, this tool allows users to draw their own SZ polygon to evaluate the potential suitability of a SZ in any environment. Since LANDFIRE vegetation heights may not be as accurate as airborne lidar, given that it is a modeled product driven by satellite imagery, it is important to quantify the effects of differing sources of vegetation height data on SSD evaluation. Accordingly, the primary objectives of this study are to: (1) introduce and describe the algorithm that underlies a new tool for calculating SSD and analyzing SZ suitability; (2) compare SSD and potential SZ suitability using different sources of vegetation height data; (3) demonstrate an example case study of the tool in a simulated wildland firefighting situation.

2. Materials and Methods

2.1. Algorithm Description

The Safe Separation Distance Evaluator (SSDE) algorithm is built and applied in Google Earth Engine (GEE), a cloud-based platform for processing and analyzing GIS and remotely sensed data, using the JavaScript application programing interface [29]. GEE was selected for few reasons: (1) it enables the production of user-facing applications that can be widely accessed by anyone with an internet connection; (2) it hosts an immense catalog of geospatial data, including datasets necessary for the analysis of SZ suitability; (3) its cloud computing capabilities provide for rapid execution of complex geospatial functions, allowing users to quickly assess SZ suitability. Accordingly, all data processing described in this section is conducted using the GEE.

The SSDE evaluates SSD in two primary ways (Figure 1). The first is per-pixel SSD, which is a representation of how far one must be from that pixel (e.g., in meters) in order to avoid burn injury (Figure 1a). This is calculated at the individual pixel level across an entire area of interest based on the vegetation height and terrain slope within each pixel, and user-defined wind speed and burn condition classes. It provides a landscape-scale view of SSD and can be used to aid in the delineation of potential SZ polygons. However, it is perhaps more important to evaluate SSD at the level of the SZ polygon, as this can help fire personnel determine the suitability of a potential SZ. Accordingly, the second way that SSDE evaluates SSD is through the analysis of proportional SSD (pSSD) within potential SZ polygons (Figure 1b). pSSD quantifies the extent to which a potential SZ polygon provides SSD from surrounding vegetation/flames, considering the average per-pixel SSD contained within a series of segments (or clusters of contiguous pixels) around the SZ polygon. Measured in percent, a pSSD of 100% or greater for a given pixel would mean that, factoring in vegetation height surrounding the polygon, slope, wind speed, and burn condition, the pixel's location should provide sufficient SSD, should fire personnel opt to use this location as a SZ. Conversely, a pixel with a pSSD of less than 100% would indicate that firefighters located within that pixel may risk injury from burning vegetation outside the boundary of the polygon. A detailed description of the computation of both SSD and pSSD follows.

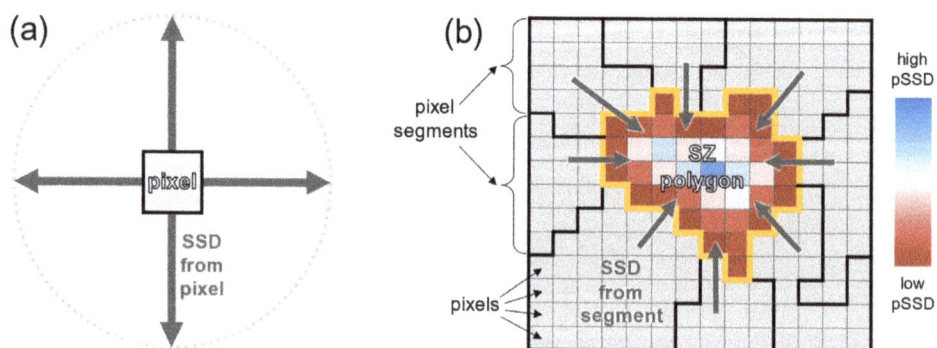

Figure 1. Conceptual depiction of the two different ways that SSDE calculates SSD, including a per-pixel SSD, representing the distance one must maintain from a pixel in order to avoid burn injury (**a**), and within-SZ polygon pSSD, representing the relative extent to which a potential SZ polygon provides SSD from surrounding flames, based on segment-level mean SSD (**b**).

Per-pixel SSD is calculated as follows. The algorithm relies on two primary input datasets, both of which are available on a nationwide basis in the US. The first is LANDFIRE

Existing Vegetation Height, version 2.0, which is a 30 m spatial resolution raster dataset where each pixel represents the average height of the dominant vegetation within that pixel (Figure 2b) [30,31]. The second is a Shuttle Radar Topography Mission (SRTM) digital elevation model (DEM), also at 30 m spatial resolution [32]. Using the SRTM DEM, terrain slope in percent is calculated (Figure 2c). This slope raster is then converted into a slope-wind factor dataset, based on user-defined wind speed and burning condition categories (Figure 2d; Table 1). Using Equation (1), SSD is calculated from the slope-wind factor and vegetation height data, producing a spatially exhaustive representation of the distance one must maintain from each pixel in the image dataset in order to avoid injury (Figure 2e). As discussed further in Sections 2.3 and 3.3, there is a user-selected bias correction option that is built into SSDE to account for underestimation of tree heights in the LANDFIRE data.

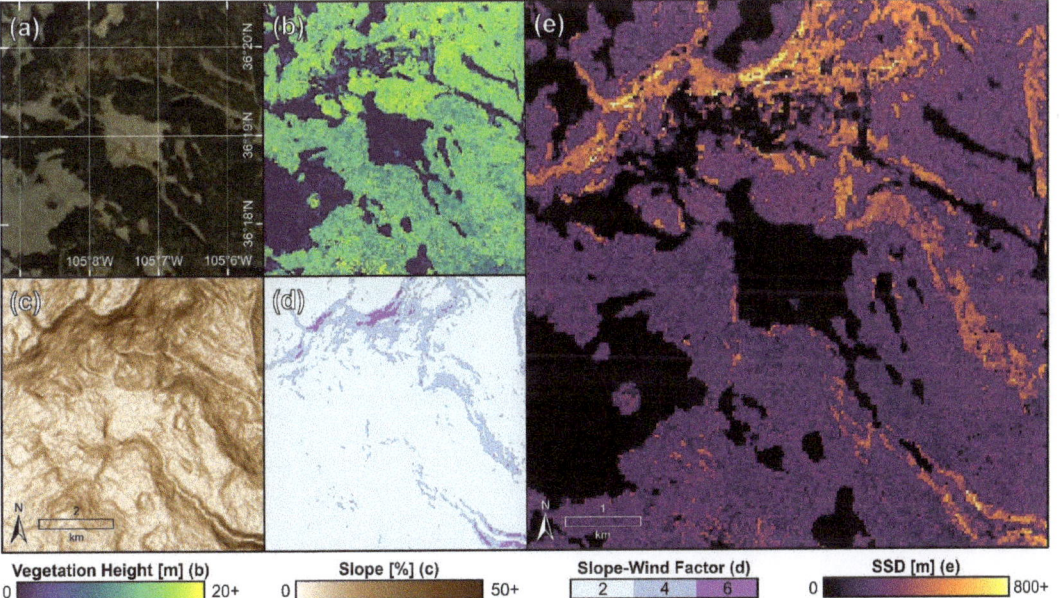

Figure 2. Illustration of the process for calculating per-pixel SSD in an example focus area in central New Mexico (a), including the vegetation height data from LANDFIRE (b), the SRTM-derived slope data (c), the slope-wind factor under moderate burning conditions and moderate wind speeds (d), and the resulting SSD raster (e). Legend labels contain parenthetical indication relevant sub-figure. Note that this figure does not illustrate the SZ selection process in a real or simulated wildland fire; it merely conveys the calculation of per-pixel SSD in a landscape in which fires could occur.

The per-pixel, landscape-wide layers representing SSD, vegetation height, and slope, as well as the high-resolution Google imagery base map, can be used to identify potential SZ on the landscape. In a realistic wildland fire scenario, this identification process would likely be undertaken by a GIS specialist working with an incident management team, who has specific knowledge of the current fire extent, projected changes in fire behavior, and crew assignments. However, given the open-access nature of SSDE, this process can likewise be undertaken by anyone with an interest in wildland firefighter safety. To calculate pSSD at the SZ polygon level, the user can define a potential SZ using the polygon drawing tools in SSDE, guided by the conditions both within the polygon and surrounding the polygon. The best SZs are those that contain no flammable material within, naturally or otherwise, so ideally this SZ polygon would be drawn in an area with low fuel loading, such as short or sparse grasses or litter. Alternatively, a SZ polygon could also be drawn in an area that has recently burned, or an area that would be targeted for fuel removal to create a SZ.

pSSD within a SZ is dependent upon the slope and vegetation height of the surrounding landscape. As discussed in the Introduction, heat transfer from flames generally increases with increasing vegetation height and terrain slope. Accordingly, a SZ in the midst of steep terrain and tall vegetation will require a larger SSD than a SZ in the midst of flat terrain and short vegetation. If we assume that the SZ itself contains little or no flammable material, then the primary concern for SZ evaluation is the area surrounding the SZ. Accordingly, to calculate pSSD within the SZ polygon, slope and vegetation height need to be evaluated within a "buffer" surrounding a SZ [20,33]. In an environment with short vegetation, a relatively small buffer surrounding a SZ needs to be evaluated, because flame heights will tend to be low and more distant vegetation will not affect the suitability of a SZ given its limited capacity for transferring heat over long distances (Figure 3a). Conversely, in the presence of tall trees and steep slopes, larger buffer areas are needed to account for the effects of more distant fuel and flames (Figure 3b). For example, a stand of 20 m tall forest in a low-slope environment with moderate wind speed and burning conditions would require an SSD of 320 m—that is, to maintain safety, firefighters would need to find a SZ that provides at least 320 m of separation from this stand. However, if the SZ evaluation procedure only included an analysis of vegetation within a small (e.g., 20 m) buffer around the SZ, then the potentially dangerous effects of the heat emitting from a crown fire in this stand could be missed, putting the firefighters at risk.

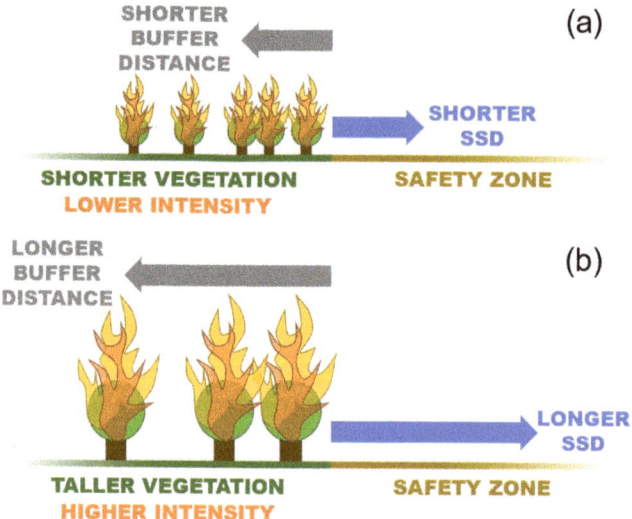

Figure 3. Relationship between the safe separation distance (SSD) defined by vegetation surrounding the safety zone and the size of the buffer needed around the safety zone to ensure that all relevant surrounding vegetation is considered in the safety zone analysis. With shorter vegetation, a smaller buffer is needed (**a**), whereas with taller vegetation, a larger buffer is needed (**b**). Note that the relationship between vegetation height and SSD is not drawn to scale.

Accordingly, to ensure that an appropriately sized buffer area is evaluated around a potential SZ polygon, the buffer size must be defined by the highest per-pixel SSD in the surrounding area. To do this, the largest theoretically possible SSD in the US (SSD_{USmax}) is first calculated as follows:

$$SSD_{USmax} = 8 \times VH_{USmax} \times \Delta_{max}, \qquad (2)$$

where VH_{USmax} is the height of the tallest tree in the US (116 m) and Δ_{max} is the maximum slope-wind factor from Table 1 (10). As a result, SSD_{USmax} is equal to 9280 m. While this is purely a theoretical SSD, it provides a useful initial buffer distance. The potential SZ polygon is buffered by this SSD_{USmax}, and then the highest SSD pixel value within the initial buffer (SSD_{SZmax}) is then identified. This value is then used to create the final buffer around a designated SZ polygon, which will serve as the basis of SSD evaluation.

The buffer area is then used to clip out a subset of the SSD raster (Figure 4a). Given that each pixel possesses its own SSD (based on its unique combination of vegetation height and slope-wind factor), it is important to consider the variation in SSD that is found throughout the area surrounding the SZ polygon. Vegetation may be taller or slopes may be steeper on one edge of a SZ polygon as compared to the other. Accounting for this spatial variability at the pixel level would greatly increase computation time, so the SSDE algorithm applies an image segmentation function on the clipped SSD raster to generate a series of segments (i.e., clusters of similar pixels) representing areas with relatively homogeneous SSD values (Figure 4a). From each segment, Euclidean distance is calculated on a continuous basis as a 30 m spatial resolution raster where each pixel represents the distance from that segment. Each segment also has a mean SSD value, aggregated from the SSD pixel values within the segment. Accordingly, a comparison between the Euclidean distance raster and the segment's mean SSD value should provide insight into whether pixels in the SZ polygon are within or beyond the SSD. To determine this on a relative basis, the Euclidean distance is then divided by the segment's mean SSD, resulting in a dataset containing pixels representing pSSD, where pixel values under 1 (or 100%) represent areas that would be unsafe if the fuel within that segment were burning under crown fire conditions (Figure 4b). Values over 1 (or 100%) represent areas that would be safe, according to the SSD guideline. This process is repeated for every segment surrounding the SZ polygon, producing a series of pSSD rasters. To reduce these rasters down to a single representation of pSSD within the SZ polygon, the minimum pixel value (representing the "worst-case" pSSD for all segments surrounding the polygon) is extracted among all of the rasters (Figure 4c). This pSSD raster enables the identification of the safest area within the potential SZ polygon, defined as the location with the highest pSSD (Figure 4d). Lastly, a threshold is applied to the pSSD to distinguish between safe areas, or areas that are equal to or greater than 100% of pSSD, and unsafe areas, or areas that are less than 100% of pSSD (Figure 4d). The size of the resulting safe areas is important for wildland firefighters, as it provides a sense for how many resources (e.g., firefighters, trucks, dozers, engines) can utilize the SZ at once. It has been estimated that approximately 5 m^2 (50 ft^2) is recommended for each firefighter, and 28 m^2 (300 ft^2) for each piece of heavy equipment [33,34]. To enable users to interact with the resulting data outside of GEE, there are options for downloading all of the resulting data layers, including the SSD raster, pSSD raster, safe/unsafe raster, the SZ polygon vector feature, and the safest point vector feature.

2.2. Vegetation Height Analysis

Given that LANDFIRE Existing Vegetation Height is a modeled product based on multispectral satellite imagery, it can be assumed that there is inherent inaccuracy in its vegetation height estimates. Conversely, airborne lidar data, which result from laser pulses emitted from an airborne platform reflecting off the ground and aboveground surfaces, provide direct measurements of vegetation height. Thus, airborne lidar is considered a reliable source of data for deriving accurate and precise canopy height models in vegetated environments [35]. In fact, the models that derive the LANDFIRE Existing Vegetation Height product are trained in part using airborne lidar data, in addition to ground-level plot measurements [36]. Inaccuracy in vegetation height results in inaccuracy in SSD, and inaccuracy in SSD could have significant effects on the safety of wildland firefighters. Accordingly, to quantify these effects, we compared LANDFIRE-driven SSD to lidar-driven SSD.

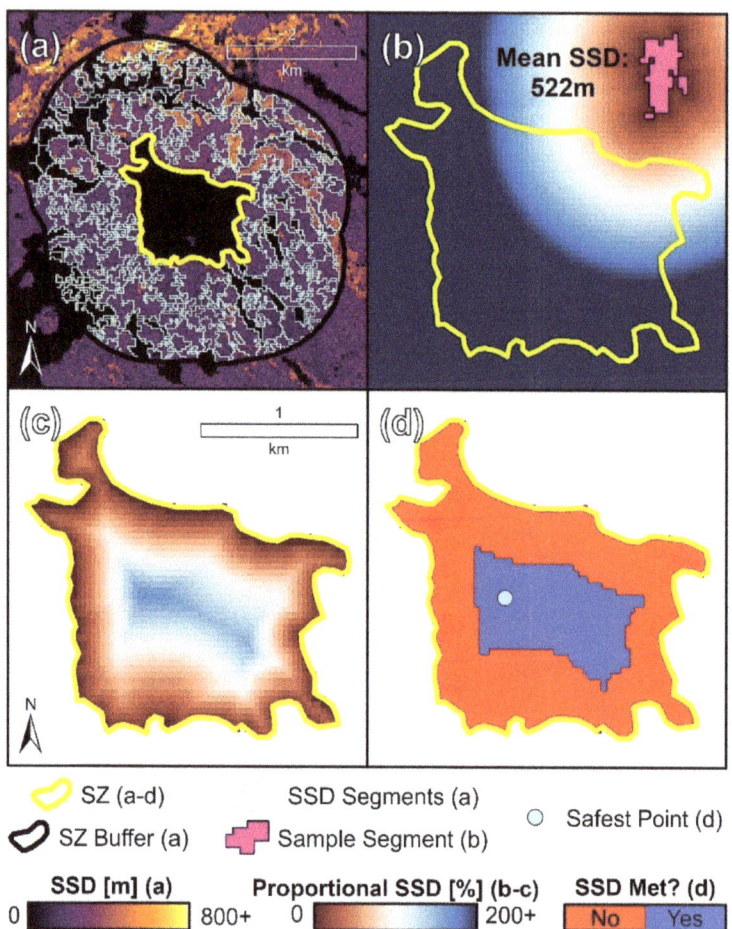

Figure 4. Illustration of the process for calculating within-safety zone (SZ) safe separation distance (SSD) in the same example focus area in central New Mexico shown in Figure 1, including the manual input of a SZ polygon, the buffering of that SZ polygon, and the generation of SSD segments (**a**), the calculation of pSSD from a single example segment (**b**), the combined minimum of pSSD from all segments (**c**), and derivative safety products, including the safest point and a binary classification of safe versus unsafe within the SZ polygon (**d**). Legend labels contain parenthetical indication relevant sub-figure(s).

Although LANDFIRE is one of the most widely used, nationwide vegetation height products in the US, particularly among wildland fire scientists, it is not the only one. A recent study by Potapov et al. introduced a new global forest height product driven by a combination of Global Ecosystem Dynamics Investigation (GEDI) data and Landsat 8 OLI imagery [37,38]. GEDI is a spaceborne lidar instrument that produces large footprint (~25 m diameter) full waveform surface elevation data. Distinct from airborne lidar, however, which produces a dense point cloud of measurements that can be used to interpolate spatially exhaustive, high-resolution height models, GEDI's sampling design is such that large gaps exist between successive and adjacent footprints, limiting the capacity to produce spatially contiguous height models. However, Potapov et al. used GEDI data to train a model to predict forest height using Landsat imagery on a global scale. Similarly to LANDFIRE, this GEDI/Landsat hybrid product is a modeled product and, as such, possesses inherent uncertainty. In the interest of basing our algorithm on the best and most reliable vegetation height data, we also compared the GEDI/Landsat-derived SSD to airborne lidar-derived SSD.

To compare LANDFIRE, GEDI/Landsat, and airborne lidar, we selected three areas in fire-prone regions of the Western US. These areas were selected to capture a range of terrain and vegetation height conditions. Importantly, each of the three areas was required to have recently collected, freely available airborne lidar data and no recent major disturbances. The 3 areas selected, each 20 × 20 km in size, are shown in Figure 5. The first area is in western New Mexico (Figure 5a), ranging in elevation from 1659 to 3032 m (mean = 2253 m), ranging in slope from 0 to 153% (mean = 26%), and is dominated by ponderosa pine forests and piñon-juniper woodlands. According to the airborne lidar data, vegetation heights range from 0 to 41 m (mean = 13 m). The second area is in western Wyoming (Figure 5b), ranging in elevation from 1741 to 3038 m (mean = 2284 m), ranging in slope from 0 to 179% (mean = 35%), and is dominated by Douglas fir and lodgepole pine forests and sagebrush shrublands. According to the airborne lidar data, vegetation heights range from 0 to 40 m (mean = 15 m). The third area is in northern California (Figure 5c), ranging in elevation from 137 to 1399 m (mean = 729 m), ranging in slope from 0 to 195% (mean = 30%), and is dominated by Douglas fir forests, black oak woodlands, and ruderal grasslands. According to the airborne lidar data, vegetation heights range from 0 to 72 m (mean = 22 m).

Airborne lidar datasets for each of these 3 areas were acquired from the USGS 3D Elevation Program (3DEP) [39]. The New Mexico data came from the USGS 3DEP dataset titled "NM_SouthCentral_2018_D19", collected in 2018 with an average point density of 6.0 pts/m^2. The Wyoming data came from the USGS 3DEP dataset titled "WY_Southwest_2020_D20", collected in 2020 with an average point density of 7.6 pts/m^2. The California data from the USGS 3DEP dataset titled "USGS_LPC_CA_NoCAL_Wildfires_B4_2018", collected in 2018 with an average point density of 10.1 pts/m^2. It should be noted that there are some temporal discrepancies between the airborne lidar data (2018–2020), LANDFIRE data (2016), and GEDI/Landsat data (2019). To avoid issues associated with major vegetation height changes that may have occurred between these time frames, these areas were selected specifically to avoid containing any fires that had occurred between 2015 and 2021. To match the training and validation data used in the development of LANDFIRE and Potapov et al.'s height products, airborne lidar canopy height models were derived at a 30 m spatial resolution using the 90th percentile of aboveground point returns within each pixel. All of the airborne lidar data processing was conducted using LAStools [40].

To assess SSD uncertainty, the algorithm described in Section 2.1 was applied to each of the three vegetation height datasets (LANDFIRE, GEDI/Landsat, and airborne lidar). However, to enable the flexibility needed for this comparative analysis, the algorithm was implemented in Python with heavy reliance on the Esri ArcPy library, rather than GEE. Given that the GEDI/Landsat product does not include vegetation heights lower than 3 m (as it is considered a "forest" height product), LANDFIRE vegetation heights were used in areas lower than 3 m in height for the GEDI/Landsat analysis. For all three datasets, the same terrain slope was used, derived from a USGS 3DEP 30 m digital elevation model. In each of the three test areas, SZ polygons were automatically generated, using a random shape generator created for this study. Briefly, it starts with a randomly located point, and then creates a shape based on a randomly defined number of equally spaced vertices that are placed at random distances from the center point. The distances are selected from a normal distribution with a mean between 100 and 1000 m and a standard deviation of 0.25 times the mean. The resulting shape is then smoothed to produce a final SZ polygon, as can be seen in Figure 5a–c. In all, 25 SZ polygons were generated for each test area (75 total). Since each polygon was evaluated independently, randomly placed polygons could overlap. For each SZ polygon, every unique combination of burning condition and wind speed was evaluated. As a result, the algorithm was run 2025 times (3 vegetation height datasets × 75 SZs × 3 burning conditions × 3 wind speeds).

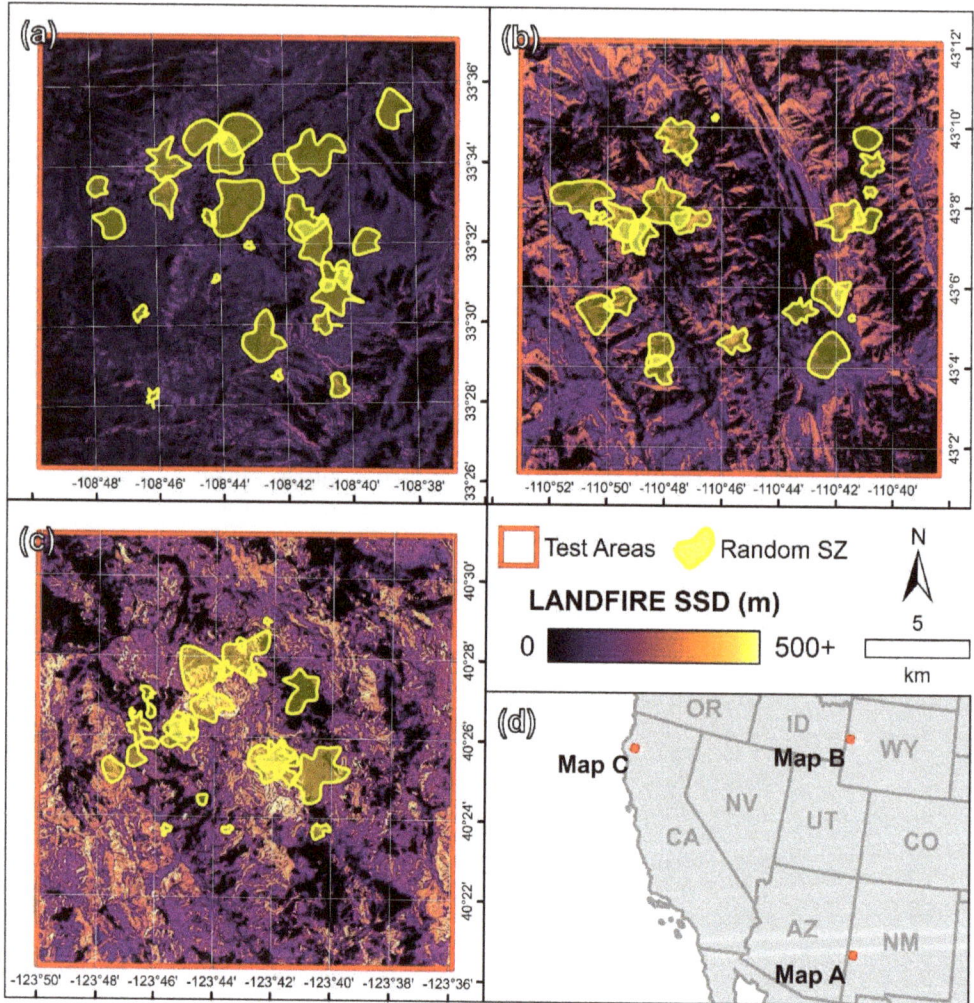

Figure 5. Three test areas used to assess SSD uncertainty, including a site in western New Mexico (**a**), western Wyoming (**b**), and northern California (**c**), as well as a locator map illustrating their broader geographic context (**d**). The LANDFIRE SSD shown represents low burning conditions and light wind. The SZs represent randomly generated polygons used for testing.

Four statistical comparisons were made between the results from the three different vegetation height datasets. The first was a per-pixel SSD comparison. For this, a set of 25 random points was generated within each combination of vegetation height dataset, burning condition, and wind speed (675 points total), and pixel values were extracted from SSD rasters. The second was the maximum pSSD found within each of the randomly generated SZ polygon for each combination of vegetation height dataset, burning condition, and wind speed. The third was the total safe area (the areas meeting or exceeding SSD) within the SZ polygon for each combination of vegetation height dataset, burning condition, and wind speed. All three of these comparisons were made using ordinary least squares regression between LANDFIRE and airborne lidar, GEDI/Landsat and airborne lidar, and LANDFIRE and GEDI/Landsat. The fourth and final statistical comparison was aimed at determining the geographic divergence between the locations identified as being the safest point within each SZ. For this, the horizontal distance between safe points was calculated between LANDFIRE and airborne lidar, GEDI/Landsat and airborne lidar, and LANDFIRE and GEDI/Landsat, from which histograms and associated descriptive statistics were derived.

2.3. LANDFIRE Bias Correction

As mentioned in Section 2.1 and discussed in greater detail in the Section 3.2, the per-pixel SSD comparison revealed that, in comparison to the airborne lidar-based SSD values, the LANDFIRE and GEDI/Landsat SSD values were underestimated. To ensure that the algorithm is not overestimating the relative safety of drawn SZs, a bias correction procedure was developed. Although both LANDFIRE and GEDI/Landsat featured a similar trend in bias, the bias correction was only tested on the LANDFIRE data, given the fact that LANDFIRE was ultimately selected as the vegetation height dataset used in SSDE, the justification for which is further discussed in Section 4. To correct for underestimation in SSD, the linear regression model resulting from the per-pixel SSD comparison was used to adjust LANDFIRE SSD pixel values through a linear transformation. To test the extent to which this bias correction procedure increased the agreement between LANDFIRE- and lidar-derived SSD values and SZ suitability metrics, the same four statistical comparisons previously described were conducted: (1) per-pixel SSD linear regression; (2) within-SZ maximum pSSD linear regression; (3) within-SZ total safe area linear regression; (4) safest point distance histogram comparison. Lastly, to enable users to select between using the raw LANDFIRE data or the bias-corrected LANDFIRE data as a basis of SSD calculation, a selection option was added to SSDE.

2.4. Use Case Demonstation

To provide an example use case for SSDE, we simulated a situation in which wildland firefighters are tasked with selecting a SZ during an active wildfire. To do this, we created a fictional fire perimeter in a ponderosa pine forest in an area of northern New Mexico that has a variety of potential SZs in the form of open grassy meadows. Two of these potential SZs are delineated and analyzed using SSDE. To test the viability of these potential SZ under different weather and fire conditions, they are each evaluated for two sets of extremes: (1) light wind and low burn condition; (2) high wind and extreme burn condition.

3. Results

3.1. GEE Application

The SSDE was developed in GEE and a free, open-access, web-based application can be viewed at https://firesafetygis.users.earthengine.app/view/ssde (accessed on 3 January 2022). Screen captures of the interface can be seen in Figures 6–8. Figure 6 represents the view that a user would have when zoomed to an area of interest, including an instructional panel on the left that explains how to use the tool (Figure 6a), the main map interface with a Google imagery base map (Figure 6b), and the polygon drawing tool that can be used to digitize a potential SZ polygon (Figure 6c).

The user then has the option to select the burning condition, wind speed, and whether or not to correct for vegetation height bias (Figure 7a). Following these selections, the user generates a per-pixel SSD map (Figure 7b). The user also has the option to download the resulting raster image (Figure 7a).

The user can then manually draw in a potential SZ polygon and evaluate a variety of suitability metrics by selecting 'Calculate SZ SSD' (Figure 8a). As a result, three new spatial layers will be added to the map, each within the extent of the SZ polygon, including the pSSD (Figure 8b), as well as a point feature representing the safest location and a safe (pixels greater than or equal to SSD) and unsafe (pixels less than SSD) raster. These layers can be toggled on and off in the 'Layers' menu (Figure 8c). The layers can also be individually downloaded (Figure 8a). Lastly, a series of tabular metrics are also computed and displayed to aid in the evaluation of SZ suitability (Figure 8d). The user can evaluate and compare multiple different wind speeds, burning conditions, and whether or not bias is corrected for.

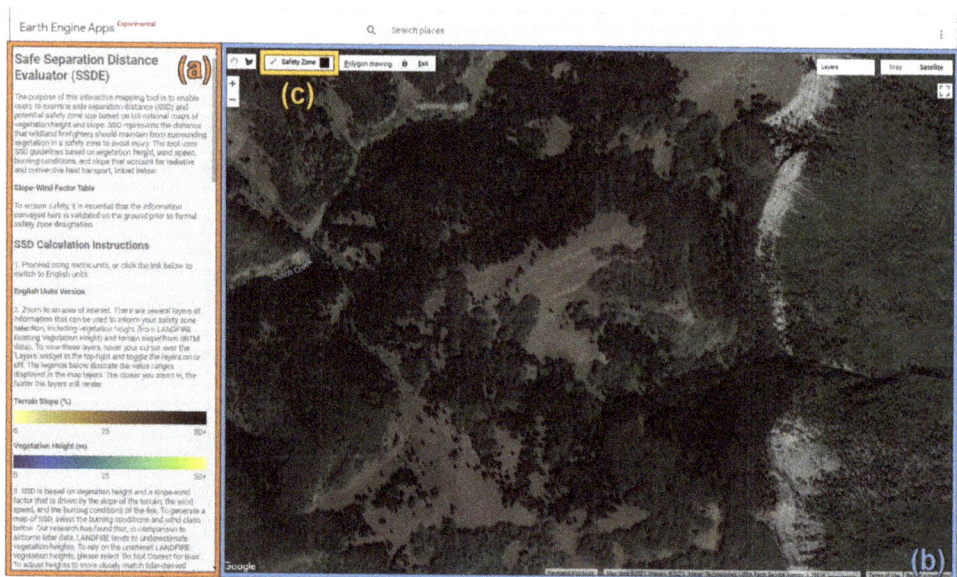

Figure 6. A screen capture of the Wildland Fire Safe Separation Distance Evaluator Google Earth Engine web application, including an instructional panel on the left that explains how to use the tool (**a**), the main map interface with a Google imagery base map (**b**), and the polygon drawing tool that can be used to digitize a potential SZ polygon (**c**).

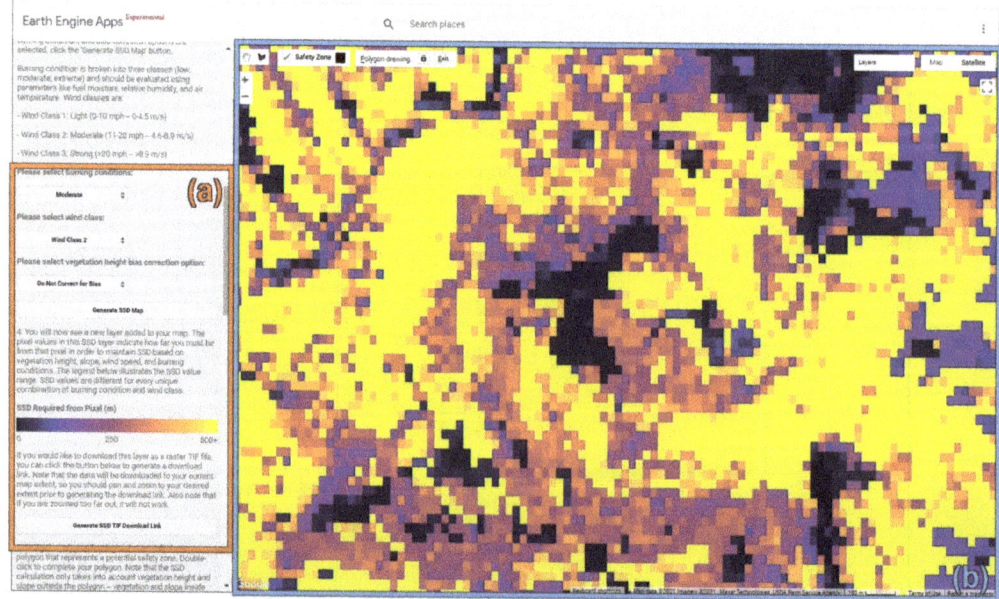

Figure 7. A screen capture of the Wildland Fire Safe Separation Distance Evaluator Google Earth Engine web application, including the user-selected options for generating and downloading a SSD map (**a**), and an example of the resulting map (**b**).

Figure 8. A screen capture of the Wildland Fire Safe Separation Distance Evaluator Google Earth Engine web application, including the widgets used for calculating SZ suitability and downloading the resulting layers (**a**), a layer representing pSSD (**b**), the 'Layers' menu for toggling on and off layer visibility (**c**), and a widget with several quantitative suitability metrics for two different wind speed-burning condition combinations (**d**).

3.2. Vegetation Height Analysis

The results of the per-pixel SSD uncertainty analysis can be seen in Figure 9. When compared to airborne lidar-derived SSD, which we assume to be the most accurate given the fact that airborne lidar vegetation heights are directly-measured rather than modeled, it becomes clear that both LANDFIRE- and GEDI/Landsat-derived SSD values, on average, are underestimated (Figure 9a,b). Since all of the other variables (slope, wind speed, burning condition) were held constant, this means LANDFIRE and GEDI/Landsat vegetation height products underestimated vegetation heights, as taller vegetation results in higher SSD. Per-pixel SSD values are still fairly strongly correlated ($R^2_{LANDFIRE-lidar} = 0.63$; $R^2_{GEDI/Landsat-lidar} = 0.68$), suggesting that the modeled products trend towards similar results as the airborne lidar-derived per-pixel SSD, but the regression slopes of less than 1 indicate that both LANDFIRE and GEDI/Landsat will underestimate per-pixel SSD, particularly on the high end. This further points towards the fact that LANDFIRE and GEDI/Landsat fail to capture the structure of very tall vegetation well, though this failure appears more pronounced in the LANDFIRE data, given the lower regression line slope. Of particular note is the presence of a number of sample points where the LANDFIRE and GEDI/Landsat per-pixel SSD values approach zero but the airborne lidar per-pixel SSD values are greater than zero. A visual assessment of the three vegetation height datasets in comparison to aerial imagery revealed that, at least in some cases, this can be attributed to the presence of standing dead vegetation such as in a post-fire environment. The airborne lidar captures this standing dead vegetation since the laser pulses interact with any above-ground objects. However, the LANDFIRE and GEDI/Landsat, both of which rely heavily on spectral data tend to suggest there is no vegetation in these areas, given the fact that the typical vegetation reflectance signal (e.g., high near infrared reflectance, low visible wavelength reflectance) is no longer present after a disturbance event. The comparison between LANDFIRE and GEDI/Landsat directly further suggests that LANDFIRE tends to underestimate vegetation heights to a greater extent than GEDI/Landsat, though they are strongly correlated ($R^2_{\text{LANDFIRE-GEDI/Landsat}} = 0.74$) (Figure 9c).

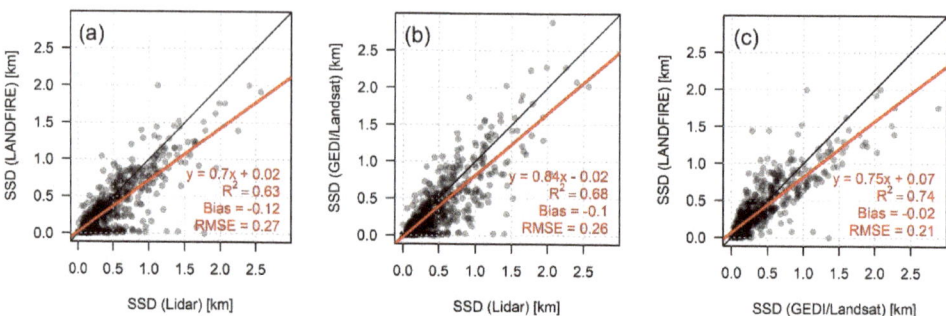

Figure 9. Results of the per-pixel SSD comparison between LANDFIRE and airborne lidar (**a**), GEDI/Landsat and airborne lidar (**b**), and LANDFIRE and GEDI/Landsat (**c**). The red line represents the ordinary least-squares regression model between the *x* and *y* variables, whereas the black line represents the 1:1 line, where *x* is equal to *y*.

The results of the SZ polygon-level maximum pSSD uncertainty analysis can be seen in Figure 10. The results follow a similar trend from the per-pixel SSD comparison, which makes sense given that within-SZ pSSD is driven by the pixel-level SSD values surrounding the SZ polygon. Both LANDFIRE and GEDI/Landsat overestimate the maximum pSSD within SZ polygons as compared to airborne lidar (Figure 10a,b). This can once again be attributed to vegetation height underestimation from these modeled products as compared to airborne lidar data. In addition, there is a much greater correlation at the SZ polygon level than at the pixel level. For example, at the pixel level, LANDFIRE and lidar SSD have a coefficient of determination (R^2) of 0.63, whereas at the SZ polygon level, LANDFIRE and lidar maximum pSSD have an R^2 of 0.92. This is likely due in part to the segmentation procedure that aggregates adjacent pixels with similar SSD values and computes a mean, which drives the pSSD calculation, thus reducing the pixel-level noise and producing a stand-level estimate of vegetation height. This high correlation also suggests that SZ polygon size is likely the major driver of within-SZ pSSD, outweighing the effects of differing vegetation height estimates. When comparing between the 2 modeled products, LANDFIRE and GEDI/Landsat produce very similar results with a high correlation ($R^2_{\text{LANDFIRE-GEDI/Landsat}} = 0.88$) and a regression line slope of close to 1 (0.94) (Figure 10c). Again, even though LANDFIRE tended to produce slightly lower per-pixel SSD values due to lower vegetation height estimation (Figure 9c), when pixels are aggregated at the segment level and when those segment-level pSSD estimates are further aggregated at the SZ polygon level, subtle differences in vegetation height bear little effect on pSSD calculation.

The results of the SZ polygon-level total safe area uncertainty analysis can be seen in Figure 11. These results mirror those seen in Figure 9, which makes sense given that safe area is a direct product of the within-SZ pSSD calculation. Both LANDFIRE and GEDI/Landsat tend to overestimate total safe area within SZ polygons as compared to airborne lidar (Figure 11a,b). While such an overestimation is potentially problematic, the most acutely problematic disagreement between the modeled products and the airborne lidar is the large number of SZ polygons where lidar indicates there is 0 ha safe area and the modeled products indicate there is greater than 0 ha safe area—in some cases as much as 100 ha of safe area. However, it is important to recognize that since the safe versus unsafe classification is based on a defined pSSD threshold, a SZ polygon that reaches 99% of SSD would still be mapped as unsafe, so small differences in vegetation heights can have large effects on these results. Once again, the LANDFIRE and GEDI/Landsat safe area estimates are both highly correlated and relatively unbiased when compared to one another (Figure 11c).

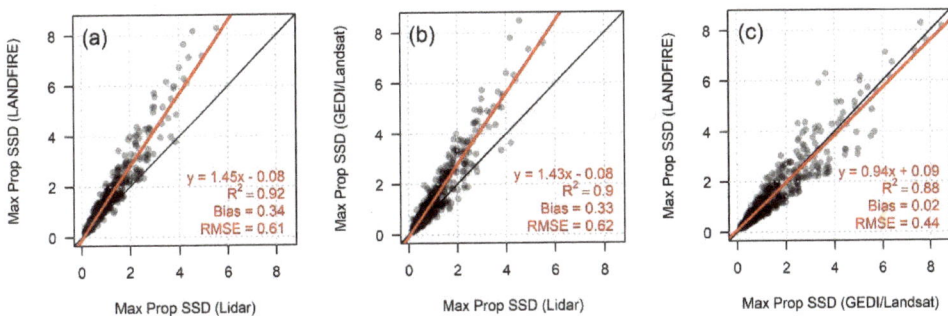

Figure 10. Results of the safety zone polygon-level maximum pSSD comparison between LANDFIRE and airborne lidar (**a**), GEDI/Landsat and airborne lidar (**b**), and LANDFIRE and GEDI/Landsat (**c**). The red line represents the ordinary least-squares regression model between the x and y variables, whereas the black line represents the 1:1 line, where x is equal to y.

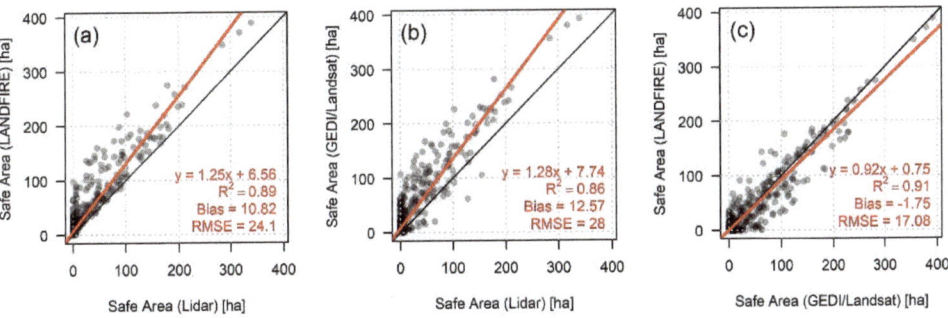

Figure 11. Results of the safety zone-level total safe area comparison between LANDFIRE and airborne lidar (**a**), GEDI/Landsat and airborne lidar (**b**), and LANDFIRE and GEDI/Landsat (**c**). The red line represents the ordinary least-squares regression model between the x and y variables, whereas the black line represents the 1:1 line, where x is equal to y.

The results of the safest point geographic displacement analysis can be seen in Figure 12. LANDFIRE and GEDI/Landsat result in safest points being on average between about 100 and 150 m from the lidar-derived safest point location (Figure 12a,b). Given the 30 m spatial resolution of the analysis, that means, on average, the safest points are within 3–5 pixels of one another. This suggests that the geography of the safest point is perhaps not as sensitive to the vegetation height as it is to the geometry of the SZ polygon. However, there are exceptions, given the right-skewed tail of the distributions. In a select few instances, safest points were found to be 800 m apart or more. From a firefighter safety perspective, this is a significant deviation, as the safest point is likely the target gathering point within the SZ for the crew and equipment. The LANDFIRE and GEDI/Landsat safest points are, on average, closer to one another than they each are to the airborne lidar-derived safest points (Figure 12c).

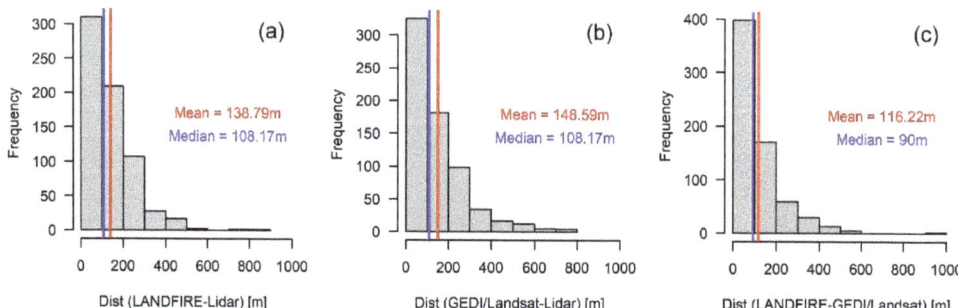

Figure 12. Results of the safety zone polygon-level total safest point location comparison between LANDFIRE and airborne lidar (**a**), GEDI/Landsat and airborne lidar (**b**), and LANDFIRE and GEDI/Landsat (**c**).

3.3. LANDFIRE Bias Correction

The results of the LANDFIRE bias correction procedure can be seen in Figure 13. The bias correction resulted in SSD and SZ suitability metrics that align more closely with those derived from airborne lidar. Specifically, in all three regression analyses, including the per-pixel SSD (Figure 13a), the within-SZ polygon maximum pSSD (Figure 13b), and the within-SZ polygon total safe area (Figure 13c), the regression slope was much closer to one than the un-corrected data. The bias correction bore little effect on the safest point location comparison, which makes sense given the fact that vegetation heights were all increased linearly, meaning the placement of the safest point would not have changed significantly (Figure 13d).

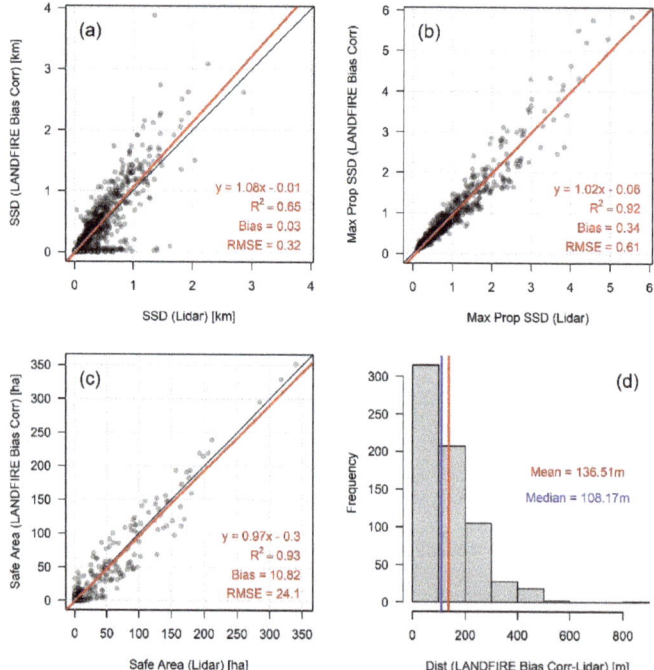

Figure 13. Results of the comparison between bias-corrected LANDFIRE data and airborne lidar data, including per-pixel SSD (**a**), within-SZ polygon maximum pSSD (**b**), within-SZ polygon total safe area (**c**), and distance between modeled safest points (**d**).

3.4. Use Case Demonstration

The results of the case study demonstration can be seen in Figure 14. The two potential SZ polygons varied in shape, size, and proximity to the fire perimeter. In a direct attack situation, a fire crew would be working in close proximity to the perimeter of the fire. Thus, having a SZ that is in close proximity to the crew is the best way to ensure that they can access the SZ quickly via a pre-planned escape route. Based on proximity alone, SZ 1 is superior to SZ 2 (Figure 14a). Indeed, provided that wind speeds were light and burn conditions were low, SZ 1 would provide sufficient SSD for the crew, providing a maximum within-SZ pSSD of 115%, and a total safe area of 0.36 ha (Figure 14b). However, under high winds and extreme burn conditions, this SZ would no longer be suitable (Figure 14c). In comparison, SZ 2 is further away, but due to its larger size, provides greater SSD, particularly under light winds and low burn conditions (Figure 14d). Even though this polygon is much larger, it still does not meet the SSD guideline under high winds and extreme burn conditions, providing a maximum within-SZ pSSD of 74% (Figure 14e). Armed with this information, however, the crew could opt to expand SZ 2 to increase its suitability through mechanical fuel removal or controlled burning.

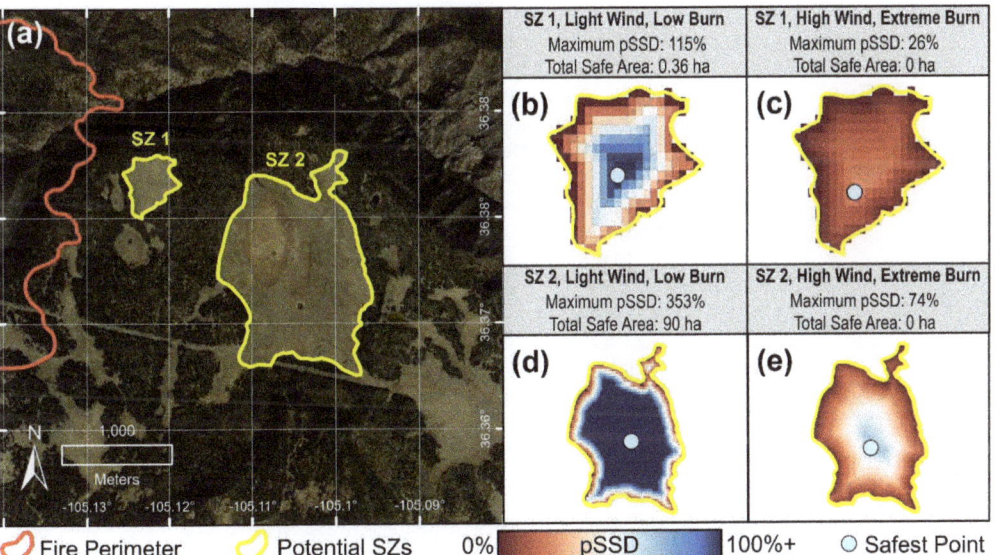

Figure 14. Results of the evaluation of two potential SZs in a simulated wildland fire scenario in northern New Mexico, including an overview map (**a**), and the suitability analysis results for SZ 1 under light wind and low burn conditions (**b**), SZ 1 under high wind and extreme burn conditions (**c**), SZ 2 under light wind and low burn conditions (**d**), and SZ 2 under high wind and extreme burn conditions (**e**).

4. Discussion

We envision the SSDE being of broad interest to the wildland fire community, from fire scientists to incident management personnel to wildland firefighters. Its open-access nature allows anyone to explore, examine, and interact with the concepts of SZs and SSD. Even if not used in an operational context, there is great value in being able to quickly and easily examine the conditions that define potential SZ suitability on a broad spatial scale. Wildland firefighters designate SZs on a daily basis as a part of their fire management duties. Built into this designation process is an inherent degree of subjectivity that can result in differences in the interpretation of SZ suitability between and among crews. By using an objective tool for SZ suitability analysis that can be broadly applied in the US, fire crews across the country can increase the consistency and reliability of the SZ evaluation process. However, given that this is a web-based platform requiring an internet connection, that also does not translate well to a mobile environment, we do not envision this as a real-time decision-making tool that firefighters could use on the ground. Instead, operational use of SSDE could be at the incident command level, for daily or more frequent evaluation of potential SZs for crews working on a fire. Given the dynamic and fast-paced nature of fire management, it is essential to be able to make rapid assessments of SZ suitability, particularly as fire conditions change. For example, cross-referencing near-real time data representing the current fire perimeter with SSDE can enable the evaluation of whether or not previously burned areas can provide SSD from nearby unburned fuel. Additionally, our SZ suitability driver analysis revealed the strong influence of wind on maximum within-SZ pSSD. This highlights the need to continually re-evaluate SZ suitability not only as the fire evolves, but as local weather conditions change as well.

There is relatively little research on the actual utilization of SZ in wildland firefighting. As discussed in the Introduction, SZ are largely designated ad hoc by firefighters, with little support from GIS and remote sensing data. As a result, there are essentially no publicly available datasets that represent SZ that were used by firefighters. If they were available, we could perform more empirically driven studies of SZ effectiveness, but as it stands, we are limited to using theoretical models to predict safety zone effectiveness based on fire physics [21,22,25–27]. While we do not know how many firefighters have suffered injuries or fatalities due specifically to the selection of inadequate SZ, we do know that burnovers and entrapments occur on an annual basis, and that these injuries and fatalities would be much less likely to occur if escape routes and SZ were properly employed. Every wildland fire safety incident is different from the last, so there is no systematic way to determine the extent to which a tool such as SSDE would have mitigated the incident. However, with the increased infusion of reliable geospatial data and modeling into wildland firefighter safety, we hope to minimize future incidents. In future research, it would be beneficial to compare the results of SSDE to actual SZ used by firefighters. This could be accomplished through the GPS collection of SZ in the field or perhaps through simulation of SZ selection in a virtual environment, to minimize risk to fire personnel while still gaining valuable, realistic insight into SZ selection [41–43].

The SSD guideline upon which SSDE is based (Equation (1), Table 1) can be quite restrictive, particularly in landscapes featuring rugged terrain and tall trees. For example, in a worst-case scenario (high winds, extreme burning conditions), trees that were 25 m tall and on slopes that were steep (>40%) would produce an SSD equal to 2 km, making SZ designation impractical, if not impossible. However, even if no SZ is found to meet guidelines, SSDE still can provide an analysis of which safety zones are closest to meeting the SSD requirement. SSDE enables evaluation of newly burned potential SZs or existing clearings that nearly meet the SSD requirement and could be enlarged through prescribed burning or mechanical fuel removal. For example, if a potential SZ polygon delineated within an existing clearing was determined through the use of SSDE to only reach a maximum pSSD of 90%, SSDE could be further used to evaluate the relative suitability of one or more potential SZ expansion options. These options could be informed by the aerial imagery, terrain, vegetation height, and SSD data built into SSDE, which could aid in the determination of the areas that might be best suited for fuel removal (i.e., removing sparse, short shrubs on flat slopes rather than dense, tall trees on steep slopes to improve the SZ).

Our comparative analysis between SZ suitability metrics derived from LANDFIRE, GEDI/Landsat, and airborne lidar revealed that the former two 30 m products tended to underestimate vegetation height and thus overestimate the suitability of potential SZs. However, given that airborne lidar data are not widely available in the Western US, and even where they are available may not reflect current vegetation structural conditions, the SSDE had to be built using either LANDFIRE or GEDI/Landsat to enable broad application. If one product or the other demonstrated a significantly closer alignment with the lidar-derived suitability metrics, then we would have built SSDE using the superior product, yet no such clear winner emerged, with LANDFIRE marginally outperforming GEDI/Landsat at the SZ level and GEDI/Landsat marginally outperforming LANDFIRE at the pixel level. Thus, we were left to rely on other factors in choosing which dataset to base the SSDE algorithm on. We chose LANDFIRE for the two primary reasons. First, LANDFIRE is a well-established interagency US government-funded program that has a lengthy history and is already well-known to the wildland fire community in the US. Secondly, LANDFIRE data are scheduled to be continually updated to reflect changes in vegetation structure over time. This is particularly important given the dynamic nature of vegetation in disturbance-prone regions of the US. SSDE could potentially be expanded to other regions of the world if adequate vegetation height data becomes available. The GEDI/Landsat forest height data used in this study are globally available but does not include vegetation below 3 m in height; thus, additional height information would be needed to fill in this lower range. Alternatively, regionally specific height models could be developed where airborne lidar is more common.

Although other variables were found to be more important predictors of maximum within-SZ pSSD than vegetation height (e.g., wind speed and slope; Appendix A), we only performed an uncertainty analysis on the vegetation height source data. The slope data in our study were derived from SRTM DEMs, which are widely regarded to be an accurate source of elevation (and, in turn, slope) data [44,45]. That being said, a previous comparison between SRTM and other DEM source datasets have revealed that SRTM-derived slope estimates tend to be comparatively high [46]. However, given the relatively coarse categorization of slope classes used to calculate SSD in our model, we believe the results would be minimally affected by using different sources of elevation data. With respect to the other variables (wind speed, burn condition, SZ geometry), these are all user-defined in SSDE. That is not to say that user definition does not come with inherent uncertainty, but the uncertainty analysis can, and indeed should, be conducted by the user of SSDE. For example, by simulating different wind speed and burn conditions, the user can gain a valuable sense of the variability of SZ suitability. There is also inherent uncertainty in the fire physics research that has defined the SSD equation (Equation (1)); however, testing that uncertainty is beyond the scope of this study. The SSDE platform is sufficiently flexible to enable future updates to the fundamental SSD equation that underlies the algorithm.

This study represents an important contribution to a growing body of literature surrounding the application of GIS and remote sensing to wildland firefighter safety. This study builds upon previous work by Dennison et al. and Campbell et al. [20,33] using airborne lidar to identify and evaluate SZ suitability but scales the application to the contiguous US. Escape routes are a second component of LCES that can be modeled, including variation in travel rates with slope [47,48], impacts of lidar-derived vegetation density and surface roughness on travel rates [49], and a landscape scale Escape Route Index [50]. Travel rates used in geospatial modeling can simulate the evacuation of a fire crew in comparison to predicted fire behavior in order to assess potential trigger points on the landscape that a fire crew could use [51]. Beyond SZs and escape routes, there is an emergence of other geospatially driven approaches for fire management that have implications for firefighter safety, such as the mapping of snag hazards [52], the spatially explicit estimation of suppression difficulty [2,53], the prediction of potential control locations [3,54], and the mapping of potential wildland fire operational delineations [55,56]. These tools are now commonly used in the US to enhance situational awareness and improve the quality of strategic decision-making on large and complex wildfire incidents [57]. Taken together, these tools stand to greatly improve the efficiency and safety with which wildland firefighters can engage in their life-saving work.

5. Conclusions

SZs are one of the most important safety tools available to wildland firefighters. The difference between a good SZ (one that provides sufficient SSD) and a bad SZ (one that does not), can be the difference between life and death. Although guidelines exist for the assessment of SSD, errors in estimating vegetation height, distances, and SSD based on anticipated slope, wind, and fire conditions could result in firefighter injury or fatality. In this study we have introduced a broadly applicable methodology and associated open-access web mapping tool for the assessment of SZ suitability based on vegetation height, terrain slope, wind speed, and burning conditions. The tool enables users to delineate their own potential SZ anywhere in the contiguous US and rapidly compute a variety of suitability metrics that are driven by free and publicly available datasets. SSDE allows for those with wide-ranging expertise to explore and examine the complexities of potential SZ suitability, and how the conditions that affect SZ suitability vary over space with vegetation height and slope.

As with all geospatially driven tools, however, it is important to consider the effects that data inaccuracy and uncertainty may have on the tool's results. This importance is greatly amplified when the tool is designed to ensure the safety of others. To that end, we compared two different modeled vegetation height datasets to reference data derived from airborne lidar data with respect to their relative agreement on the array of SZ suitability metrics produced by the tool. Although correlations between the two datasets and airborne lidar were relatively high across all metrics, they tended to underestimate vegetation heights, resulting in an overestimation of the relative safety of SZs. However, this disagreement enabled us to implement a bias correction option in the tool to enable users to perform raw and lidar-adjusted SZ evaluations. Even with the bias correction, it is of the utmost importance that all results emerging from the use of the SSDE be validated on the ground by professionals prior to the formal designation of SZs. In the future, as airborne lidar becomes more widely available and frequently updated, efforts should be made towards basing SSD on lidar, rather than modeled geospatial vegetation height products. As fire science continues to advance our understanding of the complex nature between fuel, topography, and weather, the analysis of SSD should reflect the most updated understanding of how radiative and convective heat transfer affects SZ suitability. In addition, the practical application of SSD and SZs needs to be developed, to determine how firefighters utilize the information provided by guidelines and mapping tools.

Author Contributions: Conceptualization, M.J.C., P.E.D. and B.W.B.; Methodology, M.J.C., P.E.D. and B.W.B.; Software, M.J.C.; Formal Analysis, M.J.C.; Resources, M.J.C. and P.E.D.; Data Curation, M.J.C.; Writing—Original Draft Preparation, M.J.C. and P.E.D.; Writing—Review & Editing, M.J.C., P.E.D., M.P.T. and B.W.B.; Visualization, M.J.C.; Supervision, M.J.C., P.E.D., M.P.T. and B.W.B.; Project Administration, M.J.C. and P.E.D.; Funding Acquisition, M.J.C., P.E.D. and M.P.T. All authors have read and agreed to the published version of the manuscript.

Funding: This research was funded by National Science Foundation grant number BCS-2117433 and USDA Forest Service grant number 19-JV-11221637-142.

Data Availability Statement: The Wildland Fire Safe Separation Distance Evaluator online tool can be accessed here: https://firesafetygis.users.earthengine.app/view/ssde (accessed on 3 January 2022). All data in this study come from freely- and publicly-available US Federal Government repositories.

Acknowledgments: We would like to thank Wesley Page and Alex Heeren for their feedback on the development of SSDE.

Conflicts of Interest: The authors declare no conflict of interest.

Disclaimer: The findings and conclusions in this report are those of the author(s) and should not be construed to represent any official USDA or U.S. Government determination or policy. Any use of trade, firm, or product names is for descriptive purposes only and does not imply endorsement by the U.S. government.

Appendix A

There are several factors that drive the suitability of potential SZ according to the SSDE. These include geometry of the SZ polygon (area, shape), height of the surrounding vegetation, slope of the surrounding terrain, the wind speed, and burn condition. However, these factors are not equal in terms of their influence on SZ suitability. Quantifying the relative influence of suitability controls can provide valuable insight into the relative stability (or inversely, variability) of SZ suitability under different conditions. For example, if SZ polygon size explained the vast majority of variance in maximum within-SZ pSSD, then that would suggest SZs are fairly robust to changes in surrounding landscape conditions, and weather/fire conditions. To quantify the relative influence of different drivers of pSSD, we used random forests [58]. Random forests are an ensemble machine learning method that uses iterative subsampling of data to generate a series of decision trees to make predictions and assess variable importance. For each of the 75 random SZ polygons generated for the vegetation height analysis (Section 2.2; Figure 5) we had nine different pSSD values, resulting from unique combinations of wind speed and burn condition classes, totaling 675 data points. Using the randomForest library in R [59,60], we built a model to predict maximum within-SZ polygon pSSD, based on the independent variables described in Table A1. These models were run using the default parameters and were not tuned for performance, as the goal was merely to assess variable importance and not maximize predictive accuracy. The equations for calculating shape metrics were gleaned from Lindsay [61] and computed using Esri ArcGIS. Variable importance was assessed in terms of the relative proportion of mean squared predictor error that would result from the removal of a particular variable from the model.

Table A1. Predictor variables used in the analysis of the SZ suitability driver analysis.

Predictor	Data Type	Description
Vegetation Height	Continuous	Median vegetation height among pixels within buffer area around SZ polygon
Slope	Continuous	Median slope among pixels within buffer area around SZ polygon
Wind Speed	Categorical	Wind speed class from Table 1
Burn Condition	Categorical	Burn condition class from Table 1
SZ Polygon Area	Continuous	Total area of the SZ polygon
SZ Polygon Perimeter-to-Area Ratio	Continuous	Length of the SZ polygon perimeter divided by its area
SZ Polygon Elongation Ratio	Continuous	One minus the length of the shortest polygon axis divided by the length of the longest polygon axis
SZ Polygon Related Circumscribing Circle	Continuous	One minus the area of the polygon divided by the area of the smallest encompassing circle
SZ Polygon Shape Complexity Index	Continuous	One minus the area of the polygon divided by the area of a convex hull containing the polygon

The results of the SZ polygon suitability driver analysis can be seen in Table A2. Wind speed emerged as the most important determinant of maximum within-SZ pSSD by a large margin. This stands in stark contrast to the relatively minimal effect of burn condition, which was determined to be the least important predictor. These differences can be explained through the examination of Δ values in Table 1, where increases in wind speed always result in a higher Δ, whereas increases in burn condition do not. Slope was the second most important predictor which, when taken together with the importance of wind speed, highlights the impact that convective heating, as captured by Δ, has on pSSD. SZ polygon area is the third most important predictor; however, its relatively small influence in comparison to wind speed suggests that the suitability of potential SZ polygons is not very robust to changes in wind conditions. Vegetation height is the fourth most important determinant of maximum within-SZ pSSD, meaning that SZs must take into account local vegetation conditions in order to ensure safety of wildland firefighters. The four SZ polygon shape metrics all had similar and relatively low importance, but their inclusion does have a significant effect on the amount of variance in maximum pSSD explained. For example, a random forest model built without these four metrics only explains 83% of variance in pSSD, whereas their inclusion increases that number to 94%. The ideal shape for a SZ would be approximately a circle, where SSD can be maintained from surrounding vegetation on all sides. So, a large circular SZ will possess very different suitability than an equal area linearly-shaped SZ (e.g., a road). Thus, the importance of SZ shape should not be underestimated.

Table A2. Variable importance for predicting maximum within-SZ pSSD, measured in terms of the percent increase in mean squared error that would result from the removal of that variable (%IncMSE).

Predictor	Rank	%IncMSE
Wind Speed	1	95.3
Slope	2	40.6
SZ Polygon Area	3	31.0
Vegetation Height	4	27.9
SZ Polygon Perimeter-to-Area Ratio	5	25.9
SZ Polygon Shape Complexity Index	6	17.4
SZ Polygon Elongation Ratio	7	16.7
SZ Polygon Related Circumscribing Circle	8	16.7
Burn Condition	9	15.2

References

1. Wei, Y.; Thompson, M.P.; Scott, J.H.; O'Connor, C.D.; Dunn, C.J. Designing Operationally Relevant Daily Large Fire Containment Strategies Using Risk Assessment Results. *Forests* **2019**, *10*, 311. [CrossRef]
2. Silva, F.R.Y.; O'Connor, C.D.; Thompson, M.P.; Martínez, J.R.M.; Calkin, D.E. Modelling Suppression Difficulty: Current and Future Applications. *Int. J. Wildland Fire* **2020**, *29*, 739–751. [CrossRef]
3. Connor, C.D.O.; Calkin, D.E.; Thompson, M.P. An Empirical Machine Learning Method for Predicting Potential Fire Control Locations for Pre-Fire Planning and Operational Fire Management. *Int. J. Wildland Fire* **2017**, *26*, 587–597. [CrossRef]
4. Cheney, P.; Gould, J.; McCaw, L. The Dead-Man Zone—A Neglected Area of Firefighter Safety. *Aust. For.* **2001**, *64*, 45–50. [CrossRef]
5. Page, W.G.; Butler, B.W. An Empirically Based Approach to Defining Wildland Firefighter Safety and Survival Zone Separation Distances. *Int. J. Wildland Fire* **2017**, *26*, 655–667. [CrossRef]
6. Arizona State Forestry Division. Yarnell Hill Fire: Serious Accident Investigation Report. 2013. Available online: https://dffm.az.gov/sites/default/files/YHR_Data_092813_0.pdf (accessed on 3 January 2022).
7. Butler, B.W.; Bartlette, R.A.; Bradshaw, L.S.; Cohen, J.D.; Andrews, P.L.; Putnam, T.; Mangan, R.J. *Fire Behavior Associated with the 1994 South Canyon Fire on Storm King Mountain, Colorado*; Research Paper RMRS-RP-9; U.S. Department of Agriculture, Forest Service, Rocky Mountain Research Station: Ogden, UT, USA, 1998; 82p. [CrossRef]
8. Alexander, M.E.; Taylor, S.W.; Page, W.G. Wildland firefighter safety and fire behavior prediction on the fireline. In Proceedings of the 13th International Wildland Fire Safety Summit & 4th Human Dimensions Wildland Fire Conference, Boise, ID, USA, 20–24 April 2015; pp. 20–24.
9. Butler, C.; Marsh, S.; Domitrovich, J.W.; Helmkamp, J. Wildland Firefighter Deaths in the United States: A Comparison of Existing Surveillance Systems. *J. Occup. Environ. Hyg.* **2017**, *14*, 258–270. [CrossRef] [PubMed]
10. National Wildfire Coordinating Group Glossary A-Z | NWCG. Available online: https://www.nwcg.gov/glossary/a-z (accessed on 17 February 2017).
11. Abatzoglou, J.T.; Williams, A.P. Impact of Anthropogenic Climate Change on Wildfire across Western US Forests. *Proc. Natl. Acad. Sci. USA* **2016**, *113*, 11770–11775. [CrossRef]
12. Abatzoglou, J.T.; Battisti, D.S.; Williams, A.P.; Hansen, W.D.; Harvey, B.J.; Kolden, C.A. Projected Increases in Western US Forest Fire despite Growing Fuel Constraints. *Commun. Earth Environ.* **2021**, *2*, 227. [CrossRef]
13. Dennison, P.E.; Brewer, S.C.; Arnold, J.D.; Moritz, M.A. Large Wildfire Trends in the Western United States, 1984–2011. *Geophys. Res. Lett.* **2014**, *41*, 2928–2933. [CrossRef]
14. Balch, J.K.; Bradley, B.A.; Abatzoglou, J.T.; Nagy, R.C.; Fusco, E.J.; Mahood, A.L. Human-Started Wildfires Expand the Fire Niche across the United States. *Proc. Natl. Acad. Sci. USA* **2017**, *114*, 2946–2951. [CrossRef]
15. Westerling, A.L. Increasing Western US Forest Wildfire Activity: Sensitivity to Changes in the Timing of Spring. *Phil. Trans. R. Soc. B* **2016**, *371*, 20150178. [CrossRef]
16. Morse, G.A. A Trend Analysis of Fireline "Watch Out" Situations in Seven Fire-Suppression Fatality Accidents. *Fire Manag.* **2004**, *66*, 66–69.
17. Ziegler, J.A. The Story Behind an Organizational List: A Genealogy of Wildland Firefighters' 10 Standard Fire Orders. *Commun. Monogr.* **2007**, *74*, 415–442. [CrossRef]
18. Gleason, P. Lookouts, Communications, Escape Routes, and Safety Zones. Available online: https://www.fireleadership.gov/toolbox/documents/lces_gleason.html (accessed on 17 February 2017).
19. Thorburn, R.W.; Alexander, M.E. LACES versus LCES: Adopting an "A" for "Anchor Points" to improve wildland firefighter safety. In Proceedings of the 2001 International Wildland Fire Safety Summit, Missoula, MT, USA, 6–8 November 2001.
20. Campbell, M.J.; Dennison, P.E.; Butler, B.W. Safe Separation Distance Score: A New Metric for Evaluating Wildland Firefighter Safety Zones Using Lidar. *Int. J. Geogr. Inf. Sci.* **2017**, *31*, 1448–1466. [CrossRef]
21. Butler, B.W. Wildland Firefighter Safety Zones: A Review of Past Science and Summary of Future Needs. *Int. J. Wildland Fire* **2014**, *23*, 295–308. [CrossRef]

22. Butler, B.W.; Cohen, J.D. Firefighter Safety Zones: A Theoretical Model Based on Radiative Heating. *Int. J. Wildland Fire* **1998**, *8*, 73–77. [CrossRef]
23. National Wildfire Coordinating Group. Incident Response Pocket Guide. 2014. Available online: https://www.nwcg.gov/sites/default/files/publications/pms461.pdf (accessed on 3 January 2022).
24. Dupuy, J.-L.; Maréchal, J. Slope Effect on Laboratory Fire Spread: Contribution of Radiation and Convection to Fuel Bed Preheating. *Int. J. Wildland Fire* **2011**, *20*, 289–307. [CrossRef]
25. Frankman, D.; Webb, B.W.; Butler, B.W.; Jimenez, D.; Forthofer, J.M.; Sopko, P.; Shannon, K.S.; Hiers, J.K.; Ottmar, R.D. Measurements of Convective and Radiative Heating in Wildland Fires. *Int. J. Wildland Fire* **2013**, *22*, 157–167. [CrossRef]
26. Parsons, R.; Butler, B.; Mell, W. *"Ruddy" Safety Zones and Convective Heat: Numerical Simulation of Potential Burn Injury from Heat Sources Influenced by Slopes and Winds*; Imprensa da Universidade de Coimbra: Coimbra, Portugal, 2014; ISBN 978-989-26-0884-6.
27. Page, W.G.; Butler, B.W. Fuel and Topographic Influences on Wildland Firefighter Burnover Fatalities in Southern California. *Int. J. Wildland Fire* **2018**, *27*, 141–154. [CrossRef]
28. Firefighter Safety | Missoula Fire Sciences Laboratory. Available online: https://www.firelab.org/project/firefighter-safety (accessed on 26 October 2021).
29. Gorelick, N.; Hancher, M.; Dixon, M.; Ilyushchenko, S.; Thau, D.; Moore, R. Google Earth Engine: Planetary-Scale Geospatial Analysis for Everyone. *Remote Sens. Environ.* **2017**, *202*, 18–27. [CrossRef]
30. Rollins, M.G. LANDFIRE: A Nationally Consistent Vegetation, Wildland Fire, and Fuel Assessment. *Int. J. Wildland Fire* **2009**, *18*, 235–249. [CrossRef]
31. LANDFIRE LANDFIRE Remap 2016 Existing Vegetation Height (EVH) CONUS. Available online: https://landfire.cr.usgs.gov/distmeta/servlet/gov.usgs.edc.MetaBuilder?TYPE=HTML&DATASET=FE3 (accessed on 2 November 2021).
32. Farr, T.G.; Rosen, P.A.; Caro, E.; Crippen, R.; Duren, R.; Hensley, S.; Kobrick, M.; Paller, M.; Rodriguez, E.; Roth, L.; et al. The Shuttle Radar Topography Mission. *Rev. Geophys.* **2007**, *45*, 25–36. [CrossRef]
33. Dennison, P.E.; Fryer, G.K.; Cova, T.J. Identification of Firefighter Safety Zones Using Lidar. *Environ. Model. Softw.* **2014**, *59*, 91–97. [CrossRef]
34. Andrews, P.L. Current Status and Future Needs of the BehavePlus Fire Modeling System. *Int. J. Wildland Fire* **2014**, *23*, 21–33. [CrossRef]
35. Hopkinson, C.; Chasmer, L.; Lim, K.; Treitz, P.; Creed, I. Towards a Universal Lidar Canopy Height Indicator. *Can. J. Remote Sens.* **2006**, *32*, 139–152. [CrossRef]
36. Peterson, B.; Nelson, K.J.; Seielstad, C.; Stoker, J.; Jolly, W.M.; Parsons, R. Automated Integration of Lidar into the LANDFIRE Product Suite. *Remote Sens. Lett.* **2015**, *6*, 247–256. [CrossRef]
37. Potapov, P.; Li, X.; Hernandez-Serna, A.; Tyukavina, A.; Hansen, M.C.; Kommareddy, A.; Pickens, A.; Turubanova, S.; Tang, H.; Silva, C.E.; et al. Mapping Global Forest Canopy Height through Integration of GEDI and Landsat Data. *Remote Sens. Environ.* **2021**, *253*, 112165. [CrossRef]
38. Dubayah, R.; Blair, J.B.; Goetz, S.; Fatoyinbo, L.; Hansen, M.; Healey, S.; Hofton, M.; Hurtt, G.; Kellner, J.; Luthcke, S.; et al. The Global Ecosystem Dynamics Investigation: High-Resolution Laser Ranging of the Earth's Forests and Topography. *Sci. Remote Sens.* **2020**, *1*, 100002. [CrossRef]
39. Sugarbaker, L.J.; Constance, E.W.; Heidemann, H.K.; Jason, A.L.; Lukas, V.; Saghy, D.L.; Stoker, J.M. *The 3D Elevation Program. Initiative: A Call for Action*; U.S. Geological Survey: Reston, VA, USA, 2014; p. 48.
40. Isenburg, M. *LAStools*; Rapidlasso GmbH: Gilching, Germany, 2015.
41. Yang, Y.; Xu, Z.; Wu, Y.; Wei, W.; Song, R. Virtual Fire Evacuation Drills through a Web-Based Serious Game. *Appl. Sci.* **2021**, *11*, 11284. [CrossRef]
42. Keil, J.; Edler, D.; Schmitt, T.; Dickmann, F. Creating Immersive Virtual Environments Based on Open Geospatial Data and Game Engines. *KN J. Cartogr. Geogr. Inf.* **2021**, *71*, 53–65. [CrossRef]
43. Kersten, T.; Drenkhan, D.; Deggim, S. Virtual Reality Application of the Fortress Al Zubarah in Qatar Including Performance Analysis of Real-Time Visualisation. *KN J. Cartogr. Geogr. Inf.* **2021**, *71*, 241–251. [CrossRef]
44. Szabó, G.; Singh, S.K.; Szabó, S. Slope Angle and Aspect as Influencing Factors on the Accuracy of the SRTM and the ASTER GDEM Databases. *Phys. Chem. Earth Parts A/B/C* **2015**, *83*, 137–145. [CrossRef]
45. Gorokhovich, Y.; Voustianiouk, A. Accuracy Assessment of the Processed SRTM-Based Elevation Data by CGIAR Using Field Data from USA and Thailand and Its Relation to the Terrain Characteristics. *Remote Sens. Environ.* **2006**, *104*, 409–415. [CrossRef]
46. Gesch, D.B.; Oimoen, M.J.; Evans, G.A. *Accuracy Assessment of the US Geological Survey National Elevation Dataset, and Comparison with Other Large-Area Elevation Datasets: SRTM and ASTER*; US Department of the Interior, US Geological Survey: Reston, VA, USA, 2014; Volume 1008.
47. Campbell, M.J.; Dennison, P.E.; Butler, B.W.; Page, W.G. Using Crowdsourced Fitness Tracker Data to Model the Relationship between Slope and Travel Rates. *Appl. Geogr.* **2019**, *106*, 93–107. [CrossRef]
48. Sullivan, P.R.; Campbell, M.J.; Dennison, P.E.; Brewer, S.C.; Butler, B.W. Modeling Wildland Firefighter Travel Rates by Terrain Slope: Results from GPS-Tracking of Type 1 Crew Movement. *Fire* **2020**, *3*, 52. [CrossRef]
49. Campbell, M.J.; Dennison, P.E.; Butler, B.W. A LiDAR-Based Analysis of the Effects of Slope, Vegetation Density, and Ground Surface Roughness on Travel Rates for Wildland Firefighter Escape Route Mapping. *Int. J. Wildland Fire* **2017**, *26*, 884–895. [CrossRef]

50. Campbell, M.J.; Page, W.G.; Dennison, P.E.; Butler, B.W. Escape Route Index: A Spatially-Explicit Measure of Wildland Firefighter Egress Capacity. *Fire* **2019**, *2*, 40. [CrossRef]
51. Fryer, G.K.; Dennison, P.E.; Cova, T.J. Wildland Firefighter Entrapment Avoidance: Modelling Evacuation Triggers. *Int. J. Wildland Fire* **2013**, *22*, 883–893. [CrossRef]
52. Dunn, C.J.; O'Connor, C.D.; Reilly, M.J.; Calkin, D.E.; Thompson, M.P. Spatial and Temporal Assessment of Responder Exposure to Snag Hazards in Post-Fire Environments. *For. Ecol. Manag.* **2019**, *441*, 202–214. [CrossRef]
53. Silva, F.R.Y.; Martínez, J.R.M.; González-Cabán, A. A Methodology for Determining Operational Priorities for Prevention and Suppression of Wildland Fires. *Int. J. Wildland Fire* **2014**, *23*, 544–554. [CrossRef]
54. Dunn, C.J.; O'Connor, C.D.; Abrams, J.; Thompson, M.P.; Calkin, D.E.; Johnston, J.D.; Stratton, R.; Gilbertson-Day, J. Wildfire Risk Science Facilitates Adaptation of Fire-Prone Social-Ecological Systems to the New Fire Reality. *Environ. Res. Lett.* **2020**, *15*, 25001. [CrossRef]
55. Thompson, M.P.; Bowden, P.; Brough, A.; Scott, J.H.; Gilbertson-Day, J.; Taylor, A.; Anderson, J.; Haas, J.R. Application of Wildfire Risk Assessment Results to Wildfire Response Planning in the Southern Sierra Nevada, California, USA. *Forests* **2016**, *7*, 64. [CrossRef]
56. Thompson, M.P.; Gannon, B.M.; Caggiano, M.D.; O'Connor, C.D.; Brough, A.; Gilbertson-Day, J.W.; Scott, J.H. Prototyping a Geospatial Atlas for Wildfire Planning and Management. *Forests* **2020**, *11*, 909. [CrossRef]
57. Calkin, D.E.; O'Connor, C.D.; Thompson, M.P.; Stratton, R. Strategic Wildfire Response Decision Support and the Risk Management Assistance Program. *Forests* **2021**, *12*, 1407. [CrossRef]
58. Breiman, L. Random Forests. *Mach. Learn.* **2001**, *45*, 5–32. [CrossRef]
59. Cutler, F.; Cutler, A.; Liaw, A.; Wiener, M. RandomForest: Breiman and Cutler's Random Forests for Classification and Regression. 2018. Available online: https://cran.r-project.org/web/packages/randomForest/index.html (accessed on 3 January 2022).
60. R Core Team. *R: A Language and Environment for Statistical Computing*; R Foundation for Statistical Computing: Vienna, Austria, 2018.
61. John Lindsay Patch Shape Tools—WhiteboxTools User Manual. Available online: https://www.whiteboxgeo.com/manual/wbt_book/available_tools/gis_analysis_patch_shape_tools.html (accessed on 11 November 2021).

Article

Predicting Wildfire Fuels and Hazard in a Central European Temperate Forest Using Active and Passive Remote Sensing

Johannes Heisig [1,*], Edward Olson [2] and Edzer Pebesma [1]

1 Institute for Geoinformatics, Westfälische Wilhelms-Universität Münster, Heisenbergstraße 2, 48149 Münster, Germany; edzer.pebesma@uni-muenster.de
2 Tropical Silviculture and Forest Ecology, Georg-August-Universität Göttingen, Büsgenweg 1, 37077 Göttingen, Germany; edward.olson@posteo.de
* Correspondence: jheisig@uni-muenster.de

Abstract: Climate change causes more extreme droughts and heat waves in Central Europe, affecting vegetative fuels and altering the local fire regime. Wildfire is projected to expand into the temperate zone, a region traditionally not concerned by fire. To mitigate this new threat, local forest management will require spatial fire hazard information. We present a holistic and comprehensible workflow for quantifying fuels and wildfire hazard through fire spread simulations. Surface and canopy fuels characteristics were sampled in a small managed temperate forest in Northern Germany. Custom fuel models were created for each dominant species (*Pinus sylvestris*, *Fagus sylvatica*, and *Quercus rubra*). Canopy cover, canopy height, and crown base height were directly derived from airborne LiDAR point clouds. Surface fuel types and crown bulk density (CBD) were predicted using random forest and ridge regression, respectively. Modeling was supported by 119 predictors extracted from LiDAR, Sentinel-1, and Sentinel-2 data. We simulated fire spread from random ignitions, considering eight environmental scenarios to calculate fire behavior and hazard. Fuel type classification scored an overall accuracy of 0.971 (Kappa = 0.967), whereas CBD regression performed notably weaker (RMSE = 0.069; R^2 = 0.73). Higher fire hazard was identified for strong winds, low fuel moisture, and on slopes. Fires burned fastest and most frequently on slopes in large homogeneous pine stands. These should be the focus of preventive management actions.

Keywords: fuels; wildfire; fire behavior; fire hazard; remote sensing; LiDAR; Sentinel; modeling; simulation

1. Introduction

Climate change is expected to cause more extreme drought periods and heat waves throughout Central Europe [1]. Temperate regions, which are traditionally not prone to fire, may experience more and larger wildfires [2]. In recent years, Central Europe has already been exposed to multiple consecutive extreme drought events. Forest health in Germany has declined, increasing susceptibility to biotic and abiotic disturbances [3]. A significant rise in number and extent of wildfires has been reported for some parts of the country [4].

The vast majority of the wildfires in Germany are man-made (~95%) and tend to be much smaller than fires in typical fire-prone regions, such as North America [5]. High population density and related risks require instant fire suppression measures. In particular, the wildland–urban interface bears great risk potential [6]. However, in contrast to, for example, Mediterranean countries, in Central Europe, both the vegetation firefighting capacities and society's awareness of fire hazard are in an early stage of development. This gap between the increased susceptibility to fire through rapid environmental change and lagging fire suppression capabilities may hold challenges in the near future.

Forest fire behavior is driven by the interplay of three components: topography, weather, and fuels [7]. These components need to be quantified precisely to identify locations with elevated fire hazard potential. Topography and weather data are available and easily accessible in most cases. As fuels strongly vary in space and time, their detailed characterization often demands resource-intensive field observation [8]. Remote sensing data can help in modeling fuel data and predict continuous surfaces of fuel from punctual fuel samples [9]. In particular, dense airborne light detection and ranging (LiDAR) point clouds allow for three-dimensional assessments and help describe the vertical canopy structure [10].

Flame length (FL) and the rate of spread (ROS) are the most meaningful variables originating from fire behavior modeling [2]. Considered together, they can already support decision making. Spatial fire spread simulations may additionally provide insight into the probability of an area burning, given that a fire has started [11]. By integrating these spatial modeling outputs, one can draw conclusions about wildfire hazard [6]. Forest and fire management can benefit from this knowledge and initiate preventive measures according to local priorities [12]. In light of a prevailing need for action to adapt Central European forests to a changing climate, the efficient allocation of limited resources is crucial.

Fire behavior and hazard have been assessed in various case studies, predominantly in regions with substantial wildfire history. Botequim et al. [13] derived fuel characteristics from inventory data for even-aged management units in a Maritime pine forest in central Portugal. With the resulting fire hazard calculations, they established simple discrimination rules to implement fuel treatments. Taccaliti et al. [14] calculated basic fire behavior for black pine forests in Northeastern Italy. They used low-density LiDAR and field data to model the required fuel characteristics, yet omitted the estimation of crown bulk eensity (CBD). Stockdale et al. [15] simulated the effects of different mitigation scenarios on fire hazard in wildlife conservation areas in Alberta, Canada. The study could benefit from existing fuel maps (100 m spatial resolution) and a historic wildfire occurrence database.

This research demonstrates an end-to-end fire hazard assessment in a small study area with high-quality data from an extensive field survey and dense airborne LiDAR. It involves (i) the collection of surface and canopy fuel parameters in the field, (ii) spatial predictive modeling of fuels supported by remote sensing data, and (iii) the identification of locations vulnerable to wildfire through fire behavior and hazard modeling. Thus, we address local fuel conditions as precisely as possible, as no reliable data exists for this region. The study intends to provide a complete and structured workflow to produce spatial predictions of fire hazard in an area lacking historic fire records. Data and workflow details are shared in a way that is easy to reproduce, increasing accessibility to fire hazard information. Our study thus reacts to the current and future needs in Central European forest management and planning by supporting capacity building in the wildfire hazard domain.

2. Study Area

The Haard is a managed temperate forest located at 51.7° N, 7.2° E in North Rhine-Westphalia, Germany, with an approximate extent of 60 km^2 (Figure 1). Surrounded by small cities, infrastructure, industry and agriculture, it is highly frequented for recreational activities. Other land use types comprised by the Haard include a medical facility, a holiday resort, and an abandoned mine. Major roads and water bodies confine the forest in the north and west, while one public road crosses it from north to south. The study area is mostly flat, while moderate slopes can be found in northern and southwestern parts. Three dominant species form the majority of trees in the Haard, namely *Pinus sylvestris* (Scots pine), *Fagus sylvatica* (European beech) and *Quercus rubra* (red oak). All three species are economically relevant for timber production.

Considered individually, each of the three species has a different role in the wildfire hazard context. Often found on dry and sandy soils, Scots pine is adapted to limited water availability. Its crown structure allows light to reach the forest floor, which promotes the accumulation of understory vegetation and surface fuels. Scots pine is highly abundant in Northeastern Germany, the part of the country historically most affected by wildfire. European beech is the most common deciduous tree species in Germany. As a climax community species, it naturally outcompetes other species by shading the forest floor while being shade-tolerant itself. Old beech stands produce only small surface fuel loadings dominated by litter with few woody components. Limited light availability entails higher soil moisture and sparse understory. Both conditions hinder fire spread. Young planted stands in the Haard on the other hand are very dense, row-wise plantations. Their crown base height is low, suggesting an elevated potential for flames to reach the crown via natural ladders. Red oak is native to eastern North America and became a solid silvicultural alternative to local oak species in Central Europe due to its superior growth rate. Its ability to cope with higher temperatures further makes it a popular choice for adapting forests to climate change. Regarding shading, understory, and live surface fuel loads, red oak lies in between Scots pine and European beech. Among others, these three species were reported to be strongly impacted by climatic extremes during the summers of 2003 and 2018, making them vulnerable to secondary drought effects in the following years [3]. It is conceivable that this pattern may reoccur in the future, raising their susceptibility to fire.

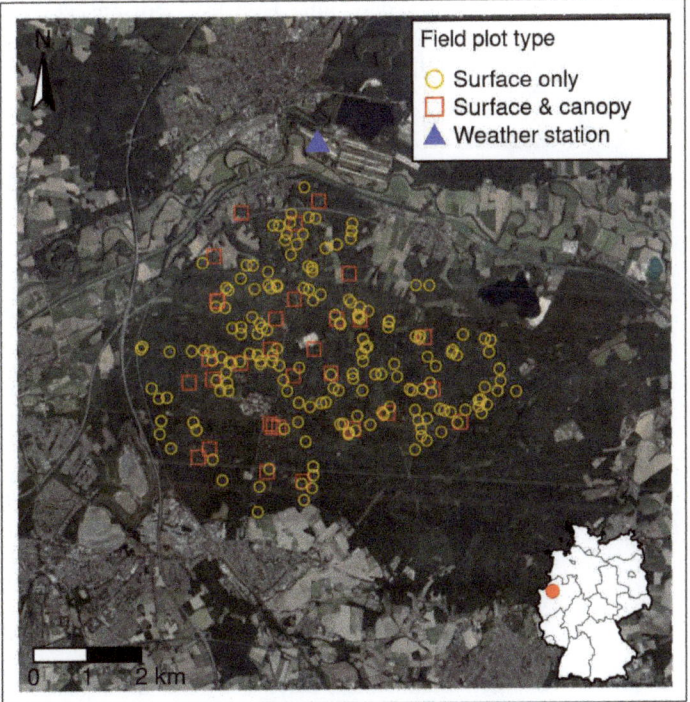

Figure 1. Our study area is a managed temperate forest dominated by Scots pine, read oak and European beech. Located at 51.7° N, 7.2° E (red dot) in densely populated North Rhine-Westphalia, Germany, it is surrounded by agriculture and industry, serving as a local recreation area. The map displays 215 surface fuel sampling locations, of which 30 were used to additionally sample canopy fuel characteristics. Wind properties were assessed at a nearby weather station.

3. Materials and Methods

3.1. Field Data

Surface and canopy fuel characteristics were sampled at points throughout the study area. The collected data enable the generation of fire behavior fuel models (FBFM) and serve as reference for predictive machine learning models.

3.1.1. Surface Fuels

Surface fuel characteristics were assessed during a field campaign in the summer of 2019. A total of 215 plots were split among the three dominant tree species found in the Haard forest (beech: 73, pine: 73, oak: 69). Locations were distributed within the study area by generating random coordinates on a 25 m grid. The process was spatially stratified by a forest inventory map containing polygons of homogeneous stands dominated by either one of the three species. Each sample plot was visited physically, and centers were mapped via GPS.

Surface fuels were sampled using a standardized field method developed by Lutes and Keane [16]. Downed woody debris, humus, leaf litter and ground vegetation were sampled, using a compound cluster-plot of intercepts, point measurements and fixed-area plots. Downed woody debris was estimated based on the planar intercept principle of Brown [17], wherein three time-lag classes (1, 10 and 100 h) are defined according to diameter. Their respective particle counts along transects are translated to loadings expressed as mass per area. Humus and leaf litter depth were measured at multiple locations within each plot, averaged and converted into loadings by their respective bulk densities [16]. Similarly, vegetation loadings were estimated based on fixed-area plot observations of percent ground cover and average height of live and dead shrubby and herbaceous vegetation within the fuel bed (i.e., below two meters above ground). Field data were aggregated for each dominant species group using the median, resulting in three custom FBFMs.

3.1.2. Canopy Fuels

Canopy fuel characteristics were assessed during a follow-up field survey in the summer of 2021. The main objective here was to quantify CBD, defined as the mass per unit volume of canopy biomass. It describes foliage and twigs with diameters less than 3 mm that would burn in a crown fire [18] and represents one of four canopy fuel parameters required for fire behavior modeling. Recorded field measurements include diameter at breast height (DBH), canopy height (CH), crown base height (CBH) and crown class. During this campaign, only 30 locations (10 per dominant species) were revisited. Plot locations were randomly selected from the existing set of surface fuels plots, stratified by dominant species. Each live individual within a 10 m radius around the plot center and a DBH greater 7 cm was considered. This led to recording a total of 695 individual trees. DBH was derived from circumference measurements, assuming tree stems to be perfectly cylindrical. Heights were approximated with a trigonometric method. It makes use of the fixed distance between the observer and stem and the inclination when targeting the top or bottom of the live canopy. Crowns were subjectively classified into categories by extent, height and growth form, as described by Lutes [19]. An individual was considered *dominant* if its height and extent were unmatched within the plot. Multiple trees forming the main canopy layer were considered *co-dominant*. The categories *intermediate* and *suppressed* were assigned to individuals that reached into the lower portion of the canopy or that were overtopped by surrounding crowns, e.g., due to competition.

Tree lists were input to FuelCalc [20] to derive plot-level CBD in m^3/kg. Canopy characteristics described above served as input to species-specific allometric equations. This workflow for deriving CBD has been applied by other studies, such as that of Erdody and Moskal [21], as direct observations require a costly destructive sampling method [22]. FuelCalc produces CBD estimates in 1-foot (~30 cm) vertical bins and summarizes the plot-level result as the maximum 5-foot (~1.5 m) running mean [19].

3.2. *Remote Sensing Data*

3.2.1. Sentinel-1 and -2

All Sentinel-1 and -2 tiles from 2019 covering the study area were processed to annual composites (Table 1), using Google Earth Engine [23]. The Sentinel-1 SAR GRD image collection (n = 120) was pre-processed following a framework by Mullissa et al. [24]. It applies border noise correction, speckle filtering and radiometric terrain normalization.

Resulting bands include polarizations VV, VH and their ratio. Optical data from Sentinel-2 Level-2A were filtered by scene cloud coverage (<5%). Clouds in the remaining tiles (n = 24) were masked using the quality assessment (QA) band, which flags pixels that may be affected by normal or cirrus clouds. Both collections were reduced in the temporal domain by calculating three percentiles: 10th, 50th (=median), and 90th. The approach attempts to better capture the intra-annual variation in backscatter or reflectance. In particular, vegetation mapping may benefit from taking seasonal shifts of land surface properties into account.

3.2.2. LiDAR

Open-access airborne laser scanning (ALS) data are available for the entire German state of North Rhine-Westphalia [25]. The latest acquisition covering the study area is from February 2020. Laser returns are declared to have a mean location error of 30 cm and a mean vertical error of 15 cm. Point cloud density is specified with 4–10 points per square meter. However, this value was found to be considerably higher in parts of the study area with a complex vertical vegetation structure. Among other meta data, each point carries its return number (first, last, nth) and classification codes for ground, non-ground and other categories, which may disqualify them for further analysis (e.g., noise).

ALS data are well suitable to extract forest structure metrics. The 3D point cloud is reduced to multiple 2D representations, describing canopy cover, height, density, or terrain. Thanks to the high point density, a partition into multiple height bins is feasible. These may have fixed widths or follow percentiles.

In total 74 metrics were derived from ALS (Table 1). Processing of the LiDAR data was facilitated by the R programming language [26] and the `lidR` package [27]. In total, 90 coherent 1 × 1 km tiles were retrieved from the administrations' web service and processed to 10 m resolution metrics. First, a digital elevation model (DEM) was computed from the raw point cloud. The simple triangulation algorithm only considers formerly classified ground points. Subsequently, slope and aspect were derived from the DEM using the *terra* package [28]. For further processing, the point cloud was normalized to eliminate the effect of varying elevation on vegetation structure calculations.

LiDAR-derived metrics produced for this analysis include several basic descriptive statistics that have been proven useful when modeling vertical structure for forestry applications [21,29–31]. A large portion of the metrics (>60%) is related to the height values of laser returns (Z-dimension). Some describe the point cloud as a whole (e.g., maximum, mean, standard deviation, or skewness). Others aim at revealing structural differences using (cumulative) height percentiles or the percentage of returns exceeding a certain threshold. Metrics describing the mean height of grasses, shrubs or trees as well as the vertical height gap follow the definitions introduced by the PROMETHEUS fuel type classification system [32].

Two slightly more complex metrics are the canopy height model (CHM) and CBH. Minor measurement errors in ALS allow both to be directly derived from LiDAR point clouds. The CHM was calculated using the pit-free algorithm [33]. It facilitates the triangulation of first returns in different height bins before assembling them. Further, it enriches each point with eight sub-circles that approximate the LiDAR footprint (~20 cm diameter) more realistically and create smoother results. CBH was calculated following a quantile-based approach proposed by Chamberlain et al. [31]. In theory, this method detects the height at which the first live branches are present, which intersect considerably more pulses than the stem below CBH.

Roughly one third of the metrics is devoted to canopy cover (CC). Their calculations follow the assumption made by Riaño [34] that vegetation cover is represented by the fraction of canopy (i.e., non-ground) returns from the total. Reversely, the fraction of ground returns can be obtained by subtracting CC from 1. Next to overall vegetation cover, a cumulative vertical profile at heights ranging from 0 to 30 m was computed. Again, the cover of grasses, shrubs or trees follow the PROMETHEUS fuel type classification system and were introduced by Novo et al. [35]. As a proxy for canopy surface roughness, the Rumple index was included. It is defined as the ratio of canopy surface and ground area [36].

Two metrics are related to LiDAR return density. Besides the total number of returns per area, the density of first returns within the canopy was computed following the method by Andersen et al. [29]. Three metrics with zero variance ($Z_{p[1,5,10]}$) were excluded prior to the following analysis.

Table 1. Remotely sensed predictor variables included in this study (n = 119). The majority of datasets are derived from airborne LiDAR point clouds (n = 74) and describe vertical forest structure and terrain. Temporal composites (t. c.) were computed from all Sentinel-1 and -2 acquisitions using the 10th, 50th and 90th percentiles.

	Variable	Name	Unit	Reference	n
LiDAR					
height					
	CHM	canopy height model	m	[27,33]	1
	CBH	crown base height	m	[31]	1
	Z_{max}	maximum height	m	[27]	1
	Z_{mean}	mean height	m	[27]	1
	Z_{sd}	height standard deviation	m	[27]	1
	Z_{cv}	height coefficient of variation	m	[27]	1
	Z_{iqr}	height inter-quartile range	m	[27]	1
	Z_{skew}	height skewness	-	[27]	1
	Z_{kurt}	height kurtosis	-	[27]	1
	$Z_{entropy}$	height entropy	-	[27]	1
	$Z_{p[15,20,...,95,99]}$	height percentiles	m	[27]	18
	$Z_{pcum[10,...,90]}$	cumulative height percentiles	m	[27]	9
	$Z_{mean_{[g,s,t]}}$	mean height grass, shrubs, trees	m	[32]	3
	Z_{gap}	vertical tree-shrub height gap	m	[32]	1
	$P_{z>z_{mean}}$	percent of returns above Z_{mean}	%	[27]	1
	$P_{z>2}$	percent of returns above 2 m	%	[27]	1
cover					
	C	vegetation cover	%	[34]	1
	P_{ground}	percent ground returns	%	[27]	1
	$C_{[0-4.5,5-9,10-30]}$	cumulative vertical profile	%	[37]	21
	$CZ_{[g,s,t]}$	cover of grass, shrubs, trees	%	[35]	3
density					
	$Rumple$	Rumple index	-	[36]	1
	N	total number of returns	-	[27]	1
	D	density 1st returns in canopy	%	[29]	1
terrain					
	DEM	elevation	m	[27]	1
	$Slope$	terrain slope	°	[28]	1
	$Aspect$	terrain aspect	°	[28]	1
Sentinel-1					
	$VV_{p[10,50,90]}$	VV polarization t. c.	dB	[24]	3
	$VH_{p[10,50,90]}$	VH polarization t. c.	dB	[24]	3
	$VV/VH_{p[10,50,90]}$	VV/VH ratio t. c.	-	[24]	3
Sentinel-2					
	$B01_{p[10,50,90]}$	ultra blue band t. c.	SR		3
	$B02_{p[10,50,90]}$	blue band t. c.	SR		3
	$B03_{p[10,50,90]}$	green band t. c.	SR		3
	$B04_{p[10,50,90]}$	red band t. c.	SR		3
	$B05_{p[10,50,90]}$	red edge 1 band t. c.	SR		3
	$B06_{p[10,50,90]}$	red edge 2 band t. c.	SR		3
	$B07_{p[10,50,90]}$	red edge 3 band t. c.	SR		3
	$B08_{p[10,50,90]}$	NIR 1 band t. c.	SR		3
	$B09_{p[10,50,90]}$	SWIR 1 band t. c.	SR		3
	$B11_{p[10,50,90]}$	SWIR 3 band t. c.	SR		3
	$B12_{p[10,50,90]}$	SWIR 4 band t. c.	SR		3
	$NDVI_{p[10,50,90]}$	vegetation index t. c.	-	[38]	3

3.3. Wind

Fire behavior models require information on wind speed and direction. Both may influence fire behavior significantly. Wind characteristics were assessed using the rdwd package [39] and hourly data from a close by weather station (Figure 1). Only observations from the months most relevant to wildfire (June, July, and August) between the years 2000 and 2021 were considered.

During the summer months, the dominant wind direction is southwest (240° N). Maximum wind speeds of up to 10.7 m/s were reached, with an average of 2.1 m/s (Figure 2).

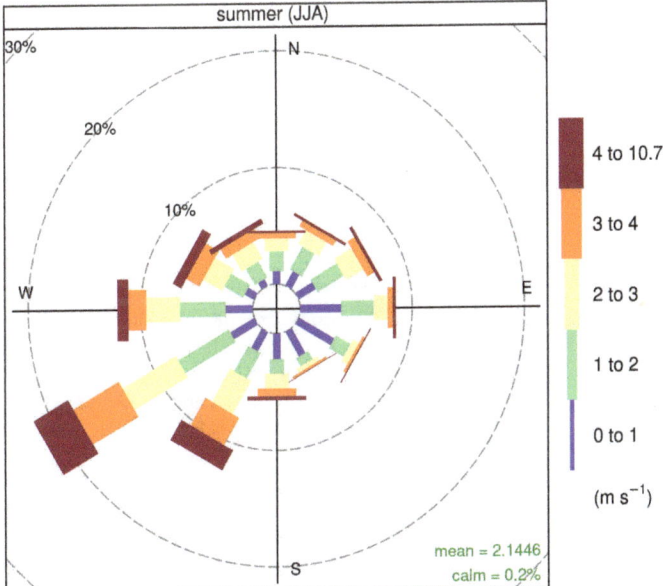

Figure 2. Wind direction and speed aggregated from hourly data in close proximity to the study area during months June, July, and August and from 2000 to 2021. The dominant wind direction was identified as southwest (240° N). Average wind speed is 2.1 m/s, whereas a maximum speed of 10.7 m/s was recorded.

3.4. Fuels Prediction
3.4.1. Surface Fuels

To relate fuel models to spatially continuous surfaces, we used remote sensing data and machine learning frameworks. Field plot locations and their respective dominant tree species label served as training points for image classification. To delimit non-relevant (non-burnable) land cover types (water, agricultural area, urban area and bare ground), 79 training points were established manually with the support of high-resolution true-color imagery.

Predictor variables comprised 119 layers at 10 m resolution from active and passive air- and space-borne sensors (Table 1). After pairing training locations with predictor variables, a random forest classification model was trained [40]. The number of trees was held constant at the default of 500. Hyperparameter *mtry* (number of candidates per split) was tuned with odd-numbered values ranging between 2 and 11. Model training was integrated in a forward feature selection (FFS) process as proposed by Meyer et al. [41]. The model is built successively, starting with the best-performing pair of predictors and adding further ones as long as the performance measure improves. This way, the resulting model is as simple as possible, which is more desirable than having tens or hundreds of terms. The FFS model was optimized using the Kappa index. Overall accuracy (OA) and the class confusion matrix were further consulted for model evaluation.

If training points are clustered in space, a traditional random k-fold cross validation (CV) comes with caveats. Spatial auto-correlation may cause an overly optimistic view on model performance. In this case, however, field sampling locations were randomly distributed over a large part of the study area. This justifies the use of a simple 5-fold random CV, which was implemented in the FFS process.

As an additional criterion, the area of applicability (AOA) was computed for the final model. It aims at estimating the extent to which a prediction model and its CV error can be applied. The underlying dissimilarity index is calculated for each pixel based on its distance to the closest training point in the multidimensional predictor space [42]. Both FFS and AOA were computed using the CAST package [43], while individual model training was facilitated by caret [44].

Finally, the model prediction yielded a four-class raster dataset. A simple majority filter (3 × 3 moving window) was applied to smooth out minor classification errors and reduce their effect on subsequent fire spread simulations. Tree species classes were assigned corresponding fuel model parameters for fire behavior analysis. The remaining class represents all non-burnable land cover types and was thus not assigned any values relevant to fire behavior.

3.4.2. Crown Bulk Density

Plot-level CBD values computed via FuelCalc [20] were subsequently used as target variable in a regression analysis. The same set of 119 predictors was used. Circular field plots (10 m radius) partially covered 4–9 pixels (10 × 10 m) of the predictor data. Therefore, all concerned values were extracted and weighted by their share of the plot area covered. Data were split into training and validation sets in a 70 to 30 ratio. Three predictor variables derived from Sentinel-2 ($B01_{p10}$; $B01_{p50}$) and ALS data (Z_{p15}) were excluded, due to their near-zero variance. The training data used in this analysis had 30 observations and 119 predictors, disqualifying them for ordinary least squares regression. Many predictors are strongly correlated. This constellation calls for a regularized regression model, which aims at making a biased estimate of regression parameters. Bias is thus traded for variance [45].

We selected a ridge regression model for this task. Its regularization parameter *lambda* is dependent on the dataset. *Lambda* was tuned using a grid of 100 values ranging from 0.1 to 100. CV yielded the optimal value by considering the root mean squared error (RMSE) as a criterion. Ridge regression analysis was facilitated by R-packages glmnet [46] and caret [44]. The training data were centered and scaled prior to model training. A 5-fold random CV was implemented for the same reason as described in the previous section. Test runs revealed a non-normal distribution of model residuals. Consequently, the target variable was transformed, using the logarithmic function. In an attempt to expand the number of training samples, 15 plots of the German National Forest Inventory were added. Their tree lists were supplemented with LiDAR-derived CH and CBH and processed in FuelCalc. Predictor variables were extracted in the same way as for the remaining training points.

3.5. Fire Behavior and Hazard Modeling

Next to fuel loadings and fuelbed depth, FBFMs further consist of several constants, such as the surface-area-to-volume ratio (SAV), moisture of extinction (MOE), and heat content [47]. Constants were selected in accordance with existing standard FBFMs [47,48]. All fuel models received 1, 10, and 100 h SAV constants of 5906, 4249, and 4593, respectively. MOE was set to 30% and heat content to 18,622 Kj/kg.

FlamMap is a quasi-empirical fire spread model, widely used among forest and fire managers in operational settings [2]. Its main requirement is the fire landscape data, which consist of three terrain variables (elevation, aspect, and slope), four forest structure variables

(CH, CBH, CC, and CBD) and a surface fuel classification [49]. Additional controls, e.g., fuel moisture, wind, or temperature, can be defined. FlamMap's objective is to calculate spatio-temporal fire behavior under a unique set of environmental conditions. This is especially interesting when assessing the effects of varying weather conditions, fuel properties, or fire-related forest management actions.

To evaluate fire hazard in the Haard under a variety of environmental conditions, we set up a grid of realistic scenarios. We selected two options each for the components of (i) fuel moisture, (ii) wind speed, and (iii) air temperature. This resulted in a total of eight individual scenarios (Table 2).

Like many other regions, Central Europe recently experienced extremely hot and dry summers. Extended heat waves and the absence of precipitation fostered an increased number of ignitions. Surface fuel moisture content dropped and caused higher rates of fire spread. To reflect variable moisture conditions and their effects on fire spread, Scott and Burgan [48] proposed a set of scenarios along with their standard FBFMs. Dead (D) and live (L) fuels can each be assigned four moisture levels, ranging from very low to low, moderate and high (1–4), which are expressed as representative moisture content values in percentage values. Dead fuel moisture scenarios are characterized for time-lag categories 1, 10 and 100 h and range from 3 to 14 percent. Live fuel moisture scenarios include values for herbaceous and woody fuels and range from 30 (fully cured) to 150 (fully green) percent. We selected fuel moisture scenarios D1L1 and D3L1, which should correspond to the effect of a longer and a shorter pre-fire drought period. Both are characterized by very low moisture levels for live herbaceous (30%) and live woody (60%) vegetation. However, they differ in dead fuel moisture levels for time-lag categories 1, 10, and 100 h with 3%, 4%, and 5%, and 9%, 10%, and 11%, respectively.

FlamMap uses either weather time series data or single value inputs for individual weather components. Wind speed and direction strongly affect fire initiation and spread [50]. Both can either be static inputs or may be adjusted dynamically using the WindNinja module [51]. This numerical micro-scale wind flow model is designed for fire modeling applications and simulates mechanical and thermal effects of terrain, vegetation, and air temperature on the flow. It scales down wind speed and direction to finer spatial resolutions.

To investigate the impact of different weather conditions on fire hazard, we considered two values each for wind speed and air temperature. We compare peak (10 m/s) and average (2 m/s) wind speed as well as extreme (35 °C) and moderate (25 °C) summer temperatures. Figure 3 displays exemplary WindNinja outputs for the study area from constant wind speed (10 m/s) and direction (240° N). Despite revealing local anomalies, down-scaled wind characteristics remained similar to their original input constants for the most part.

Table 2. Fire hazard modeling scenarios representing a range of environmental conditions. Eight combinations result from two fuel moisture scenarios (D1L1 and D3L1), two wind speed values (10 and 2 m/s), and two air temperature values (35° and 25 °C).

	S1	S2	S3	S4	S5	S6	S7	S8
FMS	D1L1	D1L1	D1L1	D1L1	D3L1	D3L1	D3L1	D3L1
Wind speed [m/s]	10	10	2	2	10	10	2	2
Air temp. [°C]	35	25	35	25	35	25	35	25

The two most critical and valuable landscape fire behavior outputs for decision makers are fire intensity (expressed as FL) and the ROS [2]. Both are calculated on a pixel basis. Considered individually, both characteristics bear a limited hazard potential. Stationary high-intensity fires or fast-spreading low-intensity fires only pose minor challenges for containment efforts. However, quickly moving high-intensity fires are difficult to manage and contain, resulting in an elevated fire hazard potential.

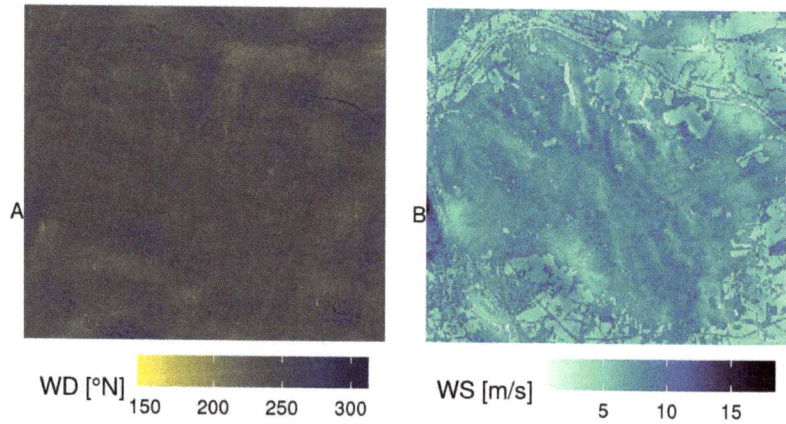

Figure 3. Wind direction (WD; (**A**)) in degrees north and wind speed (WS; (**B**)) in meters per second down-scaled and adjusted to the study area's terrain using WindNinja software. Constant input values were assessed in an exploratory analysis of weather station data. WD was fixed at 240° N, WS at 10 m/s.

In a second step, FlamMap employs the minimum travel time (MTT) algorithm to run fire spread simulations [52]. It requires a set of random or predefined ignition locations. Each fire burns under the same environmental conditions and stops after a specified duration. That makes MTT particularly useful for the analysis of effects of spatial patterns in fuels and topography [49]. Final perimeters are recorded for each simulated fire. Locations (pixels) that have burned more frequently throughout the simulation will receive higher conditional burn probabilities (CBP). This means they are more likely to burn in the case of a fire, but gives no indication of the probability of a fire starting. CBP values are scaled by their maximum and broken down into five ordinal classes, ranging from lowest to highest.

Next to CBP, MTT also produces a conditional flame length (CFL). It represents a weighted version of FL previously introduced in the context of landscape fire behavior. MTT distinguishes heading, flanking, and backing fire spread. Therefore, heading fire burns at a higher intensity than the other two. Throughout repeated fire spread simulations, every pixel can burn multiple times and with different intensities. MTT computes probabilities for 20 FL classes that sum up to 1. These are aggregated to a single value per pixel by applying probabilities as weights for the mid-point FL of each class (0.5, 1,..., 9.5, >9.5 m). The resulting CFL is considerably lower than FL from a landscape fire behavior calculation. Similar to CBP, CFL is subdivided into six standardized classes.

Integrated fire hazard (FH) is finally calculated from CFL and CBP, following the classification scheme proposed by the Interagency Fuels Treatment Decision Support System (IFTDSS) [53] (Figure 4). The scheme divides continuous fire behavior values into hazard categories, which are more intuitive to read. By combining both measures to a single metric, it offers a more holistic view on the present situation than other fire behavior variables.

A combination of high values in both categories leads to high FH, whereas high values in either one lead to medium FH at most. Equally to CBP, the five FH classes range from lowest to highest.

For each scenario, we ran fire spread simulations using the same set of 10,000 randomly distributed ignition locations. The maximum simulation time for MTT was set to 60 min. This constraint was selected to reflect a sensible reaction time for nearby fire departments. Local forest and fire management confirm that this value approximately matches reality. In addition to the non-burnable land cover class fire spread was restricted by barrier structures, such as major roads or water bodies surrounding the study area. These are represented by vector geometries, which serve as input data for the model.

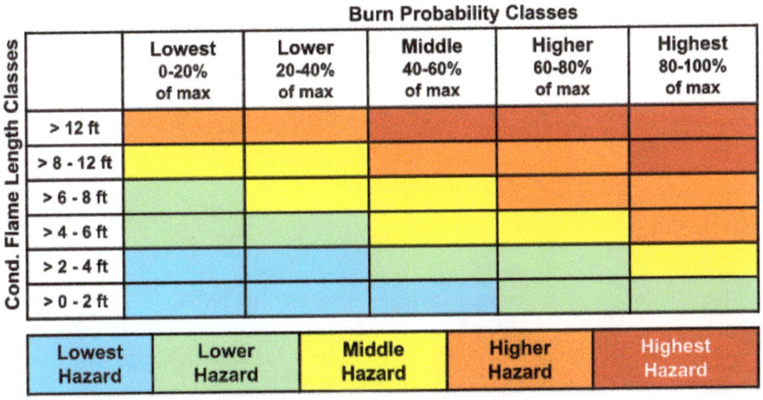

	Burn Probability Classes				
Cond. Flame Length Classes	Lowest 0-20% of max	Lower 20-40% of max	Middle 40-60% of max	Higher 60-80% of max	Highest 80-100% of max
> 12 ft					
> 8 - 12 ft					
> 6 - 8 ft					
> 4 - 6 ft					
> 2 - 4 ft					
> 0 - 2 ft					
	Lowest Hazard	Lower Hazard	Middle Hazard	Higher Hazard	Highest Hazard

Figure 4. Integrated fire hazard (FH) classification scheme proposed by the Interagency Fuels Treatment Decision Support System (IFTDSS) [53]. It categorizes continuous fire behavior variables (CFL and CBP) to present a single intuitive metric. The highest FH is only reached in places where both CFL and CBP are very high.

4. Results

4.1. Surface Fuels

Field sampling yielded surface fuel loadings for three custom FBFMs. Distributions of sampling data are displayed in Figure 5. Custom FBFMs are created from the median of each parameter (see Table 3). Significant differences between all fuel models can be observed for 1 h fuels. They are generally greater than 1 h loadings of similar standard FBFMs suggested by [48]. In turn, 10 h fuels are very similar among all groups. Red oak shows significantly more 100 h fuels with approximately twice the loading of pine or beech. Pine has the largest quantities of live fuels, especially for herbaceous vegetation. Beech, however, forming dense top canopies and shading the forest ground well, only shows small live shrub loadings and almost no live herbs. This pattern is reflected in the fuelbed depth. Pine fuelbeds are about twice as deep as the others.

Table 3. Fuel loading parameters from field sampling. Data were aggregated using the median to create fuel models for fire behavior calculation. Dead and live fuel loadings are measured in [kg/m^2], and height of the fuelbed is in [m].

Species	Fuel Loadings [kg/m^2]					Fuelbed Depth [m]
	1-h	10 h	100 h	Live Herb	Live Shrub	
Beech	1.60	0.62	0.23	0.00	0.10	0.47
Red Oak	1.41	0.62	0.40	0.06	0.42	0.52
Pine	1.17	0.58	0.20	0.20	0.43	1.06

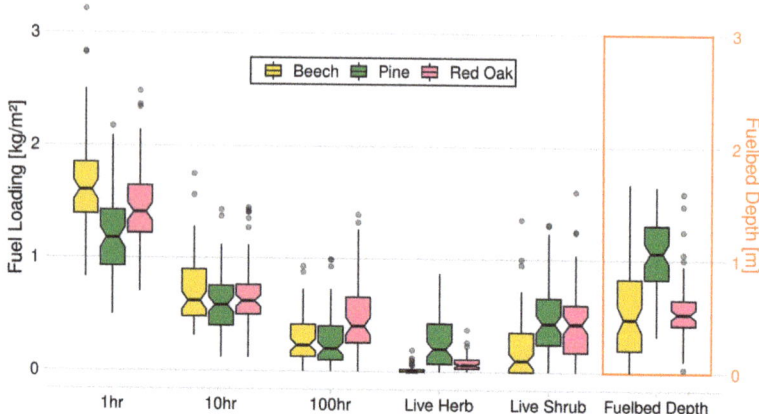

Figure 5. Boxplot of sampled surface fuel parameters for each fuel model. Dead and live fuel loadings are measured in [kg/m^2], height of the fuelbed is in [m]. Beech and red oak produce slightly more 1 and 100 h fuels, while more live herbaceous biomass is found in pine stands. Higher-growing live fuels cause pine's fuelbed depth to exceed that of the others.

The FFS process found the best performing solution among 13,924 possible combinations of predictor variables. CV selected an optimal `mtry` of 2 and reported an OA of 0.971 with a Kappa of 0.967. The final model consisted of five variables, with $B05_{p90}$ being the most important one. $B06_{p90}$ and $Rumple$ also scored high variable importance, followed by $B05_{p10}$. Finally, Z_{cov_g} was only able to add very little improvement to the model. FFS successfully reduced the number of predictors significantly (5 vs. 119) while keeping model performance high. For comparison, a classic random forest model was built, using all available predictors. Performance was still high, but both OA and Kappa decreased by 2.8% and 3.8%, respectively.

The model confusion matrix revealed a perfect classification of non-burnable areas. This may be explained by noticeable differences in the spectral signatures and vertical structure of non-burnable land cover (e.g., water bodies, urban area) compared to forest vegetation. Minor confusion was found between tree species classes. Errors ranged between 4% and 8%, with beech having slightly larger errors than pine and red oak. Previously separated validation samples (n = 90) were tested on the model prediction. All samples were classified correctly, yielding an OA and Kappa of 1.0. Overall, the fuel model classification results were very good. However, this could be expected, as it was a simple modeling task combined with a large range of powerful predictors.

The final prediction is shown in Figure 6 alongside with the AOA. The dissimilarity of predictors from the training samples was only critical in areas less relevant for fire spread. Fewer than 1% of burnable pixels fell outside the AOA and represented mainly roads, bare ground or vegetation along water bodies. These areas are often located in close proximity to the forest edge or to existing fire barriers, which makes them less relevant for fire spread modeling.

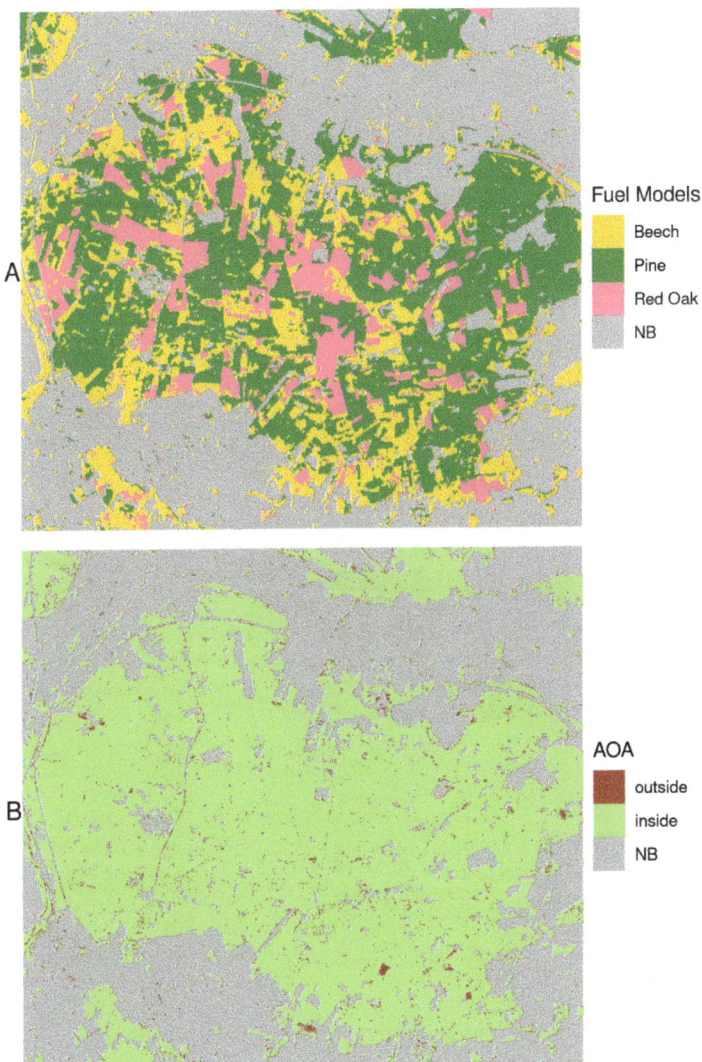

Figure 6. Surface fuel model prediction (**A**) and respective area of applicability (AOA; (**B**)). Spatial patterns originating from forest management are clearly visible. Pine is the most abundant species in the study area (51%) followed by beech (32%) and red oak (17%). Burnable pixels falling outside the AOA sum up to less than 1%.

4.2. Crown Bulk Density

Field measurements of canopy fuel characteristics show clear differences between dominant species (see Figure 7). Many beech stands in the study area are young and have small DBH. They often occur in row-wise plantations with high stem density and nearly identical CBH and CH. Red oak and pine stands are older and less dense as a result of selective logging. Their DBH and CH are two to three times greater than those of beech. CBH, the height of the lowest live branch, is generally much lower for red oak compared to pine, while the opposite is the case for CH. As a consequence, the CBD estimates for red oak are lower.

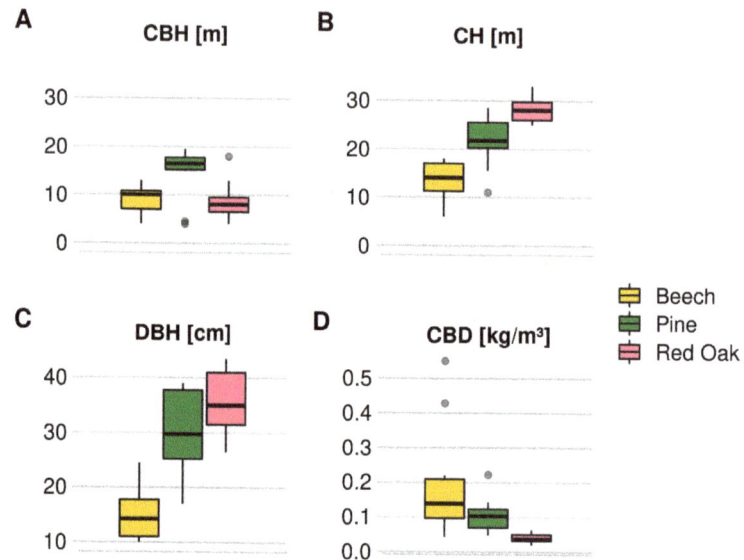

Figure 7. Boxplot of canopy fuel parameters from field sampling for each fuel model. Parameters include crown base height (CBH; (**A**)), canopy height (CH; (**B**)), diameter at breast height (DBH; (**C**)), and crown bulk density (CBD; (**D**)). CBD was calculated via FuelCalc using the other measures as inputs. Beech stands sampled in this study area are mostly young with low, yet dense canopy layers. Contrarily, red oak stands are old with open crowns and low CBH, resulting in low CBD. Pine stands are in between the others, with higher CBH leading to higher CBD than red oak.

Ridge regression analysis was applied to predict CBD for the Haard forest. CV selected an optimal regularization parameter *lambda* of 10.7. Overall, model performance was poor which was expected, considering the small number of training samples. A model-R^2 of 0.59 with a RMSE of 0.054 was reported. Independent validation samples produced a higher R^2 of 0.73, while RMSE degraded to 0.069. Although R^2 is acceptable, RMSEs are large, considering a CBD training sample mean of 0.095. Differences in model performance and validation scores indicate the introduction of bias by ridge regression.

Variable importance scores indicated strong dependence on LiDAR-derived vertical structure metrics and optical predictor data. The most relevant predictors included C_{25}, $B05_{p10}$, Z_{p20}, $NDVI_{p90}$, $B04_{p90}$, and DEM. Further, the nine next relevant variables in the ranking, Z_{iqr}, Z_{pcum80}, and $Z_{p65,\ldots,95}$, all describe vegetation structure in the upper third of the tree. This coincides with relative heights at which CBD can be found at the maximum.

We tested adding training samples from NFI plots (n = 15) but were not able to improve the model. On average, their derived CBD values were significantly smaller than the existing values based on field sampling. NFI surveys include records of species and CH among many other observations. However, they do not include CBH. Supplementing NFI tree lists, for example, with LiDAR-derived CBH at 10 m spatial resolution, is rather inaccurate, especially when considering plots with heterogeneous species composition, age, and vertical structure.

Spatial prediction and AOA for CBD are shown in Figure 8. CBD values range from 0 to 0.3. They roughly follow the tree species classification, while higher densities can be observed for pine than for beech and red oak. Anomalies in CBD within homogeneous patches dominated by a single species are related to structural differences. Considering only forested pixels, 20% fall outside the AOA. This may again be explained by the low number of training samples. A significant portion is located in areas with steeper slopes.

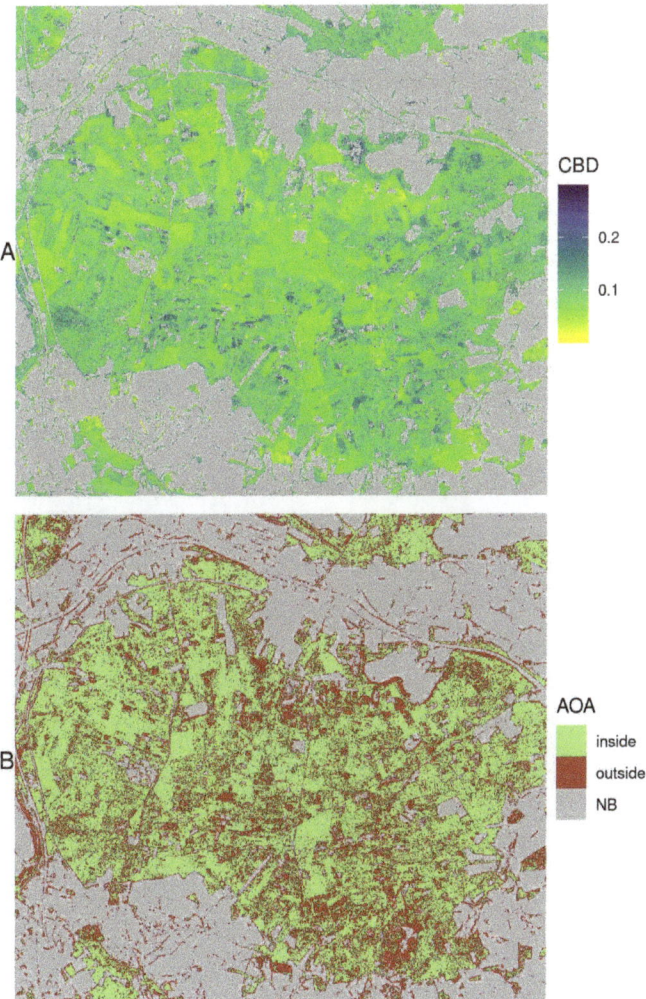

Figure 8. Spatial prediction of crown bulk density (CBD; (**A**)) and the respective area of applicability (AOA; (**B**)). Values strongly depend on the dominant tree species. Anomalies within stands are related to differences in forest structure. Pixels falling outside the AOA (20%) occur especially on slopes and are attributable to the low number of training samples for the predictive model.

4.3. Fire Behavior and Hazard

Fire behavior was computed for the Haard forest considering landscape properties, fuel metrics, and eight sets of weather conditions. Variations in fire behavior can be observed in the spatial domain. This is related to the heterogeneity of surface fuels, vertical forest structure and terrain. Further, wind speed and fuel moisture impact fire behavior, resulting in major differences among scenarios. Air temperature, however, had no significant effect on fire behavior. Therefore, only scenarios S1, 3, 5, and 7 are evaluated in the following.

Overall, the strongest fire behavior was found for S1 (Figure 9 row 1). Low fuel moisture and high wind speed led to FL exclusively larger than 2 m, sporadically even exceeding 10 m. Low to medium FL predominantly occurred in red oak and beech stands. On the contrary, pine stands in particular showed high fire intensities. The same pattern was observed for ROS. Even more so than FL, ROS was closely linked to the spatial patterns of surface fuel models. Spread rates ranging from 6 to beyond 10 m/min were common in pine-dominated stands, whereas broad-leaved stands produced significantly slower fires. Most areas in the Haard with steep slopes are populated by pine. The combined effects of fire-prone pine fuel beds and steep slopes result in the highest observations of FL and ROS within the study area.

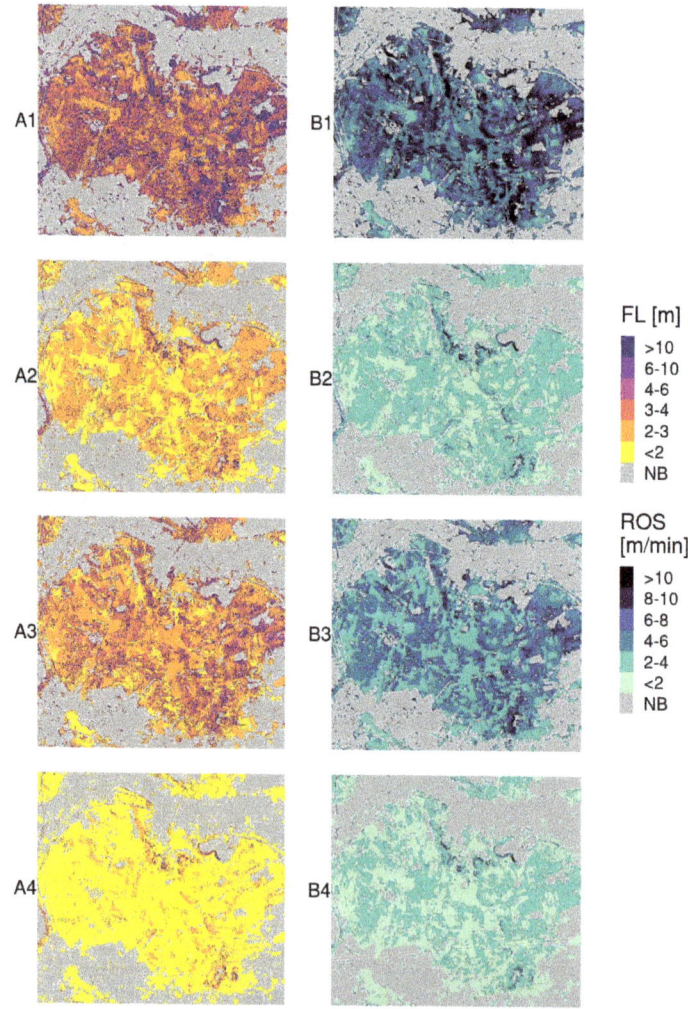

Figure 9. Landscape fire behavior outputs including flame length (FL; (**A**)) and rate of spread (ROS; (**B**)) for scenarios S1, 3, 5, and 7 (rows 1, 2, 3, and 4). Fire behavior depends more strongly on wind speed than on dead fuel moisture. Even if wind speed is low and dead fuel moisture is high, individual locations in pine stands and on steep slopes show FL > 6 m and ROS > 8 m/min. NB = Non-Burnable.

The effect of high wind speed on fire behavior becomes visible when comparing scenarios S1 with S3 (Figure 9 rows 1 and 2). Overall, a reduction in wind speed from 10 to 2 m/s cut the mean FL in half from 5.1 to 2.5 m, while ROS dropped from 6.8 to 2.6 m/min. Lower wind speed in S3 diminished fire behavior disproportionately in pine stands, which showed extremes in S1. A weaker, yet still significant, reduction occurred in beech and red oak stands. Few locations with high FL and ROS in the central northern and southwestern parts of the study area still stand out in S3 and S7 (Figure 9 rows 2 and 4). These correlate well with increased slopes and related higher wind speeds, highlighting their promoting effect on fire behavior.

Comparison of S1 and S5 (Figure 9 rows 1 and 3) emphasizes the influence of dead fuel moisture. S5 features similar spatial patterns in both FL and ROS, with extreme values consistently occurring in steep pine stands. On average, however, both are reduced significantly when facing increased dead fuel moisture levels. Mean FL dropped to 3.2 m while ROS slowed down to 4.8 m/min. Similar, yet weaker, shifts apply to average wind scenarios S3 and S7. Hence, a more substantial reduction in fire behavior was recorded for a decrease in wind speed as opposed to a decrease in dead fuel moisture.

Fire behavior outputs from MTT fire spread simulations (Figure 10) show comparable patterns as landscape fire behavior results. While strong winds in S1 and S5 result in elevated CFL for the majority of the study area, average winds only lead to a few hotspots of small extent. CFL has substantially lower values than FL, which can be attributed to the weighted calculation method.

CBP for all scenarios was scaled, using the overall maximum of 0.013. Higher and highest CBP in S1 occurred in the center and the eastern part of the Haard. Both areas are characterized by large continuous pine stands indicating an interdependence. Within the limited simulation time, fire spread fastest and formed the largest perimeters when burning in this setting. Concerned locations consequently burned more frequently, leading to higher CBP. Increased fuel moisture content in S3 yielded similar spatial patterns, yet on a lower level without exceeding medium CBP. Disregarding few exceptions, S3 and S5 only produced CBP values belonging to the lowest category. This indicates that fire spread is very limited during the absence of strong winds. Even though spatial variations in CBP exist, they are hidden by the scaled classification scheme.

Similar to the fire behavior results, extreme FH only occurred during strong winds and with low fuel moisture present (Figure 11A). In case of a fire in these conditions, large pine stands in all parts of the study area appeared to be affected more severely than other species. Within pine stands, the highest FH category coincides with areas showing increased CBD or slopes. Regardless of fuel moisture, fire spread simulations featuring strong winds (Figure 11A,C) revealed a small number of FH hotspots.

Simulations for average wind scenarios S3 and S7 (Figure 11B,D) showed clearly reduced FH predictions. Low FH prevailed under both scenarios. While medium FH still emerged in pine-dominated areas for S3, it only sums up to a negligible extent in S7. It should be noted that in both cases, high FH can be found sporadically on steep slopes, despite the absence of strong winds.

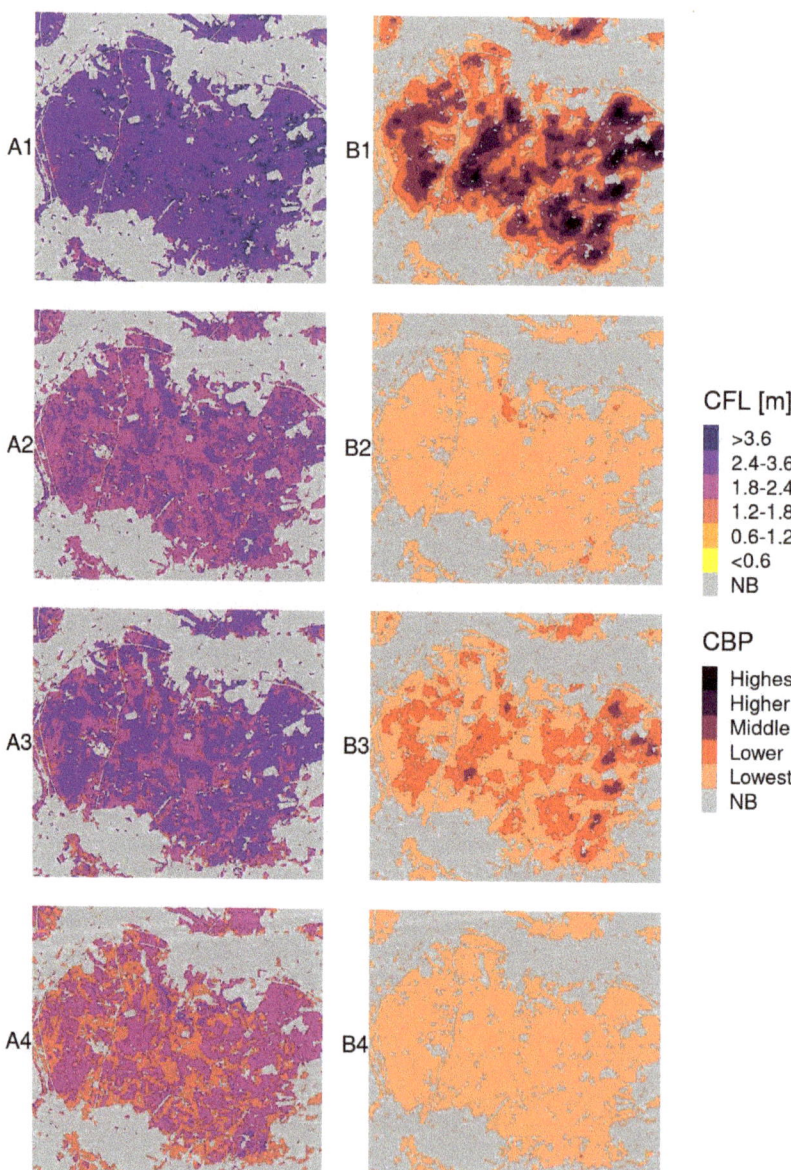

Figure 10. Conditional flame length (CFL; column (**A**)) and conditional burn probability (CBP; column (**B**)) from MTT fire spread simulations for scenarios S1, 3, 5, and 7 (rows 1, 2, 3, and 4). CFL is presented as the mid-point of 20 FL classes weighted by probabilities. CBP was scaled by its maximum (0.013) and classified into five equal bins. Large continuous areas dominated by pine show the highest probability of burning if strong winds are present. NB = Non-Burnable.

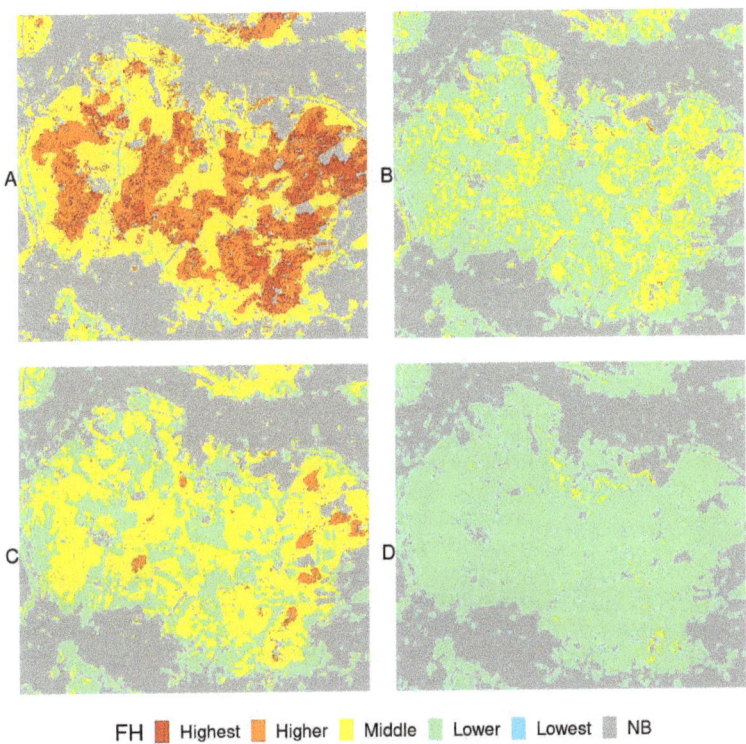

Figure 11. Integrated fire hazard (FH) as a product of conditional flame length (CFL) and conditional burn probability (CBP) for scenarios S1, 3, 5, and 7 (**A–D**). High and highest hazard throughout the study area can only be expected for strong winds and low fuel moisture. Higher fuel moisture significantly reduces high FH and limits it to large homogeneous pine stands. FH is medium to low when wind speed is low, while large pine stands and slopes bear more hazard than other areas. NB = Non-Burnable.

5. Discussion

With this paper, we demonstrate a comprehensive workflow for the assessment of wildfire hazard in a small managed temperate forest. Surface and canopy fuels were sampled in field surveys and extrapolated in space, using predictive statistical modeling methods. We combined fuel information with other variables, completing the fire landscape to feed fire behavior models, simulate the growth of several thousands fires, and derive a hazard index.

Spatial predictions of fuel characteristics were performed based on field data and statistical modeling. Sufficient training data in combination with powerful predictor variables led to a near-perfect classification result for three surface fuel models. Dense ALS point clouds allowed for directly deriving canopy fuel metrics, such as CC, CH, and CBH, due to their relatively low measurement error.

CBD, on the contrary, required regression analysis using training data from field observations. This canopy fuel variable could not be measured directly but was estimated via allometric equations, using field measurements of other tree properties. Estimates then represented the reference for statistical modeling. Through this process, training data were subject to multiple sources of errors. Due to the limited number of samples, the resulting statistical model was weak. The value ranges of field and predicted CBD were similar to previous studies by Andersen et al. [29] and Erdody and Moskal [21]. Their regression model validation produced RMSEs of 0.27 (R^2 = 0.84) and 0.024 (R^2 = 0.88) while including 101 and 57 field sampling locations, respectively. Cameron et al. [30] reported a RMSE of 0.098 (R^2 = 0.78) whilst using 52 training samples. Considering the low number of samples in the present study, a RMSE of 0.069 (R^2 = 0.73) was acceptable, especially as CBD is only one of multiple components of the FH assessment. Further, predicted CBD showed variations between stands with different dominant species, which is comprehensible from an ecological stand point. The ridge regression model introduced bias in exchange for reducing variance. This effect may have been reduced by applying a more complex tuning approach, such as nested cross validation. Alternatively, a better fit may be achieved through more flexibility, for example, by applying a penalized generalized additive model.

Errors in the prediction of all fuel variables may have been caused by the temporal gap between field (summers 2019 and 2021) and LiDAR (winter 2020) data acquisition. In addition, the collection of LiDAR data over deciduous trees during their defoliated phase is problematic. This may cause underestimation of CC and other metrics involving the number of canopy returns.

LiDAR-based predictor variables were crucial for modeling CBD and could contribute to the mapping of surface fuel models. Sentinel-2 data were able to support both efforts. In particular, bands 4, 5, and 6, and the NDVI were listed among the most important predictors. All involve wavelengths in the near-infrared, which are proven to describe vegetation characteristics. Temporal composites of the 10th and 90th percentiles were more important than the ones of the 50th percentile. They represent stages of the vegetative cycle, which reveal more differences among tree species than the annual median. The surface fuel classification model, for instance, took advantage of these differences by selecting $B05_{p10}$ and $B05_{p90}$ as two of its five predictors. Sentinel-1, on the contrary, did not play a relevant role. The SAR signal is able to partially penetrate the forest canopy and was expected to react to differences in CBD. Theses differences may have been too subtle to be captured, or were explained more precisely by other predictors.

As Papadopoulos and Pavlidou [50] stated, fire behavior and spread models require major assumptions, and their output accuracy heavily depends on the quality of input data. This study classified three FBFMs, each connected to one dominant tree species. Within stands, however, the species composition may slightly differ from observations made at sampling locations, altering surface fuel properties. Additional uncertainty in FH prediction arises from the accumulation of errors over the course of a fire growth simulation and the common problem of an unfeasible validation against real wildfire events [50]. Despite numerous sources of uncertainty, simulation-based FH is more robust and informative than simple fire behavior calculations, as it considers flanking and backing fire, not only heading fire [54]. CBP (and therein FH) gives no indication about the probability of a fire starting. As most fires are caused by humans, ignition density is likely to increase with proximity to man-made structures, such as roads, paths or recreational points of interest. Together with an accessibility assessment for fire suppression units, this information should be considered in any follow-up risk assessment.

The present analysis revealed wind speed and slope to be the most relevant factors of FH in the Haard forest. Further, low fuel moisture content can amplify FH. While high wind speed and low fuel moisture are difficult to prevent, forest management can address elevated FH on slopes. These areas could be converted to stands with low stem density, consisting of species that produce slow and weakly burning surface fuels (e.g., red oak). The same strategy could be applied to areas surrounding medical or recreational facilities, as they bear elevated risk. Other FH hotspots were identified in large pine stands. Among the three dominant species in the study area, pine's fuel bed produced the highest flame lengths and spread rates. Additionally, fires reached their largest perimeters in pine-dominated areas, resulting in the highest values for CBP. This may, in part, be related to the fact that local pine stands are larger and more contiguous than beech and oak stands, allowing fire to cover longer distances through a homogeneous and highly flammable landscape. To mitigate fire spread in large pine stands, species composition and forest structure could be diversified, creating a more heterogeneous landscape. Fire barriers in the form of planted strips of broadleaved species could be introduced perpendicular to the dominant wind direction. The Haard is confined by roads and water bodies in the west and north. However, wind and, with it, fire predominantly move toward the east. FH calculations for high wind speeds suggest a need to secure the eastern forest edge to prevent potential wildfire from affecting adjacent campsites and agriculture.

Future research in Central Europe could aim at developing detailed fuels datasets with national or continental extents. These are needed to allow for wildfire hazard and risk assessments on various spatial scales. Previous studies as well as the present one have confirmed the importance of LiDAR data for the spatial prediction of wildfire fuels. ALS data, however, have high acquisition costs and are rarely accessible. Two current trends in LiDAR remote sensing may advance the quantification of fuel variables in the near future. The growing availability of drones with lightweight laser scanning instruments enables high temporal and spatial resolution in small areas. From a global perspective, space-borne LiDAR missions, such as GEDI, will improve not only biomass and CH mapping, but also CBH or CBD estimation.

6. Conclusions

We characterized the fire landscape in a small managed temperate forest by connecting field and remote sensing data. With this baseline, we calculated fire behavior, simulated fire spread and deduced spatial wildfire hazard information. By interpreting the results of this process, we identified critical areas that should receive special attention in management actions related to wildfire safety. With fire as an emerging threat to Central European forests, interest in such information may grow. We deliberately used openly accessible data and low-cost instruments for field sampling. Processing was facilitated through freely available software without advanced computational requirements. These attributes lower the hurdle for forest management to obtain FH information, following our approach.

Author Contributions: Conceptualization, methodology, data curation, formal analysis, validation, visualization, writing—original draft preparation, project administration, J.H.; investigation, J.H. and E.O.; writing—review and editing, J.H., E.O. and E.P.; supervision, E.P. All authors have read and agreed to the published version of the manuscript.

Funding: This research received no external funding.

Institutional Review Board Statement: Not applicable.

Informed Consent Statement: Not applicable.

Data Availability Statement: Data, code, and instructions on how to reproduce this study can be accessed via https://github.com/joheisig/Haard_Wildfire_Fuels_Hazard (27 January 2022).

Acknowledgments: We thank Harald Klingebiel at RWR Ruhr Grün for sharing his experience on fire behavior and history in the Haard. Further, we are grateful to Gregory Dillon, Duncan Lutes and Charles McHugh at the Missoula Fire Sciences Laboratory for their valuable advice on fire and fuels modeling. Thanks also goes to Lukas Blickensdörfer at the Thünen Institute for his help with forest inventory data. We acknowledge support from the Open Access Publication Fund of the University of Münster.

Conflicts of Interest: The authors declare no conflict of interest.

References

1. IPCC. Climate Change 2014: Synthesis Report. In *Contribution of Working Groups I, II and III to the Fifth Assessment Report of the Intergovernmental Panel on Climate Change*; Core Writing Team, Pachauri, R.K., Meyer, L.A., Eds.; Technical Report; IPCC: Geneva, Switzerland, 2014; p. 151.
2. Cardil, A.; Monedero, S.; Schag, G.; de Miguel, S.; Tapia, M.; Stoof, C.R.; Silva, C.A.; Mohan, M.; Cardil, A.; Ramirez, J. Fire behavior modeling for operational decision-making. *Curr. Opin. Environ. Sci. Health* **2021**, *23*, 100291. https://doi.org/10.1016/j.coesh.2021.100291.

3. Schuldt, B.; Buras, A.; Arend, M.; Vitasse, Y.; Beierkuhnlein, C.; Damm, A.; Gharun, M.; Grams, T.E.; Hauck, M.; Hajek, P.; et al. A first assessment of the impact of the extreme 2018 summer drought on Central European forests. *Basic Appl. Ecol.* **2020**, *45*, 86–103. https://doi.org/10.1016/j.baae.2020.04.003.
4. BMEL. *Deutschlands Wald im Klimawandel-Eckpunkte und Maßnahmen*; Technical Report; Bundesministerium für Ernährung und Landwirtschaft: Bonn, Germany, 2019.
5. BMEL. *Waldbrandstatistik der Bundesrepublik Deutschland für das Jahr 2019*; Technical Report; Bundesministerium für Ernährung und Landwirtschaft: Bonn, Germany, 2019.
6. Dillon, G.; Menakis, J.; Fay, F. Wildland Fire Potential: A Tool for Assessing Wildfire Risk and Fuels Management Needs. In *Proceedings of the Large Wildland Fires Conference*; U.S. Department of Agriculture, Forest Service, Rocky Mountain Research Station: Missoula, MT, USA, 2015; pp. 60–76.
7. Keane, R.E.; Burgan, R.; van Wagtendonk, J. Mapping wildland fuels for fire management across multiple scales: Integrating remote sensing, GIS, and biophysical modeling. *Int. J. Wildland Fire* **2001**, *10*, 301. https://doi.org/10.1071/WF01028.
8. Keane, R.E.; Gray, K.; Bacciu, V. *Spatial Variability of Wildland Fuel Characteristics in Northern Rocky Mountain Ecosystems*; Technical Report RMRS-RP-98; U.S. Department of Agriculture, Forest Service, Rocky Mountain Research Station: Ft. Collins, CO, USA, 2012. https://doi.org/10.2737/RMRS-RP-98.
9. Chuvieco, E.; Aguado, I.; Salas, J.; García, M.; Yebra, M.; Oliva, P. Satellite Remote Sensing Contributions to Wildland Fire Science and Management. *Curr. For. Rep.* **2020**, *6*, 81–96. https://doi.org/10.1007/s40725-020-00116-5.
10. White, J.C.; Tompalski, P.; Vastaranta, M.; Wulder, M.A.; Saarinen, N.; Stepper, C.; Coops, N.C. *A Model Development and Application Guide for Generating an Enhanced Forest Inventory Using Airborne Laser Scanning Data and an Area-Based Approach*; Technical Report; Natural Resources Canada, Canadian Wood Fibre Center: Ottawa, ON, Canada, 2017.
11. Finney, M.A.; McHugh, C.W.; Grenfell, I.C.; Riley, K.L.; Short, K. A simulation of probabilistic wildfire risk components for the continental United States. *Stoch. Environ. Res. Risk Assess.* **2011**, *25*, 973–1000.
12. Oliveira, S.; Rocha, J.; Sá, A. Wildfire risk modeling. *Curr. Opin. Environ. Sci. Health* **2021**, *23*, 100274. https://doi.org/10.1016/j.coesh.2021.100274.
13. Botequim, B.; Fernandes, P.M.; Garcia-Gonzalo, J.; Silva, A. Coupling fire behaviour modelling and stand characteristics to assess and mitigate fire hazard in a maritime pine landscape in Portugal. *Eur. J. Forest Res.* **2017**, *136*, 527–542.
14. Taccaliti, F.; Venturini, L.; Marchi, N.; Lingua, E. Forest fuel assessment by LiDAR data. A case study in NE Italy. In Proceedings of the 23rd EGU General Assembly, Online, 19–30 April 2021; https://doi.org/10.5194/egusphere-egu21-12755.
15. Stockdale, C.; Barber, Q.; Saxena, A.; Parisien, M.A. Examining management scenarios to mitigate wildfire hazard to caribou conservation projects using burn probability modeling. *J. Environ. Manag.* **2019**, *233*, 238–248. https://doi.org/10.1016/j.jenvman.2018.12.035.
16. Lutes, D.C.; Keane, R.E. Fuel Load (FL). In *FIREMON: Fire Effects Monitoring and Inventory System*; Lutes, D.C., Keane, R.E., Caratti, J.F., Key, C.H., Benson, N.C., Sunderland, S., Gangi, L., Eds.; Gen. Tech. Rep. RMRS-GTR-164-CD; U.S. Department of Agriculture, Forest Service, Rocky Mountain Research Station: Fort Collins, CO, USA, 2006; pp. 1–25.
17. Brown, J.K. *Handbook for Inventorying Downed Woody Material*; Technical Report; U.S. Department of Agriculture, Forest Service, Intermountain Forest and Range Experiment Station: Ogden, UT, USA, 1974.
18. Keane, R.E.; Reinhardt, E.D.; Scott, J.; Gray, K.; Reardon, J. Estimating forest canopy bulk density using six indirect methods. *Can. J. For. Res.* **2005**, *35*, 724–739. https://doi.org/10.1139/x04-213.
19. Lutes, D.C. *FuelCalc User's Guide (Version 1.7)*; U.S. Department of Agriculture, Forest Service, Rocky Mountain Research Station: Missoula, MT, USA, 2021.
20. Reinhardt, E.; Lutes, D.C.; Scott, J.H. FuelCalc: A Method for Estimating Fuel Characteristics. In Proceedings of the Fuels Management-How to Measure Success: Conference Proceedings, Portland, OR, USA, 28–30 March 2006; p. 10.
21. Erdody, T.L.; Moskal, L.M. Fusion of LiDAR and imagery for estimating forest canopy fuels. *Remote Sens. Environ.* **2010**, *114*, 725–737.
22. Scott, J.H.; Reinhardt, E.D. Estimating canopy fuels in conifer forests. *Fire Manag. Today* **2002**, *62*, 6.
23. Gorelick, N.; Hancher, M.; Dixon, M.; Ilyushchenko, S.; Thau, D.; Moore, R. Google Earth Engine: Planetary-scale geospatial analysis for everyone. *Remote Sens. Environ.* **2017**, *202*, 18–27. https://doi.org/10.1016/j.rse.2017.06.031.
24. Mullissa, A.; Vollrath, A.; Odongo-Braun, C.; Slagter, B.; Balling, J.; Gou, Y.; Gorelick, N.; Reiche, J. Sentinel-1 SAR Backscatter Analysis Ready Data Preparation in Google Earth Engine. *Remote Sens.* **2021**, *13*, 1954. https://doi.org/10.3390/rs13101954.
25. Bezirksregierung Köln. Nutzerinformationen für Die 3D-Messdaten aus dem Laserscanning für NRW. 2020. Available online: https://www.bezreg-koeln.nrw.de/brk_internet/geobasis/hoehenmodelle/nutzerinformationen.pdf (accessed on 27 January 2022)
26. R Core Team. *R: A Language and Environment for Statistical Computing*; R Foundation for Statistical Computing: Vienna, Austria, 2021.
27. Roussel, J.R.; Auty, D.; Coops, N.C.; Tompalski, P.; Goodbody, T.R.; Meador, A.S.; Bourdon, J.F.; de Boissieu, F.; Achim, A. lidR: An R package for analysis of Airborne Laser Scanning (ALS) data. *Remote Sens. Environ.* **2020**, *251*, 112061. https://doi.org/10.1016/j.rse.2020.112061.
28. Hijmans, R.J. Terra: Spatial Data Analysis. R Package Version 1.3-22. 2021. Available online: https://CRAN.R-project.org/package=terra (accessed on 27 January 2022)

29. Andersen, H.E.; McGaughey, R.J.; Reutebuch, S.E. Estimating forest canopy fuel parameters using LIDAR data. *Remote Sens. Environ.* **2005**, *94*, 441–449.
30. Cameron, H.A.; Schroeder, D.; Beverly, J.L. Predicting black spruce fuel characteristics with Airborne Laser Scanning (ALS). *Int. J. Wildland Fire* **2021**, in press. https://doi.org/10.1071/WF21004.
31. Chamberlain, C.P.; Sánchez Meador, A.J.; Thode, A.E. Airborne lidar provides reliable estimates of canopy base height and canopy bulk density in southwestern ponderosa pine forests. *For. Ecol. Manag.* **2021**, *481*, 118695. https://doi.org/10.1016/j.foreco.2020.118695.
32. Chuvieco, E.; Riaño, D.; Van Wagtendok, J.; Morsdof, F. Fuel Loads and Fuel Type Mapping. In *Wildland Fire Danger Estimation and Mapping—The Role of Remote Sensing*; World Scientific Publishing Co. Pte. Ltd.: Singapore, Singapore, 2003; Volume 4, pp. 119–142. https://doi.org/10.1142/9789812791177_0005.
33. Khosravipour, A.; Skidmore, A.K.; Isenburg, M.; Wang, T.; Hussin, Y.A. Generating Pit-free Canopy Height Models from Airborne Lidar. *Photogramm. Eng. Remote Sens.* **2014**, *80*, 863–872. https://doi.org/10.14358/PERS.80.9.863.
34. Riaño, D. Modeling airborne laser scanning data for the spatial generation of critical forest parameters in fire behavior modeling. *Remote Sens. Environ.* **2003**, *86*, 177–186. https://doi.org/10.1016/S0034-4257(03)00098-1.
35. Novo, A.; Fariñas-Álvarez, N.; Martínez-Sánchez, J.; González-Jorge, H.; Fernández-Alonso, J.M.; Lorenzo, H. Mapping Forest Fire Risk—A Case Study in Galicia (Spain). *Remote Sens.* **2020**, *12*, 3705. https://doi.org/10.3390/rs12223705.
36. Parker, G.G.; Harmon, M.E.; Lefsky, M.A.; Chen, J.; Pelt, R.V.; Weis, S.B.; Thomas, S.C.; Winner, W.E.; Shaw, D.C.; Frankling, J.F. Three-dimensional Structure of an Old-growth Pseudotsuga-Tsuga Canopy and Its Implications for Radiation Balance, Microclimate, and Gas Exchange. *Ecosystems* **2004**, *7*, 440–453. https://doi.org/10.1007/s10021-004-0136-5.
37. Aber, J.D. Foliage-Height Profiles and Succession in Northern Hardwood Forests. *Ecology* **1979**, *60*, 18–23. https://doi.org/10.2307/1936462.
38. Rouse, J.; Hass, R.; Schell, J.; Deering, D. Monitoring vegetation systems in the great plains with ERTS. Third Earth Resour. Technol. Satell. Symp. **1973**, *1*, 309–317.
39. Boessenkool, B. rdwd: Select and Download Climate Data from 'DWD' (German Weather Service). R Package Version 1.5.0. 2021. Available online: https://CRAN.R-project.org/package=rdwd (accessed on 27 January 2022)
40. Breiman, L. Random Forests. *Mach. Learn.* **2001**, *45*, 5–32. https://doi.org/10.1023/A:1010933404324.
41. Meyer, H.; Reudenbach, C.; Hengl, T.; Katurji, M.; Nauss, T. Improving performance of spatio-temporal machine learning models using forward feature selection and target-oriented validation. *Environ. Model. Softw.* **2018**, *101*, 1–9. https://doi.org/10.1016/j.envsoft.2017.12.001.
42. Meyer, H.; Pebesma, E. Predicting into unknown space? Estimating the area of applicability of spatial prediction models. *Methods Ecol. Evol.* **2021**, *12*, 1620–1633. https://doi.org/10.1111/2041-210X.13650.
43. Meyer, H. CAST: 'caret' Applications for Spatial-Temporal Models. R Package Version 0.5.1. 2021. Available online: https://CRAN.R-project.org/package=CAST (accessed on 27 January 2022)
44. Kuhn, M. caret: Classification and Regression Training. R Package Version 6.0-88. 2021. Available online: https://CRAN.R-project.org/package=caret (accessed on 27 January 2022)
45. James, G.; Witten, D.; Hastie, T.; Tibshirani, R. (Eds). *An Introduction to Statistical Learning: With Applications in R*; Number 103 in Springer Texts in Statistics; Springer: New York, NY, USA, 2013.
46. Friedman, J.; Hastie, T.; Tibshirani, R. Regularization Paths for Generalized Linear Models via Coordinate Descent. *J. Stat. Softw.* **2010**, *33*, 1–22.
47. Anderson, H.E. *Aids to Determining Fuel Models for Estimating Fire Behavior*; Technical Report INT-GTR-122; U.S. Department of Agriculture, Forest Service, Intermountain Forest and Range Experiment Station: Ogden, UT, USA, 1982; https://doi.org/10.2737/INT-GTR-122.
48. Scott, J.H.; Burgan, R.E. *Standard Fire Behavior Fuel Models: A Comprehensive Set for Use with Rothermel's Surface Fire Spread Model*; Technical Report RMRS-GTR-153; U.S. Department of Agriculture, Forest Service, Rocky Mountain Research Station: Ft. Collins, CO, USA, 2005.
49. Finney, M.A. An overview of FlamMap fire modeling capabilities. In Proceedings of the Fuels Management-How to Measure Success: Conference Proceedings, Proceedings RMRS-P-41, Portland, OR, USA, 28–30 March 2006; Andrews Patricia L., Butler Bret W., Eds.; US Department of Agriculture, Forest Service, Rocky Mountain Research Station: Fort Collins, CO, USA, 2006; Volume 41, pp. 213–220.
50. Papadopoulos, G.D.; Pavlidou, F.N. A Comparative Review on Wildfire Simulators. *IEEE Syst. J.* **2011**, *5*, 233–243. https://doi.org/10.1109/JSYST.2011.2125230.
51. Forthofer, J.M.; Butler, B.W.; Wagenbrenner, N.S. A comparison of three approaches for simulating fine-scale surface winds in support of wildland fire management. Part I. Model formulation and comparison against measurements. *Int. J. Wildland Fire* **2014**, *23*, 969. https://doi.org/10.1071/WF12089.
52. Finney, M.A. Fire growth using minimum travel time methods. *Can. J. For. Res.* **2002**, *32*, 1420–1424. https://doi.org/10.1139/x02-068.

53. US Department of the Interior & US Department of Agriculture. Interagency Fuels Treatment Decision Support System (IFTDSS) (Version 3.4.1.3). 2021. Available online: https://iftdss.firenet.gov/ (accessed on 27 January 2022)
54. Calkin, D.E.; Ager, A.A.; Gilbertson-Day, J. *Wildfire Risk and Hazard: Procedures for the First Approximation*; Technical Report RMRS-GTR-235; U.S. Department of Agriculture, Forest Service, Rocky Mountain Research Station: Ft. Collins, CO, USA, 2010; https://doi.org/10.2737/RMRS-GTR-235.

 *fire*

Article

Considerations for Categorizing and Visualizing Numerical Information: A Case Study of Fire Occurrence Prediction Models in the Province of Ontario, Canada

Den Boychuk [1,*], Colin B. McFayden [2], Douglas G. Woolford [3], Mike Wotton [4,5], Aaron Stacey [6], Jordan Evens [4], Chelene C. Hanes [4] and Melanie Wheatley [5]

1. Aviation Forest Fire and Emergency Services, Ministry of Northern Development, Mines, Natural Resources and Forestry, Sault Sainte Marie, ON P6A 6V5, Canada
2. Aviation Forest Fire and Emergency Services, Ministry of Northern Development, Mines, Natural Resources and Forestry, Dryden, ON P8N 2Z5, Canada; colin.mcfayden@ontario.ca
3. Department of Statistical and Actuarial Sciences, University of Western Ontario, London, ON N6A 5B7, Canada; dwoolfor@uwo.ca
4. Great Lakes Forestry Centre, Canadian Forest Service, Natural Resources Canada, Sault Sainte Marie, ON P6A 2E5, Canada; mike.wotton@utoronto.ca (M.W.); jordan.evens@canada.ca (J.E.); chelene.hanes@canada.ca (C.C.H.)
5. Graduate Department of Forestry, University of Toronto, Toronto, ON M5S 3B3, Canada; melanie.wheatley@mail.utoronto.ca
6. Aviation Forest Fire and Emergency Services, Ministry of Northern Development, Mines, Natural Resources and Forestry, Peterborough, ON K9J 3C7, Canada; aaron.stacey@ontario.ca
* Correspondence: den.boychuk@ontario.ca

Citation: Boychuk, D.; McFayden, C.B.; Woolford, D.G.; Wotton, M.; Stacey, A.; Evens, J.; Hanes, C.C.; Wheatley, M. Considerations for Categorizing and Visualizing Numerical Information: A Case Study of Fire Occurrence Prediction Models in the Province of Ontario, Canada. *Fire* **2021**, *4*, 50. https://doi.org/10.3390/fire4030050

Academic Editor: James R. Meldrum

Received: 18 June 2021
Accepted: 12 August 2021
Published: 18 August 2021

Publisher's Note: MDPI stays neutral with regard to jurisdictional claims in published maps and institutional affiliations.

Copyright: © 2021 by the authors. Licensee MDPI, Basel, Switzerland. This article is an open access article distributed under the terms and conditions of the Creative Commons Attribution (CC BY) license (https://creativecommons.org/licenses/by/4.0/).

Abstract: Wildland fire management decision-makers need to quickly understand large amounts of quantitative information under stressful conditions. Categorization and visualization "schemes" have long been used to help, but how they are done affects the speed and accuracy of interpretation. Using traditional fire management schemes can unduly restrict the design of new products. Our design process for Ontario's fine-scale, spatially explicit, daily fire occurrence prediction (FOP) models led us to develop guidance for designing new schemes. We show selected historical fire management schemes and describe our method. It includes specifying goals and requirements, exploring design options and making trade-offs. The design options include gradient continuity, hue selection, range completeness and scale linearity. We apply our method to a case study on designing the scheme for Ontario's FOP models. We arrived at a smooth, nonlinear scale that accommodates data spanning many orders of magnitude. The colouring draws attention according to levels of concern, reveals meaningful spatial patterns and accommodates some colour vision deficiencies. Our method seems simple now but reconciles complex considerations and is useful for mapping many other datasets. Our method improved the clarity and ease of interpretation of several information products used by fire management decision-makers.

Keywords: colour coding; communication; forest fire; ordinal categorization; palette; risk; wildfire

1. Introduction

Situational awareness and decision-making for operational wildland fire management is supported by a large amount of complex, numerical information, often covering large areas and sometimes spanning multi-day forecasts. Comprehending and interpreting that quantity of information under time-limited and stressful conditions is challenging. Among other ways, this task is commonly made faster and easier by categorizing and visualizing the numerical information. There are many ways to do so, but how it is done can help or hinder interpretation, highlight or obscure valuable information and accurately portray or distort the data.

The categorization of numeric indicators of potential fire activity has a long history in Canadian and other fire management agencies. Established categories are deeply integrated into fire operations and culture. Although categorization is useful, conforming to traditional schemes, such as a four-category blue–green–yellow–red sequence from low to extreme, may be less than ideal for new information products. This issue arose during our implementation of fine-scale, spatially explicit fire occurrence prediction (FOP) models. Our design process for a new scheme is the basis for this paper.

FOP is one of the pillars of situational awareness, a requirement for daily and multi-day preparedness planning [1] and a key component for modelling risk [2]. Ontario's fire management agency has a lightning-caused FOP model [3] and, more recently, a human-caused FOP model [1] that was developed in collaboration between the agency's science specialists and decision-makers and external researchers.

Our objective is to provide guidance for the design of the categorization and visualization of complex information used in fire management. We begin with an overview of selected historical categorization and colouring schemes used in fire management. We then describe our method in steps including (1) specifying design goals and requirements, (2) exploring design options and seeing how they interact and (3) making trade-offs. We present a case study on designing the scheme for displaying the output of Ontario's FOP models. Although our method arose from this work, the considerations and general principles employed can be useful in many other situations. The Supplementary Material includes other applications of our method and extensions to some other considerations for the display of data. Note that we use the terms "category" and "categorization" synonymously with "class" and "classification", the latter set being conventions in fire management.

Overview of Selected Historical Categorization and Colouring Schemes

Knowing the historical origin of the schemes aids us to understand their limitations and engender improvements. Prime categorization examples are those for the outputs of two Canadian Forest Fire Danger Rating System (CFFDRS) [4] subsystems: the Fire Weather Index (FWI) System [5] and the Fire Behaviour Prediction (FBP) System [6].

The FWI System accounts for effects of past and present weather on fuel ignitability and fire behaviour and has six main numeric outputs. Three track moisture in different depths or sizes of fuel: Fine Fuel Moisture Code (FFMC), Duff Moisture Code (DMC) and Drought Code (DC). The other three indicate potential fire behaviour: Buildup Index (BUI), Initial Spread Index (ISI) and Fire Weather Index (FWI). FWI indicates the potential fireline intensity in a standard pine (genus *Pinus*) stand [5]. FWI is also used in Ontario as a general indicator of fire hazard or danger and is mapped into ordinal adjective classes: Low–Moderate–High–Extreme. The classes accompanied the introduction of the FWI System [5] and were implemented early on by Ontario [7]. Fire danger schemes are intended to "sound an alarm" about potential extreme behaviour and difficulty of control [8]. The class boundaries were set so that Extreme covered the worst 2% of historical days, and the remaining lower boundaries were set by a geometric progression [5]. For a more detailed history of the FWI classification methods in Ontario, see [9]. Other fire agencies in Canada calculated different boundaries according to their data [10,11]. The FWI System and component classifications have also been applied outside North America. For examples, see [12,13]. The current danger classes and colours used in Ontario for the FWI System's main outputs are outlined in Table 1. Examples of these are shown for roadside signs for public alerts about daily fire hazard (Figure 1a) and also for operational maps (Figure 1b). In mapping products such as Figure 1b, the FWI System values are calculated for weather station locations and are interpolated [14] across Ontario's fire management area, which is almost 50% larger than France.

Table 1. Current danger classes and colour scheme [15] used in Ontario for each of the six Fire Weather Index System components: Fine Fuel Moisture Code (FFMC), Duff Moisture Code (DMC), Drought Code (DC), Buildup Index (BUI), Initial Spread Index (ISI) and Fire Weather Index (FWI).

Class	Colour	FFMC	DMC	DC	ISI	BUI	FWI
Low		0–80	0–15	0–140	0–2.2	0–20	0–3
Moderate		81–86	16–30	141–240	2.3–5.0	12–36	4–10
High		87–90	31–50	241–340	5.1–10.0	37–60	11–22
Extreme		≥91	≥51	≥340	≥10	≥61	≥23

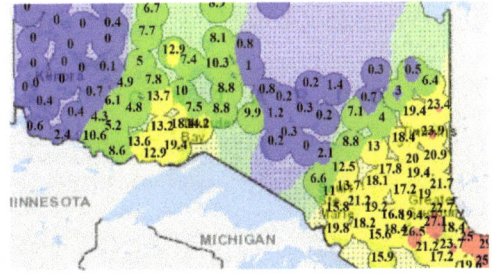

Figure 1. Examples of different displays of the Fire Weather Index (FWI) class used by the Ontario's fire agency: (**a**) a public roadside sign; (**b**) an agency operational map showing interpolated, colour-coded FWI and the raw FWI values for each weather station.

The other major CFFDRS subsystem that has outputs commonly communicated using classes is the FBP System, which provides quantitative estimates of fire behaviour outputs [6]. A primary output is fire intensity, the rate of energy or heat release per unit time per unit length of a spreading fire front [16], which ranges from 1 to ~100,000 kW/m in Canadian conditions. Fire intensity is also commonly categorized into fire intensity classes (ICs). Rather than adjective classes (i.e., low–extreme) they were given five numeric labels (IC 1–5 [17]), and later six (IC 1–6 [18]). Higher IC numbers correspond with the higher intensity values, but the boundaries are not evenly spaced (Table 2). The most commonly used IC boundaries in Canada delineate distinct differences in fire type characteristics (for example, surface, torching or crowning) in mature jack pine (*Pinus banksiana* Lamb.) stands and the corresponding general effectiveness of different types of fire suppression activities (for example, hand tools, pumps and hose, airtankers). These ICs are described in the Field Guide to the FBP System [19] and the ICs are also used to map fire intensity by many fire management agencies, for various purposes (Table 2).

Table 2 shows some of the different intensity values for higher-end class boundaries (i.e., additional thresholds beyond IC 6). The choice of colour when mapping can be operationally significant because colours covey information rapidly and have strong associations with levels of alarm—for example, red for danger [20] and green for calm [21]. Such psychological factors are not always considered in the visualization, however, which could lead to misinterpretation. For example, IC 4 in Table 2, which is associated with the upper limit of direct fire suppression effectiveness [17,22], is variously coloured a calming light green, a cautionary yellow or a warning orange. In Table 2 there are also cases where the same colour refers to different IC classes—for example, calming light green is used for IC 2 in Ontario, IC 3 in Alberta and IC 4 nationally.

Table 2. Examples of the diverse classification of fire intensity and colouring used in Canada. The Ontario, Alberta and national schemes are used in daily maps. The British Columbia (BC) scheme is used in a static map of the 90th percentile of historical fire intensity. The Field Guide to the Fire Behaviour Prediction System (Field Guide) scheme is used in printed tables. Intensity classes IC 1–IC 6 are as defined in [19]; the higher classes are informal. The colours are approximate.

Intensity Class (IC)	Range (kW/m) [1]	Ontario Map	Alberta Map [23]	Field Guide [2] [19]	National Map [24]	BC 90th Percentile Map [3] [25]	
IC 1	<10					<1 k	
IC 2	10–500					<1 k	
IC 3	500–2 k					<1 k	1 k–2 k
IC 4	2 k–4 k						
IC 5	4 k–10 k	>4 k				4 k–6 k	6 k–10 k
IC 6	>10 k					10 k–18 k	
7					10 k–30 k	18 k–30 k	
8					>30 k	30 k–60 k	
9						60 k–100 k	
10						>100 k	

[1] Range applies to the full row except where overridden by the text in the cell. [2] The colours are limited by the printing ink used. [3] BC's IC 1–5 corresponds to BC's colour progression rather than the row label.

An additional caveat in the classification is that simplifying numerical information by aggregation into few classes has a cost. For FWI and fire intensity, the wide range within some classes is operationally significant for some decisions—for example, FWIs of 11–22 become "High". Furthermore, for interpolated maps, neighbouring points that are displayed as different classes will not have operationally meaningful differences. To compensate, maps often include the raw point values that were used for interpolation (Figure 1b). The numeric information cannot, however, be read and interpreted as quickly, thus reducing the benefit of categorization. The categorizing of data for spatial application such as FWI and FBP is common, and there are many recognized considerations (for example, [26]) and built-in solutions in geographic information systems. However, using a built-in classification option without a deep understanding may be unsuitable because potential distortions can lead to radically different interpretations, as others have noted [27]. Consequently, there is a strong need to use schemes that convey information accurately.

2. Methods

We propose five steps to categorize and visualize model outputs for use in fire situational awareness and decision-making:

1. Understanding and scoping the data
2. Understanding the decision-making uses of the information
3. Specifying the design goals and requirements
4. Designing the categorization and visualization scheme
5. Evaluating and revising the scheme

We first describe these steps in general, below, and then with further detail on their application, in our case study. Although the method is described in a linear sequence, the work is partly concurrent and highly iterative, especially within Step 4.

2.1. Step 1: Understanding and Scoping the Data

Our method applies to data with a continuous numerical scale of measure (real numbers). With minor modifications, it can also apply to data with a discrete numerical scale of measure (integers) and to ordinal categorical data (for example, Very Low, Low, Low–Moderate, ...). The modification is that colouring with continuous gradients (described below) does not apply unless there are a great many discrete values or ordinal categories.

The technical details of the raw model output data may be straightforward, but unfamiliar units, scaling, storage or other conditions can lead to misinterpretation. The following need to be understood by the designers:

- Units, including any scaling and transformations
- Data storage type (for example, 64-bit floating point, signed long integer, string) and associated considerations (for example, storing a scaled real value as an integer, which truncates the precision)
- The data's range or anticipated range if using a static scale for all future maps
- The frequency distribution of historical data
    - It may or may not be useful to have more categories where there was a lot of data, and it may or may not be pointless to have multiple categories where there was little or no data (see examples in Section 2.2).
- The data's precision of measurement and storage and the data's accuracy of measurement or estimation
    - The stored precision may not correspond with the accuracy. Measured data such as weather observations have low precision (for example, to 0.1 °C), but calculated data such as FWI System values should be calculated and may be stored with full machine precision (~16 decimal digits), which is well beyond the accuracy of typical fire management data.
- If the data are generated by a model, then the model's meaning, structure, assumptions, limitations, precision and accuracy

Understanding data precision and accuracy is necessary for presenting information accurately and having decision-makers understand it easily and correctly. Regarding data storage and all subsequent calculations using data, full machine precision should be maintained to avoid accumulating rounding errors. Regarding the numbers displayed for decision-makers, the displayed precision should not exceed the data's accuracy, because that could be misleading. The numbers displayed for decision-makers should ideally have the lowest precision that is operationally significant to minimize unnecessary mental processing. Further discussion and examples are given in the Supplementary Material.

2.2. Step 2: Understanding the Decision-Making Uses of the Information

The purpose of the information is to support decision-making, so it is necessary to know who is using the information and how it is used. Working directly with fire management staff to understand their needs is necessary for ensuring that new model outputs are effectively integrated into the decision-making process [28,29].

There are key questions to consider. What decisions are being supported? Are certain parts of the range more important, needing higher attention? Is a higher resolution (smaller class size) needed in some parts of the range rather than others? For example, consider the categorizing and mapping of the accumulated 24-h rainfall from a precipitation radar [30]. For differentiating the degree and duration of the reduced fire behaviour potential, a high resolution is useful at the low end but not at the high end. Conversely, for differentiating the degree and duration of flood potential, a high resolution is useful at the high end but not at the low end. Moreover, a higher top category is appropriate.

2.3. Step 3: Specifying the Design Goals and Requirements

As stated in the introduction, the high-level goals of categorization and its visualization are simply to show the information completely and accurately and to have the information be understood quickly and easily. These goals are elaborated into criteria as follows.

- Regarding the complete and accurate display of information:
    - Are the magnitudes shown with the original or reduced precision?
    - Are the magnitudes undistorted or distorted by categorization, scale nonlinearity or truncation?
    - Are the relative magnitudes evident by the colouring without or with referral to the legend?

- Regarding the quick and easy understanding of information:
    - Can the colouring be easily matched to the legend's magnitude numbers?
    - Does the colouring draw attention and convey a suitable psychological meaning for the degree of alarm?
    - What is the overall ease of understanding?

Possible design requirements include the accommodation of colour vision deficiencies and other technical considerations such as the adequate appearance on low-quality displays, colour printers or standard photocopiers.

2.4. Step 4: Designing the Categorization and Visualization Scheme

There are four design options for categorizing and colouring the values in the scale (Figure 2):

1. Gradient continuity: whether to use the original values unaltered or categorized
2. Hue selection: the number and choice of colours and design of gradients
3. Range completeness: whether to show the full range or truncate the top or bottom of the range
4. Scale linearity: whether to have a linear or nonlinear scale or progression of category boundaries, and whether to have colour gradients that are linearly or nonlinearly proportional to the data magnitudes

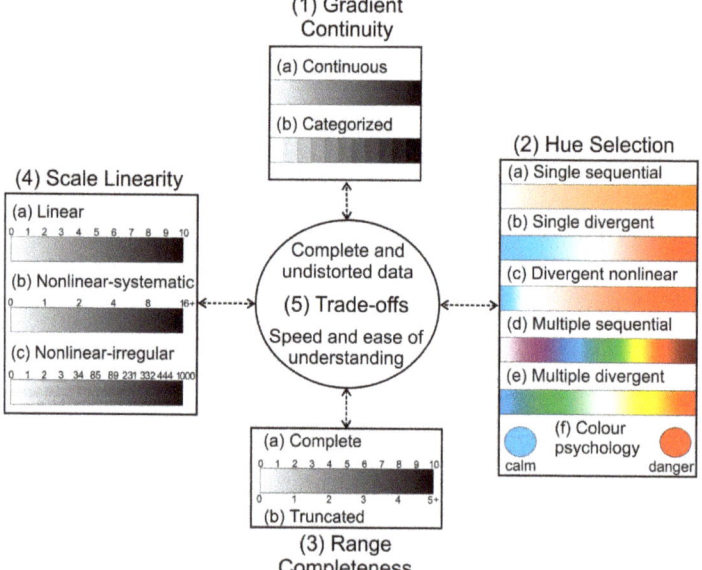

Figure 2. Illustration of the design options (rectangles), which are described in the text, and trade-offs (circle) that are made when categorizing and colouring data that are on a numerical scale of measure.

2.4.1. Gradient Continuity

The alternatives for gradient continuity are either to use the original values or categorize them (Figure 2, parts 1a and 1b). For a continuous gradient, the percentage of colour saturation is proportional to the raw datum magnitude. Continuous gradients are therefore precise and accurate but harder to interpret using the legend compared to categorized gradients.

2.4.2. Hue Selection

Colours have strong psychological associations that affect the inferred meaning of information and its speed and ease of interpretation. Blue and green are associated with relaxation, calm and hope [21], and red is associated with danger [20] (Figure 2, part 2f). Historically, variations of a blue–green–yellow–orange–red sequence (which alludes to water, growing vegetation, dried vegetation and flame) have been used to represent escalating fire danger (see examples in Table 2). Accommodating colour vision deficiencies reduces the colour combination choices, particularly most of those in the traditional fire danger sequence [31]. Tools are available to assist with the testing of colour palettes for accessibility [31,32]. The design task is to choose colours appropriate for the implications of the data magnitudes, particularly the degree of attention or alarm.

Regarding the gradient design, there are a few distinct alternatives [33] (Figure 2, parts 2a–2e). A single sequential gradient is for data ranging over a meaning of zero or neutral to bad or good. A divergent sequential gradient is for data ranging over a meaning of good through neutral to bad. In Figure 2, parts 2b and 2c grade from a calming blue through white to an alarming red. The left and right ends each have a single hue. Part 2b is neutral in the middle, whereas part 2c has compressed and expanded ends. Part 2d, multiple sequential, is analogous to part 2a except that part 2d has multiple hues, which are in a rainbow spectrum in the example. Compared to a single hue, multiple hues provide more contrast over the range, making it easier to match the legend and signal levels of attention or alarm. Part 2e, multiple divergent, is analogous to part 2b except that part 2e has multiple hues for each side. The R software [34] package "inlmisc" [35,36] is useful for constructing continuous or categorized gradients.

A key concern is how these many design alternatives support or oppose the design goals. Only the single sequential and single divergent gradients (Figure 2, parts 2a and 2b) have the accuracy of continuous gradient continuity (Figure 2, part 1a), but they have a difficult interpretability. The remaining gradient alternatives require matching the legend to identify the magnitudes, but this can become quick and easy to interpret with familiarity and an effective use of colour psychology.

2.4.3. Range Completeness and Scale Linearity

These are described together because an incomplete range is an extreme form of nonlinearity. The alternatives for range completeness (Figure 2, part 3) are whether to show the full range of the data or truncate the top or bottom and group the truncated data. All resolution is lost beyond the truncation points. The alternatives for scale linearity (Figure 2, part 4) are for two components, independently:

1. Numeric scale: whether to have a linear (part 4a) or nonlinear numeric scale (parts 4b and 4c) or progression of category boundaries (if applicable)
   - For categorized gradient continuity, a nonlinear scale is used to vary the resolution over the range
2. Colour gradient: whether to have colour gradients that are linearly or nonlinearly proportional to the data magnitudes
   - For categorized gradient continuity, a nonlinear colour gradient is used to communicate the varying meaning or importance of the information over the range

For a continuous gradient, the same result can be achieved from nonlinearity in either of the above two components.

Nonlinear-systematic (part 4b) methods use a smooth function such as log or power to transform the output, while nonlinear-irregular methods use a non-smooth progression such as Jenks [37]. Nonlinear colouring requires the referral to the legend to understand the magnitudes. This is a trade-off between the goals of drawing attention to where it is needed and improving the speed and ease of understanding.

2.4.4. The Design Process

There are copious settings and combinations of alternatives for the four design options. Getting to a result is an iterative process, with analysts and subject matter experts trying alternatives and making trade-offs, hopefully avoiding the anchoring to tradition or early trials. We cannot recommend a path through the four design options other than saying it is iterative and concurrent. We do, however, recommend a starting point or baseline, which is the extreme of displaying all the data completely and accurately and ignoring the goals of quick and easy understanding. The baseline has continuous gradient continuity, an achromatic colour gradient of white through greys to black and a linear scale with no truncation (Table 3).

Table 3. The baseline case alternatives for the design options. This is a starting point that presents complete and accurate information, while ignoring the goals of a quick and easy understanding of the information.

Design Option	Alternative	Description
1. Gradient continuity	Continuous	No categorization; the percentage of grey saturation is proportional to the datum magnitude
2. Hue selection	Single sequential	Achromatic gradient of white through greys to black; no chromatic psychological colour associations
3. Range completeness	Complete	Full range of data magnitude is shown; no truncation or aggregation at the top or bottom
4. Scale linearity	Linear	No distortion; the percentage of grey saturation equals the datum's position within its range

If categories are used, determining the number and their boundaries are fundamental design decisions [38]. The number of categories is a trade-off of accuracy (requiring more) and speed and ease of understanding (requiring fewer). Ideally, individual categories have no operationally significant physical differences, while adjacent categories do. In practice, all considerations require compromise. An example of determining categories for FWI System outputs based on physical differences is given by [9].

2.5. Step 5: Evaluating and Revising the Scheme

Once the design process is done and implemented, an essential further step is the ongoing work with decision-makers to evaluate the outputs and revise the design as necessary.

3. Case Study: Designing the Scheme for Ontario's FOP Models

We now describe the application of the above method to categorizing and visualizing FOP data.

3.1. Case Study—Step 1: Understanding and Scoping the Data

The data are outputs from process and statistical FOP models for Ontario, so we begin with their description. The lightning- and the human-caused FOP models have been used operationally since the mid-2000s and 2015, respectively. The daily lightning-caused fire occurrence is modelled as two separate processes [3]: the probability of a lightning strike will lead to a holdover ignition and the probability that an existing ignition "arrives" (is reported). The ignition model is mainly driven by the forest floor organic layer's moisture content, which determines the sustainability of smouldering and the survivability of the ignition. Additional factors are other moisture indicators, ecoregional modifiers and lightning strike polarity. The "arrival" model, which is conditional on a holdover ignition being present, is influenced by the surface litter moisture, organic layer moisture, wind speed and ecoregional differences. Ontario's human-caused fire prediction system uses a set of logistic generalized additive models to model inherently nonlinear relationships with key drivers of human-caused fire occurrence, including seasonal and spatial patterns, fuel moisture and the characteristics of human land use [1]. The models are stratified regionally and by cause categories to account for different seasonal patterns in fire occurrence.

Both the lightning- and human-caused FOP models produce outputs for each of the 2574 cells in a grid that spans the province's approximately 91.9 million ha wildland fire management area (Figure 3). Most of the cells are about 20 km × 20 km or 40,000 ha, with some fractional cells at the boundaries. The units of the FOP output data are interpreted as the expected number of fires per cell, regardless of the cell size. The model calculations and outputs have a 64-bit floating point precision, but the data are transferred to the mapping software via a text file holding up to 12 significant digits.

Figure 3. The approximately 91.9 million ha extent of Ontario's fire management area and the 20 km × 20 km resolution of the fire occurrence prediction (FOP) grid for Ontario.

Regarding the range and frequency distribution, we analysed historical FOP data for each cell for each day from 15 May to 31 August; 2016–2018 for human-caused and 1992–2006 for lightning-caused fires. For this analysis, the start and end dates in each fire season were chosen to avoid the variability in spring and fall snow-free conditions, when the models are not making predictions for the entire province. The lower limit of the range is zero; there is no theoretical upper limit. Table 4 presents summary statistics for the data, with zeros excluded to characterize the important data more clearly. Most of the distribution statistics of the human- and lightning-caused data differ by about an order of magnitude.

Table 4. Summary statistics of historical fire occurrence prediction model data (expected number of fires/cell) generated from 15 May–31 August for 2016–2018 (human-caused) and 1992–2006 (lightning-caused), with zeros removed.

	Number of Observations	First Quartile	Median	Mean	Third Quartile	Maximum
Human	607,383	0.00003	0.00013	0.00107	0.00048	1.37168
Lightning	1,026,860	0.00090	0.00260	0.00955	0.00750	3.89129

Figure 4 shows the empirical probability distributions of the non-zero data. Those data were mostly clustered close to zero in both models, so we log-transformed the data for illustration (untransformed data are required for operational use). The magnitudes of data between the lower tails of the two distributions differ by orders of magnitude. This presents a challenge for categorizing and colouring the data on a common scale for mapping.

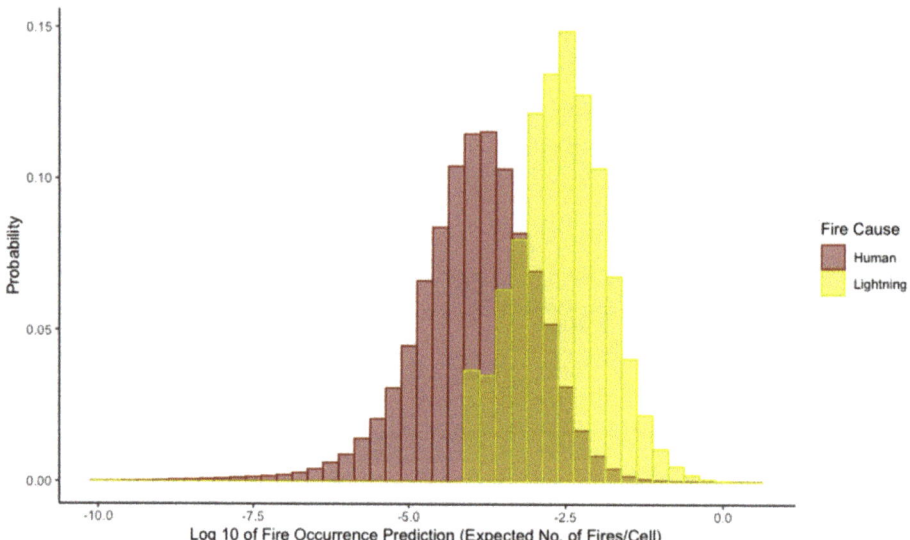

Figure 4. Empirical probability distributions of the non-zero, human- and lightning-caused fire occurrence predictions (FOPs) for Ontario, 15 May–31 August, for 2016–2018 (human-caused) and 1992–2006 (lightning-caused). FOPs for the two causes have distinctly different ranges and central tendencies because of the spatio-temporal processes of ignition. The data are transformed by log 10 for visualization here, but untransformed data are required for operational use.

3.2. Case Study—Step 2: Understanding the Decision-Making Uses of the Information

We used a variety of methods to understand the FOP information needed and how it is used for daily decision-making: reviewing documentation, observing operational decision-making and (for some) working part-time in operational functions where FOP information is used. Most importantly, we held a series of engagements with fire management agency personnel. For example, we hosted a workshop in 2017 attended by agency personnel including regional and provincial Fire Intelligence Officers, external researchers and students. The workshop's topics included the purposes and methods of subjective FOPs by experts.

To understand the agency's use of FOP information, it is necessary to outline the agency's hierarchical structure and the responsibilities of each level. Ontario's fire management area (Figure 3) has two main parts, the Northeast and Northwest Regions, each of which is divided into six or seven Fire Response Sectors. There is also a Provincial level. Each level has a sole or shared responsibility for various decisions; most are made with consultation or coordination between adjacent levels. The Regions are primarily responsible for strategic fire response decisions and management, and the Sectors are primarily responsible for tactical fire response decisions and operations. The Province is primarily responsible for adjusting the near-term (1- to 21-day) capacity according to the demand via temporary commercial hiring and inter-provincial and international resource sharing.

Several decisions are directly dependent on the potential number of fires anticipated in various parts of the province. These decisions are made at the stated levels:

- Prevention: for example, escalated and targeted messaging, temporary fire bans; made by the Province, Regions and Sectors
- Preparedness: types, numbers, locations and readiness alert levels of firefighting resources (crews, helicopters, airtankers and engines); made by the Regions and Sectors
- Detection: numbers, routes and times of aerial detection patrols described in [39]; made by the Regions, jointly
- Dispatch: different resources may be sent now if more threatening fires are anticipated later; made by the Regions and Sectors
- Inter-provincial and international resource sharing; made by the Province and Regions

The various FOP-dependent decisions have diverse needs in terms of the spatial extent and resolution of FOP information. For example, detection route designers can use relatively fine resolution information on the order of kilometres, while the Province needs only aspatial, numeric FOPs by region for resource-sharing decision-making.

3.3. Case Study—Step 3: Specifying the Design Goals and Requirements

The primary and conflicting goals are of course to display complete and accurate information and have the information quickly and easily understood. In twice-daily briefings, decision-makers have limited time (minutes) to view, interpret and absorb each of many information items regarding, for example, weather values, FWI and FBP System outputs, FOP, active fires, logistics and personnel. Completeness and accuracy are important because the information supports the many decisions described above, which are made under uncertainty and have potentially significant consequences.

Regarding specific information requirements:

- There is a need for both maps and numeric subtotals and totals of fire occurrence by cause and location (i.e., Sectors, Regions, Province)
- All the maps need to use the same categories and colours for FOP magnitudes
- The FOP models' output is the expected or average occurrence, but the actual occurrence varies around the average, so an indication of the variability is needed.

Decision-makers expressed strong preferences for the number of categories, ranging from three to many categories, and they desired a familiar colour sequence (blue–green–yellow–red). There was also a strong preference for integers for all numbers related to FOP. Finally, we wished to accommodate colour vision deficiencies.

3.4. Case Study—Step 4: Designing the Categorization and Visualization Scheme

Design has been described as a messy process with a tidy outcome. We do not detail our circuitous journey but show and describe some key alternatives, stages and considerations. Figure 5a illustrates the baseline alternative (Table 3) applied to the human-caused FOP and actual fire arrivals for a selected day. The range of the colour gradient is 0–3 fires/cell, the upper limit of which is between the maximum human- and lightning-caused FOP (Table 4). The map area looks mostly white, with three small, pale grey patches; there is little useful information, especially considering the six actual fires that day, which is a low-to-moderately busy day for this cause. Adding more hues alone would make no meaningful difference because the data are clustered very near zero. Categorizing at this stage would make it worse. Truncating the upper limit to a low value somewhat close to zero would add resolution and colour to the human-caused FOP here, but such truncation would lose all resolution of the rarer but critically important high lightning-caused FOP.

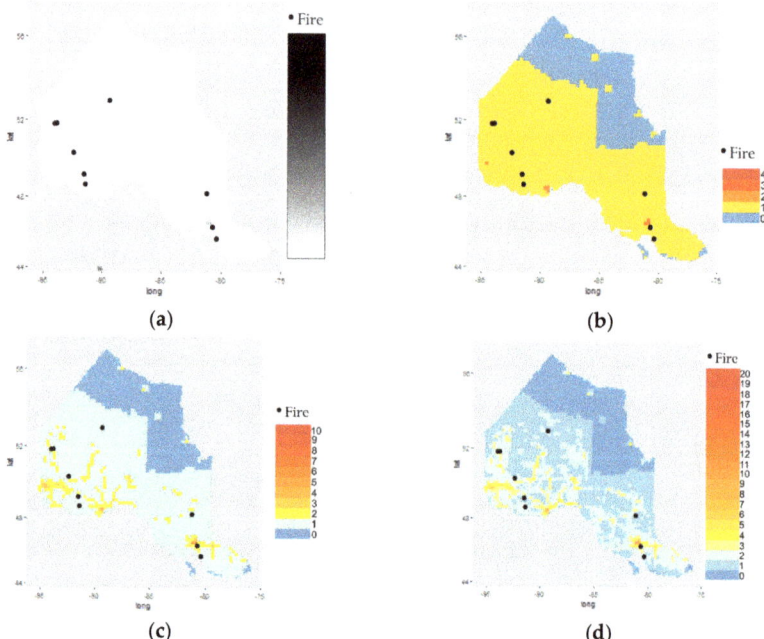

Figure 5. The human-caused fire occurrence prediction of 10 June 2018, mapped using the categorization and colouring schemes indicated in the legends. The black dots are the human-caused fires reported later that day. The scales range from zero to 3 fires/cell: (**a**) the baseline scheme (Table 3), which shows little useful information; (**b**) 4 + 1 categories; (**c**) 10 + 1 categories; (**d**) 20 + 1 categories. The additional category in (**b**–**d**) are for a "no forecast model" or true zero. Using more categories and hues greatly increases the information portrayed but makes matching the colour to the legend more difficult. The spatial pattern in (**d**) corresponds with roads and settlements.

Our original solution was to use separate scales for the two fire occurrence causes (Figure 6). The lightning-caused FOP scale was linear. For the human-caused FOP scale, we led subject matter experts through a scenario for an area of Ontario that has a relatively high occurrence. When the FFMC (a strong indicator of sustainable ignition) was 90, that area was considered to have an elevated concern suitable for a High classification. We used the corresponding FOP magnitude (0.4 fires/cell) as the upper limit for the then 10-category scale. We determined Moderate similarly and interpolated with equal linear steps. In Figure 6, this scale is shown by the blue line, except that categories 1 to 10 in the original have been mapped to a 0 to 20 scale for comparability. Scaling the original total-fires map required a creative logic. The maps by individual cause were acceptable to decision-makers, but the inconsistency between the causes was ultimately unacceptable (and motivated the present work).

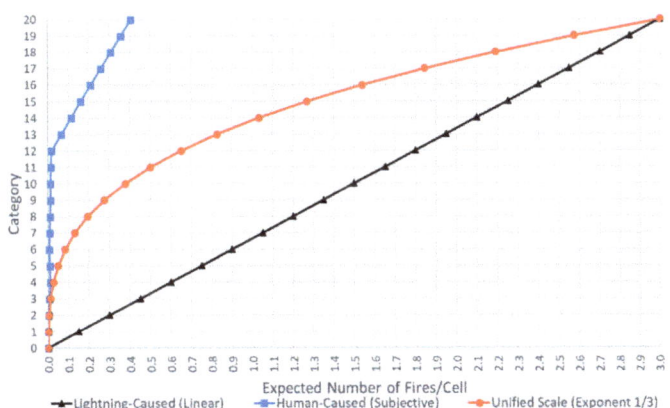

Figure 6. The original and unified fire occurrence prediction (FOP) classification scales. Because of order-of-magnitude differences in FOP, a separate scale was originally used for each cause: linear for lightning (black) and subjectively determined, irregular, nonlinear for human (blue). The unified scale (red) is a systematic, nonlinear one generated by a power function; a cube root in this example.

Several alternatives were considered for a unified scale that accommodated the conflicting needs of fine resolution at the low end and a high upper limit. Discussions with decision-makers indicated that concern increases relatively quickly as the likelihood of fire rises from zero. Providing a high resolution at the low end while retaining a high upper limit would require a great many colours (for example, like those of precipitation radar maps). That would be unfamiliar and confusing and would not correspond with psychological colour associations nor accommodate colour vision deficiencies. Piecewise linear and irregular scales were explored, but their abrupt changes made them difficult to interpret. We wanted a smooth, systematic progression of category boundaries and tested logarithmic and power functions. Those functions can be made to match fairly closely, but the power function had a more suitable shape at the low end. A power function takes the general form $f(x) = ax^b$, with the shape controlled by parameters a and b. Our desired behaviour for the scale to increase quickly for low FOP but then increase progressively more slowly is provided when $a > 0$ and $0 < b < 1$, since this family of power functions is monotonically increasing and concave down. The FOP scale is x, and the category scale is $f(x)$. We developed a parametric scaling tool with three inputs to generate and plot boundaries: shape parameter, $1/b$; the upper limit of the FOP scale, $FOPMax$; and the number of categories, $NumCat$. The boundary for the top of category Cat is

$$Boundary_{Cat} = FOPMax \cdot \left(\frac{Cat}{NumCat}\right)^{1/b}, \quad Cat = 1, 2, 3, \ldots, NumCat. \quad (1)$$

Any FOP > $FOPMax$ stays in the highest category. A category for true zero can be added if required. For convenience, we parameterized $1/b$, for which we tested values in the range of 1.5 to 4.5; for example, 2 yields a square root shape. a works out to be $\frac{NumCat}{FOPMax^b}$. The red curve in Figure 6 illustrates the boundaries for $1/b = 3$, $FOPMax = 3$ and $NumCat = 20$; for example, the boundary for $Cat = 11$ (the top of category 11) is ~0.5. The tool lists the boundaries and graphs them as in Figure 6.

While working directly with subject matter experts, the tool facilitated the joint testing of alternatives for the number of categories, truncation and nonlinearity design options. These alternatives plus colouring needed to be adjusted simultaneously when trading off the goals because their effects interact. The FOP outputs for a set of representative days were mapped using R [35] for candidate sets of boundaries and colouring. Table 5 lists ways in which the alternatives for the number of categories, the amount of truncation and nonlinearity and number of hues generally interact in affecting several attributes related to completeness and accuracy of information or speed and ease of understanding. There are exceptions for some combinations and edge conditions. Every alternative for the design options improves some attributes and worsens others. The trade-off behaviour is more straightforward to work with than it may seem because the tool shows most of the trade-offs immediately. The difficulty lies in subjectively assessing the results and compromising on the attributes and goals.

Table 5. Tabulation of how alternatives for the number of categories, amount of truncation and nonlinearity and number of hues generally interact in affecting several attributes related to completeness and accuracy of information or speed and ease of understanding. There are exceptions for some combinations and boundary conditions. Every alternative improves some attributes (blue-grey shading) and worsens others (orange-tan shading).

Goal	Attribute	Alternative for Design Options			
		More Categories	More Truncation at High end	More Nonlinearity	More Hues
Complete and accurate information	More resolution at low end	Better	Better	Better	No effect
	More resolution at high end	Better	Better in range NONE beyond truncation	Worse	No effect
	Accurate, Undistorted	Better	Better in range ABSENT beyond truncation	Worse	No effect
Fast and easy to understand	Need to refer to legend	Worse	No effect in range	Worse	Worse
	Ease of matching legend	Worse	Better	Worse	Better
	Attention drawn to important data	No effect	No effect in range ABSENT beyond truncation	No effect	Better
	Has suitable psychological meaning	No effect	No effect	No effect	Better if using colour psychology

3.5. Case Study—Step 4: Results of the Design Process

We state our current category and colouring design and give the rationale for the trade-offs made. The pressure to have a small number of categories and colours (~4) could not accommodate the need for fine resolution at the low end; we used 20 categories (plus a true zero if needed). Figure 5b–d show the same FOP data as Figure 5a but with 4, 10 and 20 categories, respectively. Only the largest number of categories reveals the meaningful network pattern of lines and nodes that corresponds roughly with roads and settlements.

In addition, a highly nonlinear scale was needed to show the patterns in the data. We used a nonlinear-systematic scale with boundaries obtained using Equation (1) with parameters $1/b = 3$, $FOPMax = 3$ and $NumCat = 20$. The boundaries are given in Table 6, stated in units of fires/cell and cells/fire. The final shape of the nonlinear scaling corresponds to parameter settings of $a = 20 \cdot 3^{-1/3}$ and $b = 1/3$.

$$f(x) = \begin{cases} 20 \cdot 3^{-1/3} x^{1/3} , & 0 \leq x < 3 \\ 20 , & x \geq 3 \end{cases}, \tag{2}$$

which is illustrated by the red curve in Figure 6.

Table 6. Categories and colours used for mapping fire occurrence prediction model outputs. Note that the map legend has a highly simplified integer scale and broad adjective categories.

Category Number	Category Upper Bound of Expected Number		Colour	Adjective Category	Map Legend
	(Fires/Cell)	(Cells/Fire)			
20	≥3.00	≤0.333		Extreme	
19	2.57	0.389			
18	2.19	0.457			
17	1.84	0.543		Very High	
16	1.54	0.651			
15	1.27	0.790			
14	1.03	0.972			
13	0.824	1.21			
12	0.648	1.54		High	
11	0.499	2.00			
10	0.375	2.67			
9	0.273	3.66			
8	0.192	5.21			
7	0.129	7.77		Moderate	
6	8.10×10^{-2}	12.3			
5	4.69×10^{-2}	21.3			
4	2.40×10^{-2}	41.7		Low	
3	1.01×10^{-2}	98.8			
2	3.00×10^{-3}	333		Very Low	
1	3.75×10^{-4}	2670			
0	0	∞		Nil	

Regarding truncation, we considered the rarity and operational importance of extreme FOP magnitudes. The 20th category ends at 3 fires/cell according to Equation (1), but that category is used for all higher FOP magnitudes. The maximum FOP in Table 4 is ≈3.9 fires/cell.

Regarding colouring, we used a nonlinear, divergent scale with one hue for the low end and multiple hues for the high end (Table 6). The hues transition from light blue to yellow through orange to red, which mostly follows the traditional blue-to-red progression. Avoiding green in that sequence accommodates some types of colour vision deficiency [31]. Even though there are 20 categories, having four main colours is easy to interpret and consistent with other CFFDRS outputs (for example, Table 1). The gradient within each main colour is difficult to match with the legend, but nonetheless reveals meaningful spatial patterns in the maps. Figure 7a–c show the final categorization and colouring scheme in example daily FOP maps of human- and lightning-caused fires and total fires, respectively. We intentionally show maps as formatted for operational use. They are intended for display on large monitors, but these reduced versions still show the colouring and categorization results adequately. Larger versions are provided in the Supplementary Material.

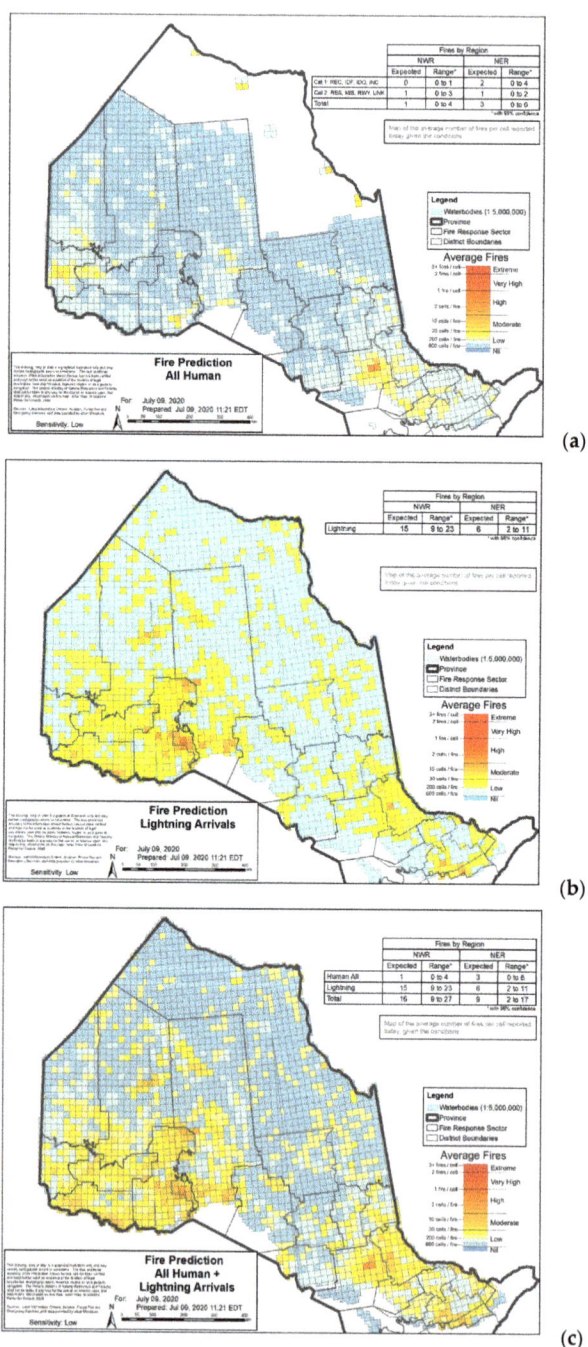

Figure 7. Examples of fire occurrence prediction maps that use the categorization and colouring in Table 6: (**a**) human-caused; (**b**) lightning-caused; (**c**) total. The legend uses units of fires/cell and cells/fire to make the magnitudes integer. Larger versions are in the Supplementary Material.

Note that a sequential scale is logical for FOP because any non-zero FOP is "bad" in this context. But the calming colours are assigned to very low magnitudes, and this provides a slightly greater distinction between the remaining colours. A much greater distinction could easily be achieved by adding more colours—for example, magenta–purple–black for the highest categories, which would also draw more attention to the critical extremes. This, however, would not accommodate colour vision deficiencies.

Broad adjective categories were added to emulate the familiar four- or five-category pattern and simplify interpretation. Those boundaries fit the general association between Ontario's fire arrival density and fire situation severity. The three-significant-digit category boundaries were replaced in the legend by integers for selected category midpoints or boundaries (Table 6). Note that the units change from fires/cell to cells/fire to show integer magnitudes, which are far more meaningful than fractions.

3.6. Case Study—Step 5: Evaluating and Revising the Scheme

Several significant revisions were done in arriving at the current design in Figure 7. The first lightning-caused FOP maps had a linear scale with four equal-interval categories using traditional blue–green–yellow–red. When human-caused FOP was added, the scale was changed to 10 categories using a smooth green–yellow–orange–red transitional gradient and later to a special blue–yellow–orange–red to accommodate some colour vision deficiencies. As stated above, the inconsistent subjective scales were replaced completely in early 2020 per our method. The maps originally showed the expected number of fires/cell but were changed to show the density of the expected number of fires/unit area because of fractional cells. Note that the map legends intentionally omit this complication; density seems to be the automatic, intuitive interpretation. Additional revisions are planned, and further evaluation is always ongoing.

4. Discussion

While implementing the FOP models for operational evaluation, we arrived at and applied this method, which may seem well structured and straightforward now, but it was far from that during the design work. The method emerged as a by-product, having evolved during iterative deliberations. As illustrated in the introduction and identified in [9], the categorization of the FWI System outputs could benefit from this approach. We have since applied the method to categorize and colour outputs from other models for operational display [39–41], examples of which are in the Supplementary Material.

The many considerations discussed in this paper need to be addressed to ensure that the models are interpreted and used appropriately. We emphasize that any schemes including those built into software applications need this careful consideration. Incorporating the outputs from scientific models into operational decision-making is not straightforward. The key for a successful application is that researchers and practitioners work together closely throughout the process, from problem identification through to implementation and evaluation [1]. Through this collaborative approach, outcomes tend to have a higher acceptance and usefulness.

An additional factor not addressed by our method is that categories need to be meaningful for more than just map design, because there is a tendency to extend the use categories to guidelines and standard operating procedures and vice versa [9]. The classifications and their boundaries can also be used as unwritten mental shortcuts or heuristics in the place of a more deliberative consideration of complex information. Well-designed, science-based classifications can be consistent with less time-constrained situational analyses, while poor classifications may lead to suboptimal decision-making.

Design considerations for presenting quantitative data for decision support go beyond the categorization and colouring of numerical scales. Also important are options for spatial resolution and the use of simulated three-dimensional displays, examples of which are given in the Supplementary Material [42].

We presented a method of designing classification and colouring schemes that consider many complex factors, interactions and trade-offs. These satisfied the ultimate goals of showing decision-makers complete and accurate quantitative information that was understood quickly and easily.

Supplementary Materials: The following are available online at https://www.mdpi.com/article/10.3390/fire4030050/s1: other applications of the model (Figures S1 and S2), classification of numbers, additional considerations for the spatial display of data (Figures S3–S5) and larger versions of Figure 7a–c (Figures S6–S8).

Author Contributions: Conceptualization: D.B., J.E., C.C.H., C.B.M., D.G.W. and M.W. (Mike Wotton); methodology: D.B., C.B.M. and D.G.W.; software: J.E., A.S. and D.G.W.; formal analysis: D.B., C.C.H., C.B.M. and M.W. (Melanie Wheatley), D.G.W. and M.W. (Mike Wotton); investigation: D.B., J.E., C.C.H., C.B.M., A.S. and M.W. (Melanie Wheatley), D.G.W. and M.W. (Mike Wotton); writing—original draft preparation: D.B., C.C.H., C.B.M., M.W. (Melanie Wheatley) and D.G.W.; writing—review and editing: D.B., J.E., C.C.H., C.B.M., A.S., M.W. (Mike Wotton), D.G.W. and M.W. (Melanie Wheatley); visualization: D.B., J.E., C.C.H., C.B.M., M.W. (Melanie Wheatley), D.G.W., M.W. (Mike Wotton) and A.S.; project administration: D.B., C.B.M. and D.G.W. All authors have read and agreed to the published version of the manuscript.

Funding: We acknowledge the support of the Natural Sciences and Engineering Research Council of Canada (NSERC) [RGPIN-2015-04221].

Acknowledgments: Thank you to the Ministry of Northern Development, Mines, Natural Resources and Forestry, Aviation Forest Fire and Emergency Services staff for important discussions and management for support. Thank you to Barry Graham, Dan Johnston, Dan Leonard and David Martell for insightful advice, Darren McLarty for the review, Jim Caputo, Natasha Jurko and Benito Russo for technical support and the Fire Intelligence Officer cadre for their regular engagement and feedback. A special thank you goes to Darryl Pajunen for supporting the model's development. We thank the anonymous reviewers for their careful reading and helpful comments that improved the manuscript.

Conflicts of Interest: The authors declare no conflict of interest.

References

1. Woolford, D.G.; Martell, D.L.; McFayden, C.B.; Evens, J.; Stacey, A.; Wotton, B.M.; Boychuk, D. The Development and Implementation of a Human-Caused Wildland Fire Occurrence Prediction System for the Province of Ontario, Canada. *Can. J. For. Res.* **2021**, *51*, 303–325. [CrossRef]
2. Johnston, L.M.; Wang, X.; Erni, S.; Taylor, S.W.; McFayden, C.B.; Oliver, J.A.; Stockdale, C.; Christianson, A.; Boulanger, Y.; Gauthier, S.; et al. Wildland Fire Risk Research in Canada. *Environ. Rev.* **2020**, *28*, 164–186. [CrossRef]
3. Wotton, B.M.; Martell, D.L. A Lightning Fire Occurrence Model for Ontario. *Can. J. For. Res.* **2005**, *35*, 1389–1401. [CrossRef]
4. Stocks, B.J.; Lynham, T.J.; Lawson, B.D.; Alexander, M.E.; Van Wagner, C.E.; McAlpine, R.S.; Dubé, D.E. The Canadian Forest Fire Danger Rating System: An Overview. *For. Chron.* **1989**, *65*, 450–457. [CrossRef]
5. Van Wagner, C.E. *Development and Structure of the Canadian Forest Fire Weather Index System*; Forestry Technical Report; Canada Communication Group Publication: Ottawa, ON, Canada, 1987; ISBN 9780662151982.
6. Forestry Canada Fire Danger Group. *Development and Structure of the Canadian Forest Fire Behavior Prediction System*; Information Report ST-X-3; Forestry Canada, Science and Sustainable Development Directorate: Ottawa, ON, Canada, 1992; ISBN 9780662198123.
7. Stocks, B.J. *Wildfires and the Fire Weather Index System in Ontario*; Information Report O-X-213; Canadian Forestry Service, Great Lakes Forestry Centre: Sault Ste. Marie, ON, Canada, 1974.
8. Taylor, S.W.; Alexander, M.E. Science, Technology, and Human Factors in Fire Danger Rating: The Canadian Experience. *Int. J. Wildland Fire* **2006**, *15*, 121. [CrossRef]
9. Hanes, C.; Wotton, B.M.; McFayden, C.; Jurko, N. *An Approach for Defining Physically Based Fire Weather Index System Classes for Ontario*. Natural Resources Canada; Information Report GLC-X-30; Canadian Forest Service: Sault Ste. Marie, ON, Canada, 2021; in press.
10. Kiil, A.D.; Lieskovsky, R.J.; Grigel, J.E. *Fire Hazard Classification for Prince Albert National Park, Saskatchewan*; Environ; Information Report NOR-X-S8; Environment Canada, Forestry Service, Northern Forest Research Centre: Edmonton, AB, Canada, 1973.
11. Kiil, A.D.; Miyagawa, R.S.; Quintilio, D. *Calibration and Performance of the CANADIAN Fire Weather Index in Alberta*; Information Report NOR-X-173; Environment Canada, Canadian Forestry Service, Northern Forest Research Centre: Edmonton, AB, Canada, 1977.

12. Alexander, M.E. *Proposed Revision of Fire Danger Class Criteria for Forest and Rural Areas in New Zealand*, 2nd ed.; National Rural Fire Authority: Wellington, New Zealand; Scion Rural Fire Research Group: Christchurch, New Zealand, 2008; 63p.
13. De Groot, W.J.; Field, R.D.; Brady, M.A.; Roswintiarti, O.; Mohamad, M. Development of the Indonesian and Malaysian Fire Danger Rating Systems. *Mitig. Adapt. Strat. Glob. Chang.* **2007**, *12*, 165. [CrossRef]
14. Flannigan, M.D.; Wotton, B.M. A Study of Interpolation Methods for Forest Fire Danger Rating in Canada. *Can. J. For. Res.* **1989**, *19*, 1059–1066. [CrossRef]
15. Ontario Forest Fire Information Map. Available online: https://www.lioapplications.lrc.gov.on.ca/ForestFireInformationMap/index.html?viewer=FFIM.FFIM (accessed on 17 June 2021).
16. Byram, G.M. Combustion of Forest Fuels. In *Forest Fire: Control and Use*; Davis, K.P., Ed.; McGraw-Hill: New York, NY, USA, 1959; pp. 61–89.
17. Alexander, M.E.; De Groot, W.J. *Fire Behavior in Jack Pine Stands as Related to the Canadian Forest Fire Weather Index (FWI) System*; Poster with Text; Canadian Forest Service, Northern Forestry Centre: Edmonton, AB, Canada, 1988.
18. Taylor, S.W.; Pike, R.G.; Alexander, M.E. *Field Guide to the Canadian Forest Fire Behavior Prediction (FBP) System*; Natural Resources Canada, Canadian Forest Service, Northern Forestry Centre and British Columbia Ministry of Forests, Research Branch: Victoria, BC, Canada, 1996.
19. Taylor, S.W.; Alexander, M.E. *A Field Guide to the Canadian Forest Fire Behavior Prediction (FBP) System*, 3rd ed.; Natural Resources Canada, Canadian Forest Service, Northern Forestry Centre: Edmonton, AB, Canada, 2018.
20. Pravossoudovitch, K.; Cury, F.; Young, S.G.; Elliot, A.J. Is Red the Colour of Danger? Testing an Implicit Red–Danger Association. *Ergonomics* **2014**, *57*, 503–510. [CrossRef]
21. Kaya, N.; Epps, H.H. Relationship between Color and Emotion: A Study of College Students. *Coll. Stud. J.* **2004**, *38*, 396.
22. Hirsch, K.; Martell, D. A Review of Initial Attack Fire Crew Productivity and Effectiveness. *Int. J. Wildland Fire* **1996**, *6*, 199. [CrossRef]
23. Alberta Wildfire, Danger Forecast, Headfire Intensity. Available online: https://wildfire.alberta.ca/files/ahfi.gif (accessed on 17 June 2021).
24. Canadian Wildland Fire Information System Fire Behaviour Maps. Available online: https://cwfis.cfs.nrcan.gc.ca/maps/fb (accessed on 17 June 2021).
25. BC Wildfire PSTA Head Fire Intensity. Available online: https://catalogue.data.gov.bc.ca/dataset/bc-wildfire-psta-head-fire-intensity (accessed on 9 September 2020).
26. Evans, I.S. The Selection of Class Intervals. *Trans. Inst. Br. Geogr.* **1977**, *2*, 98–124. [CrossRef]
27. Beverly, J.L.; McLoughlin, N. Burn Probability Simulation and Subsequent Wildland Fire Activity in Alberta, Canada—Implications for Risk Assessment and Strategic Planning. *For. Ecol. Manag.* **2019**, *451*, 117490. [CrossRef]
28. Martell, D. The Development and Implementation of Forest and Wildland fire Management Decision Support Systems: Reflections on Past Practices and Emerging Needs and Challenges. *Math. Comput. For. Nat. Resour. Sci.* **2011**, *3*, 18.
29. Noble, P.; Paveglio, T.B. Exploring Adoption of the Wildland Fire Decision Support System: End User Perspectives. *J. For.* **2020**, *118*, 154–171. [CrossRef]
30. Hanes, C.C.; Jain, P.; Flannigan, M.D.; Fortin, V.; Roy, G. Evaluation of the Canadian Precipitation Analysis (CaPA) to Improve Forest Fire Danger Rating. *Int. J. Wildland Fire* **2017**, *26*, 509. [CrossRef]
31. Jenny, B.; Kelso, N.V. Color Design for the Color Vision Impaired. *Cartogr. Perspect.* **2007**, 61–67. [CrossRef]
32. Harrower, M.; Brewer, C.A. ColorBrewer.org: An Online Tool for Selecting Colour Schemes for Maps. *Cartogr. J.* **2003**, *40*, 27–37. [CrossRef]
33. Zeileis, A.; Hornik, K.; Murrell, P. Escaping RGBland: Selecting Colors for Statistical Graphics. *Comput. Stat. Data Anal.* **2009**, *53*, 3259–3270. [CrossRef]
34. The R Project for Statistical Computing. Available online: https://www.r-project.org/ (accessed on 28 August 2020).
35. inlmisc: Miscellaneous Functions for the USGS INL Project Office. Available online: https://cran.r-project.org/web/packages/inlmisc/index.html (accessed on 4 September 2020).
36. TOL Color Schemes. Available online: https://waterdata.usgs.gov/blog/tolcolors/ (accessed on 4 September 2020).
37. Jenks, G.F.; Caspall, F.C. Error on Choroplethic Maps: Definition, Measurement, ReductioN. *Ann. Assoc. Am. Geogr.* **1971**, *61*, 217–244. [CrossRef]
38. Geospatial Analysis 6th Edition, 2020 Update. Available online: https://www.spatialanalysisonline.com/HTML/index.html (accessed on 4 June 2021).
39. McFayden, C.B.; Woolford, D.G.; Stacey, A.; Boychuk, D.; Johnston, J.M.; Wheatley, M.J.; Martell, D. L Risk Assessment for Wildland Fire Aerial Detection Patrol Route Planning in Ontario, Canada. *Int. J. Wildland Fire.* **2020**, *29*, 28–41. [CrossRef]
40. McFayden, C.B.; Boychuk, D.; Woolford, D.G.; Wheatley, M.J.; Johnston, L. Impacts of Wildland Fire Effects on Resources and Assets through Expert Elicitation to Support Fire Response Decisions. *Int. J. Wildland Fire* **2019**, *28*, 885–900. [CrossRef]
41. McFayden, C.; Boychuk, D.; Stacey, A.; Woolford, D.G.; McLarty, D.; Leonard, D.; Evens, J.; Johnston, D.; McAlpine, R.S. Daily Spatial Response Objective and Speed Weight Indicators to Assist in Daily Wildland Fire Preparedness Planning for ONTARIO, Canada. (in prep).
42. Albert-Green, A.; Braun, W.J.; Martell, D.L.; Woolford, D.G. Visualization Tools for Assessing the Markov Property: Sojourn Times in the Forest Fire Weather Index in Ontario. *Environmetrics* **2014**, *25*, 417–430. [CrossRef]

A Surrogate Model for Rapidly Assessing the Size of a Wildfire over Time

Ujjwal K. C. [1,*], Jagannath Aryal [2], James Hilton [3] and Saurabh Garg [1]

1. Discipline of ICT, University of Tasmania, Hobart, TAS 7005, Australia; Saurabh.Garg@utas.edu.au
2. Department of Infrastructure Engineering, Faculty of Engineering and Information Technology, University of Melbourne, Parkville, VIC 3052 Australia; Jagannath.Aryal@unimelb.edu.au
3. Data61, CSIRO, Clayton, Melbourne, VIC 3168, Australia; James.Hilton@csiro.au
* Correspondence: Ujjwal.KC@utas.edu.au

Abstract: Rapid estimates of the risk from potential wildfires are necessary for operational management and mitigation efforts. Computational models can provide risk metrics, but are typically deterministic and may neglect uncertainties inherent in factors driving the fire. Modeling these uncertainties can more accurately predict risks associated with a particular wildfire, but requires a large number of simulations with a corresponding increase in required computational time. Surrogate models provide a means to rapidly estimate the outcome of a particular model based on implicit uncertainties within the model and are very computationally efficient. In this paper, we detail the development of a surrogate model for the growth of a wildfire based on initial meteorological conditions: temperature, relative humidity, and wind speed. Multiple simulated fires under different conditions are used to develop the surrogate model based on the relationship between the area burnt by the fire and each meteorological variable. The results from nine bio-regions in Tasmania show that the surrogate model can closely represent the change in the size of a wildfire over time. The model could be used for a rapid initial estimate of likely fire risk for operational wildfire management.

Keywords: fire spread models; surrogate modeling; sensitivity analysis; global sensitivity analysis

1. Introduction

Wildfires are a major threat in fire-prone areas such as Australia, Mediterranean regions in Europe, and the United States. These can destroy homes and infrastructure causing millions of dollars of damage [1,2] as well as threaten lives and ecologically sensitive areas. Consequently, fire suppression and mitigation costs have significantly increased over the years [3]. For disasters such as wildfires, where the fire conditions rapidly change with time, any models that can give realistic fire predictions in as little time as possible can be critical for effective preparedness, early warning, fire suppression, and evacuation efforts. Computational wildfire simulations using different fire propagation models are an effective means of predicting the wildfire behaviors and risk [4]. However, such models are typically deterministic and may neglect uncertainties inherent in factors driving the fire [5]. Modeling these uncertainties can more accurately predict risks associated with a particular wildfire, but require a large number of simulations with a corresponding increase in computation time. Under operational constraints, this computation time may be longer than the time window available to prepare and respond to the wildfire. Consequently, computationally cheaper models that can rapidly estimate risks could be very useful for operational wildfire management.

Wildfire behavior models typically fall into one of three categories [6]: those based on physical processes [7–10], empirical models [11–13], or simulation and mathematical models [14–16]. Mathematical concepts like Ordinary Differential Equations (ODEs), polynomial chaos, Gaussian process, and spatial process-based models like Cellular Automata, and Level Set methods have been used to model the spread of wildfires [17–22]. Currently, operational fire behavior tools such as Spark [14], FARSITE [15], and Phoenix [23] are used to predict the wildfire behavior. These tools predict fire behavior by simulating the complex relationships between different contributing factors in a one-to-one deterministic manner without incorporating uncertainty in the prediction results [24]. In reality, there are uncertainties for each of the contributing factors which consequently introduce uncertainty in the resulting model output.

Quantifying such uncertainties in operational fire models requires identifying, classifying, and assessing their sources and influence on the model output [25]. Kaschek et al. [26] highlighted the importance of parameter value estimation for the experimental data in understanding the model dynamics for any permissible value of the parameters and subsequent uncertainty quantification. However, such uncertainties have not been well quantified in current operational fire models. As such, Sensitivity Analysis (SA) is a currently active area of research, in which the spatial and temporal variability in the model outputs is quantified as a function of the variation in the input parameters for an optimized model [27]. Note that uncertainty and sensitivity analysis are different as uncertainty analysis assesses the uncertainty in the model output due to uncertainties in input parameters while sensitivity analysis assesses the contribution of input parameters on the uncertainty of the model output [28]. Sensitivity analyses of input parameters for different fire models have been carried out in several studies [5,29–33], where the model outputs have been considered for a fixed period. Sensitivity analyses may also help quantify such uncertainties and determine the influence of each input parameter on fire progression but such studies for operational fire models, to our knowledge, have not been well-explored.

On the other hand, the model performance and the effectiveness of risk management achieved with such models is determined to a large degree by how well such uncertainties are understood and communicated [34]. Therefore, calculating more accurate risk metrics quantifying all these uncertainties can be a complex task [24,29]. Risk calculation can require a large number of simulations to be run under different scenarios, which can take several hours to days to complete [35]. Such time-consuming analyses are currently impractical for operational management. Consequently, developing computationally efficient methods, such as surrogate models, can be an alternative to such models for operational wildfire management.

Surrogate models, also known as metamodels, are the models of the outcomes that mimic the behavior of the simulation model as closely as possible while being computationally inexpensive. Surrogate models have been extensively used to replace, or as an alternative to, computationally expensive models, as they represent input–output relationships derived from complex models. These have been used in natural hazard models including flooding [36,37] and storms [38,39]) as well as engineering design problems [40–42] to rapidly estimate the associated risks and concerns. The surrogate model introduced in this study can facilitate rapid estimation of wildfire risks based on the data from high-fidelity operational fire simulation models.

Several studies have shown that not all the parameters in wildfire models have significant influence on the spread rate, and consequently fire behaviors can be explained by prioritizing highly significant parameters [43,44]. Sharples et al. [44] demonstrated that the spread of wildfires could be expressed in a "universal" fire spread model based on only temperature, relative humidity, and wind speed input parameters. Taking this as a starting point, we use these input parameters to construct a computationally efficient surrogate model. Next, we present a one-at-a-time (OAT) sensitivity analysis of input parameters for the variability of fire dynamics in the wildfire prediction tool Spark, whereby the influence of a parameter on the model output is assessed one at a time instead of multiple parameters simultaneously. Based on the results obtained from the analyses, the nature of the contribution of the parameters is determined and used to define the functional form for the surrogate model. Finally, we use a data fitting technique to determine the values of unknown factors in the functional form to define a complete surrogate model to explain fire behaviors based on initial weather conditions. We test the efficacy of the approach by applying it to the entire Tasmanian region and its nine different bio-regions. The specific contributions of this study are as follows:

1. An investigative time-based OAT sensitivity analysis for quantifying the influence of meteorological input on fire dynamics.
2. Surrogate modeling of fire simulations based on initial conditions (temperature, relative humidity, and wind speed).

The rest of the paper is organized as follows. Section 2 explains the general methods for the mathematical formulation of the surrogate model and time-based sensitivity analysis. Section 3 discusses the investigative results obtained in our experiments. Section 4 concludes the paper and discusses possible future work.

2. Methodology

2.1. Study Area

We chose Tasmania for our study due to the frequent occurrences of wildfires in the region, high-quality land data sets as maintained by Tasmania Fire Service (TFS) and State Emergency Service (SES), and a well-studied and systematic grid configuration for possible fire start locations in the region. Tasmania has a total area of 68,401 sq. km. The grid configuration for fire simulation, as maintained by TFS, places each possible fire start location at an interval of 1 sq. km irrespective of land classification. Any start locations falling on water bodies are shifted to the nearest land locations. There is a total of 68,048 possible fire start locations in the grid configuration for the entire Tasmanian region. We followed Interim Biogeographic Regionalization for Australia Version 7 (IBRA 7) [45] for nine different bio-regions (see Figure 1) to run fire simulations.

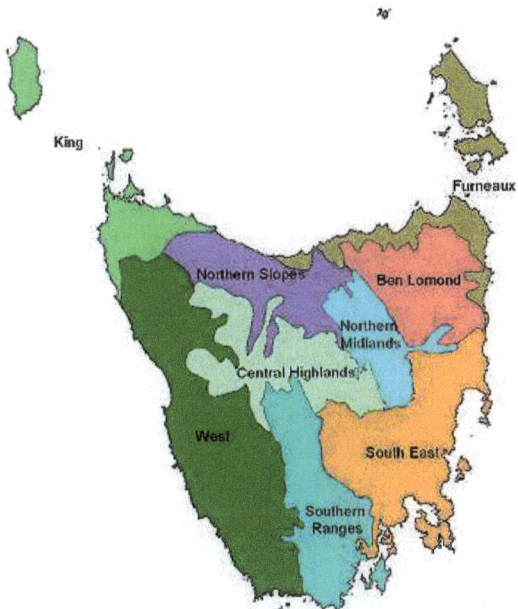

Figure 1. Nine bio-regions of Tasmania. Sourced from the work in [46].

2.2. Fire Simulations—Spark

Spark [14] is a computational wildfire modeling framework that predicts the spread and potential impact of wildfires based on different behavior fire models. The framework simulates the progression of the fire over time in different fire conditions and fuels, resulting in different rates-of-spread (RoS). The fire simulations in Spark require many input data sets, including the information on land classification, fuel type, topography, and meteorological data. The simulations can be run for any number of distinct fire perimeters and can model factors such as firebreaks, spot fires, and coalescence between different parts of the fire over time. The simulations use several different empirical fire models for fuels found in Tasmania. Different fuel types found in Tasmania are mapped from the TasVeg [47] vegetation types to several empirical fire spread models. These include models for eucalypt forest [48,49], buttongrass moorland [50], heathland [51], and grasslands [52]. A detailed description of these models is included in the Appendix A.

The models require meteorological data inputs. For this study, only temperature, relative humidity, and wind speed were used [44]. Other input parameters in the fire simulation such as fuel load and land topography were used as per the configuration and records maintained by TFS and Tasmanian Government [53]. The fuel load was based on a Tasmania fire history data set allowing the time since the fire to be calculated and populated using Olson fuel load curves [54–56]. Our current study covered only the westerly winds as they are the most common winds in Tasmania [57,58]. The configuration and input files can be made available from the authors upon request. Based on the historical records and as considered in [5], the ranges for the data inputs are fixed as follows:

- Temperature 10–40 °C
- Relative Humidity 10–90%
- Wind Speed 10–60 km/h

The fire simulations were started at locations in each of the nine bio-regions ensuring minimal obstructions (lakes/water bodies) irrespective of land classification locations, as shown in Figure 2. The simulations were run for a total of five hours and the area of each simulation was recorded every half-hour. The simulation data used in this study are a subset of the data available in [59].

Figure 2. Fire start locations for different bio-regions in Tasmania. The locations are chosen ensuring minimal obstructions due to water bodies irrespective of land classification.

2.3. Sensitivity Analysis

To quantify the interactions of meteorological inputs on fire dynamics, we focus on the parametric sensitivity analysis of fire simulations over time. For SA, we follow an OAT approach where the influence of a parameter is assessed by running the fire simulations for different values of the parameter within the permissible range while keeping the other parameters constant as explained in the work [60]. The influence of a parameter on the fire area is assessed by comparing the fire values obtained for the maximum and minimum values of the parameter (with minimum and maximum values of other parameters, respectively) against the fire areas obtained for the maximum (*all-high*) and minimum values (*all-low*) of all the parameters. The variability of the output of the fire simulations can be analyzed at different time steps by considering the model outputs in all these time steps to determine the scale of influence of that input parameter on the fire area. However, for our analysis to determine the cumulative influence of each parameter on the fire size, we use the total fire size obtained at the end of the total simulation time. The quantification of the variability in the fire area is carried out for each bio-region and the entirety of Tasmania as well. Similar analyses can be done to understand the influence of each parameter on the fire spread during a particular duration of time by assessing the values of the fire area.

2.4. Surrogate Modeling

For surrogate modeling of fire simulations, we first establish a mathematical model based on how fires spread over time when they ignite under some initial environmental conditions. Then, unknown values in the mathematical model are determined by fitting the simulation data by considering the results obtained from the sensitivity analysis. The mathematical relationship and simulation data fitting are discussed in detail as follows.

2.4.1. Mathematical Foundation

For a fire starting at a particular location, the shape of the fire is ideally assumed to be elliptical [61]. Consequently, the total area burnt by the fire, represented by A, is directly proportional to the square of the time (t) for which the fire is burning [61]. We focus our mathematical foundation on the same principle but replace the squared power of the time with a variable n to account for any dependency on fire geometry (ideally, the value of n in our mathematical foundation must be less than or equal to 2). As such, we mathematically express A as

$$A = \alpha t^n \tag{1}$$

where α is a fitting factor that depends on the initial values of temperature (T_0), relative humidity (R_0), and wind speed (W_0). This fitting factor does not account for factors such as daily temperature changes or wind changes during the progression of the fire. However, such factors are accounted for by analyzing the fire sizes for different sets (original and changed) of initial conditions and assessing the difference. The value of α is determined by defining a functional form for the three input parameters based on the results of sensitivity analysis. The values of n and α are determined by fitting simulation data.

2.4.2. Simulation Data Fitting

The unknown factors in Equation (1) can be determined by fitting a curve through experimental analyses. We use nonlinear least squares method [62] for data fitting. For experimental analyses, the fire simulations are run for five hours over a set of points in all nine different bio-regions in Tasmania. The simulation of fire behavior in Spark is graphically represented in Figure 3, which shows that the propagation of the fire over time can be different for different bio-regions. The steps for surrogate modeling are carried out initially for the entire Tasmanian region and then to specific bio-regions one at a time.

 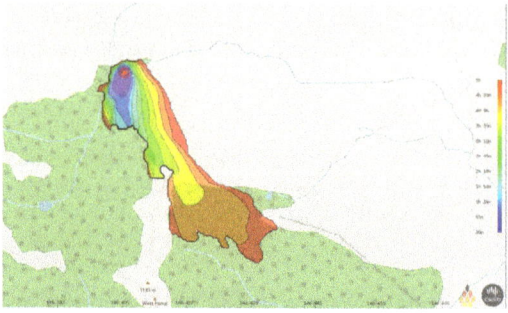

(a) Southern Ranges　　　　　　　　　　　　　　　　(b) West

Figure 3. Illustration of simulation in Fire dynamics in Spark ($T = 40, R = 90, W = 60$). The wildfire propagates differently based on the fuel load and land topography and consequently, the total area burnt by the fire varies based on the bio-region. Different colors represent the area burnt by the fire over time in half hourly steps. The purple color represents the area burnt by the fire in the first half hour, while the red color represents the area burnt from 4.5–5 h mark.

2.5. Surrogate Model Validation

For the validation of the dynamic model constructed in this study, we use the hold-out cross-validation method as explained in [63]. This method splits the simulation data into 70% training set and 30% test set for cross-validation. Then, to quantify the accuracy of the dynamic model, Pearson's correlation coefficient [64] is calculated between the model predicted values and the actual values obtained from the simulations. The value of the correlation coefficient closer to one (1) represents a more accurate model.

3. Results and Discussions

3.1. Sensitivity Analysis

We analyzed the effect of each input parameter on the fire area size at different instants of time as given by the fire simulations in Spark run over different bio-regions in Tasmania. A plot of the variation in fire area for different bio-regions caused by variability in temperature (°C) is shown in Figure 4a. The fire area generally grows bigger over time for the maximum value of the temperature. In the analysis, the fire burns a greater area at $T = 40$ °C and minimum values of other parameters. For $T = 40$ °C, the increase in the area burnt by the fire is steady until 2 h, but after that, the increase is exponential to reach a maximum value of 4121.190 ha at the five-hour mark. For a lower value of T, the increase in the fire area over time is steady and close to a linear relationship. For the entire Tasmania, the variability in burnt fire area caused by the variation of temperature is as high as 4115.340 ha. The respective variations in the fire area for different bio-regions are listed in Table 1.

The variation of the fire area with the variability of relative humidity (%) is shown in Figure 4b. It is clear from the plots that relative humidity has a significant influence on the variation of the fire area. The maximum values of the area burnt by the fire are achieved with the value of relative humidity at 10% and maximum values of other parameters for all the locations. For this particular combination of parameter values, the increase in the fire area size with time is brisk. There is a sharp increase in the rate of increase in the fire area after 2 h in general for most of the bio-regions. The maximum area burnt by fire over 5 h is 28,682.6 ha. The variation in the fire area caused by the variability of relative humidity for the entire Tasmanian region is as high as 28,680.08 ha, which is significantly higher and signifies the greater impact of the parameter. The respective variations in the fire area for different bio-regions are listed in Table 1.

The variation in the fire area caused by the variability in the wind speed (kmh^{-1}) is shown in Figure 4c. The area burnt by the fire is the greatest at $W = 60$ kmh^{-1} and minimum values of other parameters for all the locations. The fire areas for other cases are relatively low. The increase in the fire area with time is steady until 2 h after which the fire area starts to increase at a higher rate. The area burnt by the fire over 5 h reaches 11,543.2 ha. The variation in the fire area caused by the variability of wind speed is approximately 11,539.96 ha for the entirety of Tasmania, which indicates the significant influence of the wind on the fire area. The respective variations in the fire area for different bio-regions are listed in Table 1.

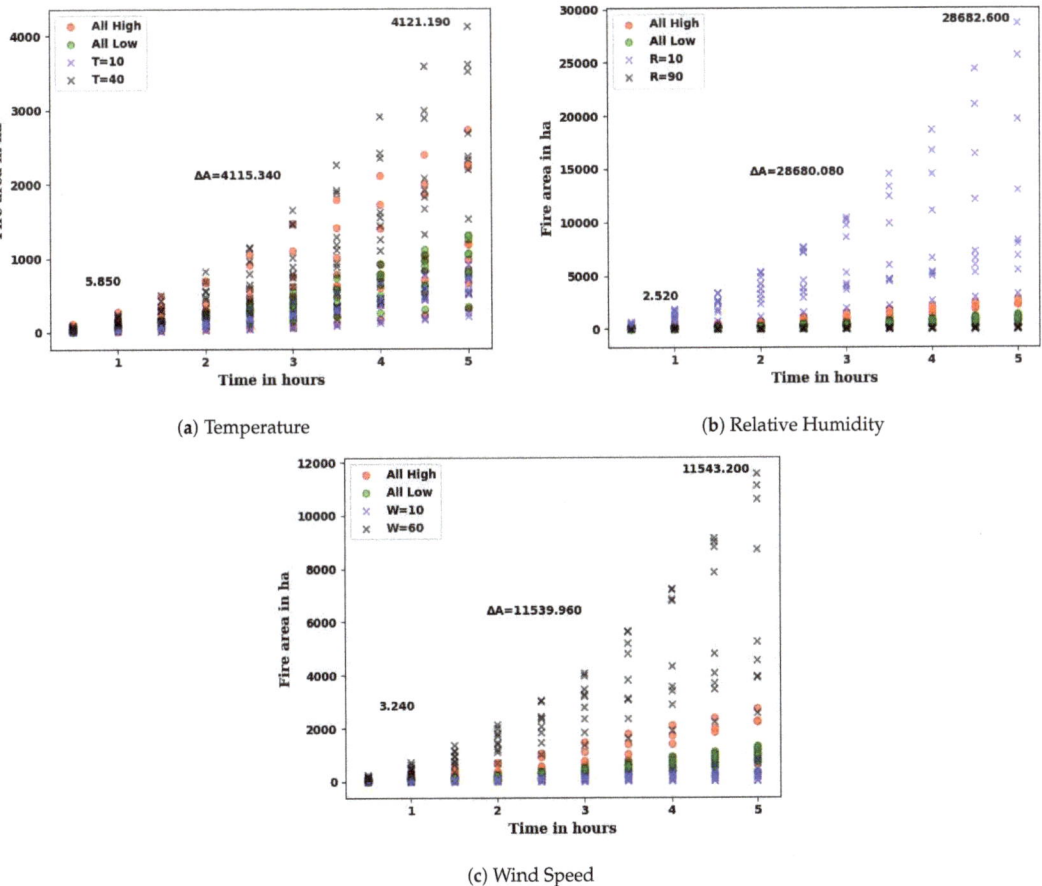

Figure 4. Variation of fire area (in *ha*) with the variability of different input parameters, shown as ΔA on the plots. *All high* is the plot for fire area obtained with maximum values of all parameters, while *all low* is with minimum values. The variation of fire area caused by relative humidity is the highest while that of the temperature is the lowest.

In the analyses, we derived the upper and lower bounds for the values of the fire area for each input parameter considering the values of other parameters as well. The minimum value of an input parameter did not necessarily signify the lower bound for the fire area given by the parameter, and vice versa. For the entire Tasmanian region, the variability in fire area brought about by extreme values of temperature for extreme values of other parameters is ~4115.340 ha in five hours (see Figure 4a). The variability in fire area in the same duration caused by relative humidity is ~28,680.08 ha (see Figure 4a), while the variability caused by wind speed stands at 11,539.96 ha (see Figure 4a). As such, the extent of variability in fire area caused by relative humidity and wind speed is approximately 7 and 3 times greater than the same caused by temperature, respectively. Thus, it can be concluded that relative humidity has the highest effect on the area burnt by the fire at different instants of time for the entire Tasmanian region given by the fire simulations by Spark while the temperature has the least effect. For Northern Slopes, the variations in fire area caused by temperature and wind speed are comparable, which points out the impact caused by fuel load and land topography. Similar analyses can be done to understand the influence of each parameter on the fire spread during a particular duration of time. For example, during the second hour of a wildfire (from 1 to 2-h mark), the variation caused in the fire area by the variability in the values of temperature, relative humidity, and wind speed in the entire Tasmanian region are 739.26 ha, 5378.58 ha, and 2095.47 ha, respectively. Such information about the influence of the input parameters on the model output with consideration of the time can lead to a better quantification of the feedback of meteorological inputs on the fire dynamics.

We used an OAT method to measure the influence of each input parameter on the output of the fire simulations. To understand the variation in fire area in different bio-regions, we carried out the analyses for all the regions independently. In our analyses, we derived the upper and lower bounds for the values of the fire area for each input parameter considering the values of other parameters as well. For all the bio-regions, the variation in the fire area caused by the variability of relative humidity is the highest, while the variation caused by that of the temperature is the lowest. The variations in fire area caused by the variability of relative humidity and wind speed are as high as about 11 and 5 times more when compared to the same caused by the variability of temperature, respectively. For Northern Slopes, the variations in fire area caused by temperature and wind speed are comparable, which points out the impact caused by fuel load and land topography. The quantitative measure of the variability in fire area brought by each parameter during the OAT method is conclusive. Our findings established relative humidity as the parameter with the greatest impact on the spread of the fire over time and temperature as the one with the least impact. This finding can be attributed to the fire models in Spark that are highly sensitive to fuel moisture content, which in turn is highly sensitive to relative humidity compared to temperature (for example, the Dry Eucalypt model equation for fuel moisture content, MC, has a greater dependency on relative humidity than temperature). However, the extent of the influence of each parameter on the variability of the fire area size is not necessarily the same for all the bio-regions even though they have similar relative comparisons.

Table 1. Variability in fire area caused by variability in input parameters.

Bio-Regions	Variation in Area burnt by fire (ΔA) in ha		
	Temperature	Rel. Humidity	Wind
Southern Ranges	2655.18	28,679.9	11,531.23
South East	2360.97	25,645.92	11,114.86
West	1143.36	3308.58	2554.11
Central Highlands	2290.32	8328.78	4524.47
Northern Slopes	4092.21	5543.11	3910.05
Northern Midlands	2177.55	7995.96	5239.71
Ben Lomond	3573.72	19,713.78	10,601.4
Furneaux	3491.91	12,972.73	8702.5
King	1497.33	6908.67	3943.89
Tasmania	4115.34	28,680.08	11,539.96

3.2. Surrogate Modeling

Following the methods defined for surrogate modeling (Equation (1)), we first incorporated the meteorological influence and adapted a functional form for temperature, relative humidity, and wind speed to determine the value of α. The functional form was developed by further analyzing the results of the sensitivity analysis in which the growth of fire area was investigated at different time instants (every half-hour mark) against the highest values of the input parameters (Figure 5). As can be seen in the figure, the growth of the fire area is close to linear over time for temperature and relative humidity, whereas an exponential fit is more suitable for wind speed. As such, the functional form for input parameter to define the factor in our surrogate model (Equation (1)) is

$$\alpha = k_1 T_0 + k_2 R_0 + e^{k_3} W_0 \qquad (2)$$

After fitting the simulation data obtained from all the Tasmanian regions, the surrogate model is defined as follows:

$$A(t) = (16.72 T_0 - 26.80 R_0 + e^{2.03} W_0) \, t^{1.76} \qquad (3)$$

Note that the functional form used for the surrogate model is slightly different from the formulation in the work of Sharples et al. [44] as the surrogate model is based on the findings from the uncertainty quantification.

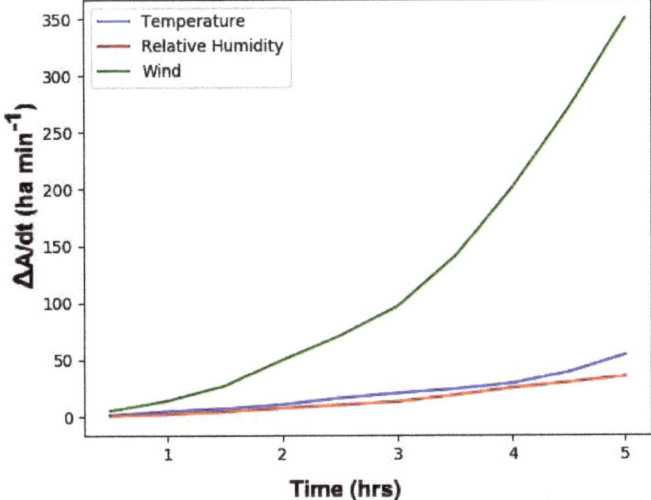

Figure 5. Rate of increase of fire area with time for Southern Ranges bio-region (The rate of increase in the fire area is in hectares per minute (ha min^{-1}). The contributions in the rate of change of fire area made by temperature (blue line) and relative humidity (red) are steady over time while the same made by the wind speed (green) increases exponentially over time).

The simulation and predicted data are shown in Figure 6. The value of the correlation coefficient is 0.897. Interestingly, the value of power to which the time is raised in the model is close to 2 (1.76), indicating fire growth scales approximately as the expected square of the time. Despite the high value of the correlation coefficient, the model over fits the fire area for some extreme values of the input parameters. Although not ideal, an overfitted result produces a more conservative estimate of larger fire sizes, rather than a possibly dangerous underestimate. The overfitting in the model is possibly due to the significant impact of the fuel loads, which can be different for different bio-regions, and the interaction between the input parameters, which are not considered in the study. The fitting parameters for each individual region in Tasmania are given in Table 2 and plotted in Figure 7.

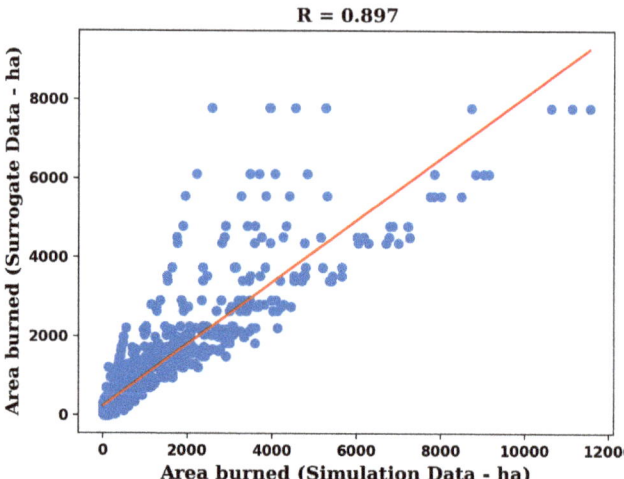

Figure 6. Simulation Data vs. model predicted data for entire Tasmania (The area burnt by the fire is in hectares (*ha*). The red line represents the ideal line where the scatter plots should align for the most accurate surrogate prediction data. The calculated value of correlation coefficient is 0.897, which indicates that the surrogate model closely predicted the actual simulation data.)

Table 2. Fitting parameters in the surrogate model for nine bio-regions in Tasmania.

Bio-Regions	k_1	k_2	k_3	n
Southern Ranges	17.26	−41.98	2.09	1.91
South East	35.56	−137.33	2.38	1.81
West	2.49	−5.56	1.46	1.49
Central Highlands	14.95	−16.54	1.87	1.76
Northern Slopes	12.11	−39.69	2.21	1.68
Northern Midlands	30.38	−19.13	1.95	1.72
Ben Lomond	23.09	−35.29	2.34	1.81
Furneaux	26.09	−53.85	2.43	1.73
King	11.21	−9.80	1.78	1.60
Tasmania	16.72	−26.80	2.03	1.76

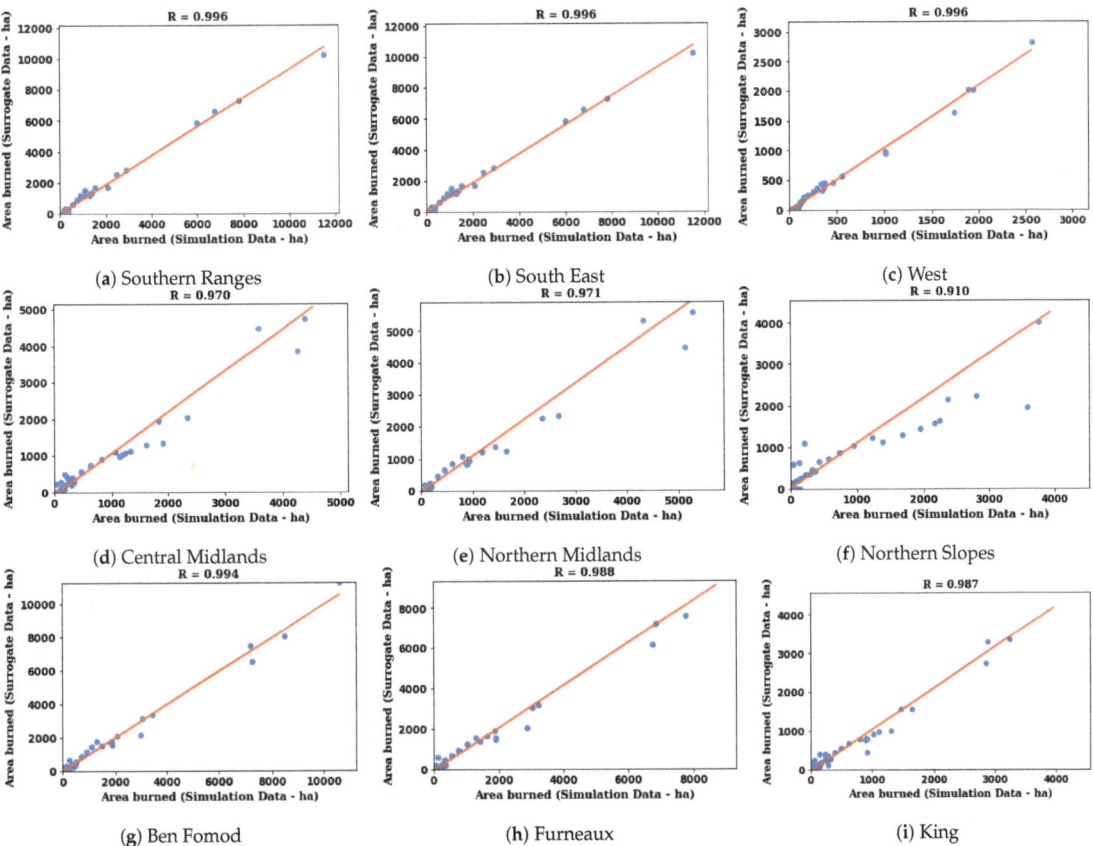

Figure 7. Bio-region-specific surrogate model predicted data compared against simulation data. The fire area is in hectares (ha). The red line represents the ideal line where the scatter plots should align for the most accurate surrogate prediction data. The surrogate model constructed for the Southern Ranges bio-region has the highest value of correlation coefficient of 0.996 while that of Northern Slopes is the lowest with a value of 0.910.

Our surrogate model based on fire simulations has reasonable success with promising findings. The mathematical equation defined in the surrogate model in terms of the initial values of different meteorological data gave satisfactory information about the progression of the fire over time as the value of the calculated correlation coefficient for the entire Tasmanian region is about 0.897. This finding indicates the fact that constructing a surrogate model to represent all the bio-regions with a diverse range of fuel loads and land topography may not be so efficient. Consequently, we constructed a single surrogate model for each bio-region and calculated the correlation coefficient between the simulation data and the model predicted data, as shown in Figure 7. Moreover, as listed in Table 2, the values of unknown parameters also represent the respective influence of the parameters in the forest area burnt by the fire. The findings of the surrogate models are consistent with the results obtained from our sensitivity analysis carried out by considering different instants of time. The negative values of k_2 indicate that the fire grows rapidly under lower values of relative humidity when compared to the higher values, while the higher values of k_2 reflect the strong influence of relative humidity on the fire area. The negative value of k_2 as determined in our study is quite consistent with the functional form defined by Shaples et al. in [44] for spread index, where the relative humidity in the denominator indicates a higher spread rate (consequently higher fire area) at lower values when compared to the same at higher values. Further, the surrogate model constructed to represent the dynamics of fire in terms of initial environmental conditions closely represents the actual behavior of the fire over time as the values of the correlation coefficient are well over 0.987. The surrogate model incorrectly identifies small fires, possibly due to the presence of a lake or other water bodies near the fire start location. However, the model shows a good match for larger fires in most cases. This is useful for operational fire management, as recent studies have shown the opposite trend for fire simulations [65], where fire predictions using forecast weather in fire simulation tools under-predicted the fire size in extreme weather conditions. Considering other location-specific information in the model can improve the closeness of the model estimation with the actual values. For the **Southern Ranges** bio-region, the model has a correlation coefficient of 0.996, which signifies the fact that the model closely matches the fire dynamics from the simulations in the region. Such a good prediction by the model appears to be due to the region being relatively free of obstructions (lakes/built-up areas) where the fire would otherwise be stopped. This finding indicates that surrogate models, like the one built here in the study, can be helpful for quickly estimating fires under extreme fire weather conditions for large open areas. The lowest value of the correlation coefficient for the model is 0.910 in the **Northern Slopes** region, which is likely due to the presence of wetlands in the region and due to nonlinear interactions between the input parameters. The nonlinear interactions between the input parameters are not considered for this study. This particular aspect of the surrogate model will be strengthened in the future to improve the closeness of the data predicted by the model with the actual simulation data.

The constructed surrogate model can represent the change in the size of the fire with time-based on the initial values of meteorological inputs (temperature, relative humidity, and wind speed). This could be used to rapidly estimate the wildfire risks, although it should be emphasized that the model can only give an estimate of the size of the fire rather than the shape, growth, or possible impact on certain areas. The functional form to estimate the fire size obtained in this study can be an alternative form to the one explained in [44]. Additionally, our attempt to investigate the path of fire growth has shown that in practical scenarios, the fire area does not follow an ideal t^2 relationship as the values of n as obtained in our experimental analyses for all nine bio-regions are less than 2.

4. Conclusions and Future Works

In this paper, we analyzed the contribution of different parameters on the area (geographical extent) burnt by the fire over time in a conventional one-at-a-time (OAT) fashion by considering nine different bio-regions in Tasmania. Relative Humidity was found to be the parameter with the greatest impact on the spread of fire over time, while Temperature was found to have the least effect. Moreover, our analyses were time-dependent, allowing new ways to quantify the feedback of meteorological inputs on the fire dynamics, and vice versa. Information about the sensitivity of parameters in wildfire models could facilitate new operational approaches for risk management by excluding less sensitive parameters from the models allowing risk to be calculated in a shorter time-frame. We used simulations to construct a computationally-efficient surrogate model for the area of a fire over time in terms of initial starting conditions. The surrogate model considered the nature and the extent of the influence of each meteorological input on the fire dynamics. The region-specific surrogate models constructed in this study represented the dynamic fire behaviors except for some extreme cases. Such surrogate models may be helpful in operational wildfire management during emergencies to quickly make informed decisions for response activities by providing a rapid estimate of potential fire size.

This study is one of the preliminary works to explore sensitivity analyses at different instants of time and how results of such analyses can be integrated into natural hazard modeling systems. In the future, we will take the nonlinear interactions between the input parameters into account to strengthen the surrogate model. The study can be extended for other regions as well with sufficient fire simulation data in hand.

Author Contributions: Conceptualization, U.K. and J.A.; Formal analysis, U.K.; Investigation, U.K.; Methodology, U.K. and J.A.; Software, J.H.; Supervision, J.A., J.H. and S.G.; Validation, U.K. and J.H.; Visualization, U.K.; Writing—original draft, U.K. and J.H.; Writing—review and editing, U.K., J.A., J.H., and S.G. All authors have read and agreed to the published version of the manuscript.

Funding: This research received no external funding.

Institutional Review Board Statement: Not applicable.

Informed Consent Statement: Not applicable.

Data Availability Statement: Not applicable.

Acknowledgments: The authors would like to thank our academic colleagues who helped in improving the quality of the manuscript at various stages of the work.

Conflicts of Interest: The authors declare no conflicts of interest.

Appendix A

Appendix A.1. McArthur Grassland Fire Danger Meters

McArthur [48] described his research into grassland fire behavior in terms of the Grassland Fire Danger Index (GFDI), based on which the rate of spread was calculated. The equations for GFDI and rate of spread (R) are listed as follows:

$$GFDI = 2exp(-23.6 + 5.01lc(C) + 0.0281T - 0.226\sqrt{RH} + 0.633\sqrt{U_{10}})$$

C is the degree of curing (%), T is the air temperature ($^\circ C$), RH is the relative humidity (%), and U_10 is the wind speed (km/h) measured at a height of 10-m in the open.

$$R = 0.13GFDI$$

R is the headfire rate of spread (km/h).

Appendix A.2. Dry Eucalypt Model (Cheney et al.)

This wildfire model is used for predicting the spread of fire behaviors in dry eucalypt forest. The model was developed from a sequence of experiments called "project Vesta" [49] carried out in south-western Australia. The mathematical equations for the model are listed as follows:

$$R = \begin{cases} 30 \; \phi M_f & , U_{10} \leq 5 \text{ km/h} \\ [30 + 1.531(U_{10} - 5)^{0.858} FHS_s^{0.93} (FHS_{ns} H_{ns})^{0.637} B_1] \phi M_f & , U_{10} > 5 \text{ km/h} \end{cases}$$

Note that the term B_1 is the model correction for bias and is taken as 1.03 for this. FHS_{ns} is the near-surface fuel hazard score, FHS_s is the surface fuel hazard score, and H_{ns} is near-surface height and are derived from fuel age using the tall shrubs regression equations as explained in [66]. R is the rate of spread (m/h), U_{10} is the average 10-m open wind speed (km/h), and ϕM_f is the fuel moisture function.

$$\phi M_f = 18.35 \; MC^{-1.495},$$

MC is the moisture content (%) and taken from [67] and [68].

$$MC = 2.76 + 0.124 \; RH - 0.0187 \; T$$

T is the air temperature ($^\circ C$) and RH is the relative humidity (%).

Appendix A.3. CSIRO Grassland Model

Based on the experimental results obtained from the burning project in the Northern Territory of Australia, Cheney et al. [52] described a relationship for the rate of fire spread on grasslands. This model was developed to settle down the confusion created after the calculation of different Grassland Fire Danger Index (GFDI) values under the same conditions thereby questioning the actual effect of fuel load on fire spread rate. The mathematical equations in the model are explained as follows.

$$R_{cu} = \frac{1}{3.6} \begin{cases} (0.054 + 0.209 \ U_{10}) \ \phi M \phi C & , U_{10} \leq 5 \text{ km/h} \\ (1.1 + 0.715 \ (U_{10} - 5)^{0.844}) \ \phi M \phi C & , U_{10} > 5 \text{ km/h} \end{cases}$$

R_{cu} is the cut/grazed rate of spreed in m/s, U_{10} is the 10-m open wind speed (km/h), ϕM is the fuel moisture coefficient, and ϕC is the curing coefficient.

$$\phi M = \begin{cases} exp(-0.108 \ MC) & , MC < 12\% \\ 0.684 - 0.0342 \ MC & , MC \geq 12\%, \ U_{10} < 10 \text{ km/h} \\ 0.547 - 0.0228 \ MC & , MC \geq 12\%, \ U_{10} \geq 10 \text{ km/h} \end{cases}$$

MC is the dead fuel moisture content (% oven-dry weight basis).

$$MC = 9.58 - 0.205 \ T + 0.138 \ RH$$

T is the temperature (°C and RH is the relative humidity (%).

$$\phi C = \begin{cases} \frac{1.036}{1+103.99 exp(-0.0996(C-20))} & , C > 20\% \\ 0 & , C \leq 20 \end{cases}$$

C is the degree of grass curing (%).

Note that the original equation for ϕC proposed by Cheney et al. was modified by Cruz et al. as the curing level can be as low as 20% and the damping effects of the fuel in grassland is less.

Appendix A.4. Marsden–Smedley and Catchpole Buttongrass Model

Marsden–Smedly and Catchpole [50] described the fire behaviors for buttongrass moorlands in Tasmania for fire danger rating and fire behavior prediction. The mathematical equations for the models are explained as follows:

$$R = 0.678 U_2^{1.312} exp(-0.0243 MC)(1 - exp(-0.116 AGE))$$

whee U_2 is the wind speed (km/h) measured at a 2-m height, MC is the dead fuel moisture content (%), and AGE is the time since the last fire (years).

$$MC = exp(1.66 + 0.0214 RH - 0.0292 T_{dew})$$

T_{dew} is the dew point temperature (°C) and RH is the relative humidity (%).

Appendix A.5. Anderson et al. Heathland Model

Anderson et al. [51] used a dataset that covered a wider range of heathland and shrubland species to develop a model for healthland as an extension to the work [69]. The equations for the model are given as follows:

$$R = \begin{cases} [R_o + 0.2(5.67(5WF)^0.91 - R_o)U_{10}]H^{0.22} exp(-0.076 MC) & , U_{10} < 5 \\ 5.67(WFU_{10}^0.91)H^{0.22} exp(-0.076 MC) & , U_{10} \geq 5 \end{cases}$$

where R is the rate of spread (m/min) with vegetation height and without live fuel moisture content, U_{10} is the 10-m open wind speed (km/h), H is the average vegetation height (m), WF is the wind adjustment factor, and MC is the dead fuel moisture content (%).

$$MC = 4.37 + 0.161RH - 0.1(T-25) - \Delta 0.027RH$$

$\Delta = 1$ for sunny days from 12:00 to 17:00 from October to March and 0 elsewhere. T is the ambient air temperature ($°C$ and RH is the relative humidity (%).

References

1. National Inter-Agency Fire Center. Available online: https://www.nifc.gov/fireInfo/fireInfo_statistics.html (accessed on 12 May 2019).
2. Munich RE. Available online: https://www.munichre.com/australia/australia-natural-hazards/bushfires/economic-impacts/index.html (accessed on 15 May 2019).
3. North, M.; Stephens, S.; Collins, B.; Agee, J.; Aplet, G.; Franklin, J.; Fule, P.Z. Reform forest fire management. *Science* **2015**, *349*, 1280–1281. [CrossRef] [PubMed]
4. Whelan, R.J. *The Ecology of Fire*; Cambridge University Press: Cambridge, UK, 1995.
5. Ujjwal, K.; Garg, S.; Hilton, J.; Aryal, J. A cloud-based framework for sensitivity analysis of natural hazard models. *Environ. Model. Softw.* **2020**, *134*, 104800.
6. Sullivan, A.L. Wildland surface fire spread modelling, 1990–2007. 3: Simulation and mathematical analogue models. *Int. J. Wildland Fire* **2009**, *18*, 387–403. [CrossRef]
7. Grishin, A. Mathematical Modeling of Forest Fires and New Methods of Fighting Them, edited by FA Albini Publishing House of the Tomsk University. *Tomsk. Russ.* **1997**, *29*, 917–919.
8. Mell, W.; Jenkins, M.A.; Gould, J.; Cheney, P. A physics-based approach to modelling grassland fires. *Int. J. Wildland Fire* **2007**, *16*, 1–22. [CrossRef]
9. Morvan, D. Physical phenomena and length scales governing the behaviour of wildfires: a case for physical modelling. *Fire Technol.* **2011**, *47*, 437–460. [CrossRef]
10. Morvan, D.; Dupuy, J. Modeling the propagation of a wildfire through a Mediterranean shrub using a multiphase formulation. *Combust. Flame* **2004**, *138*, 199–210. [CrossRef]
11. Gould, J.S.; McCaw, W.; Cheney, N.; Ellis, P.; Knight, I.; Sullivan, A. *Project Vesta: Fire in Dry Eucalypt Forest: Fuel Structure, Fuel Dynamics and Fire Behaviour*; Csiro Publishing: Collingwood, Australia, 2008.
12. Tanskanen, H.; Granström, A.; Larjavaara, M.; Puttonen, P. Experimental fire behaviour in managed Pinus sylvestris and Picea abies stands of Finland. *Int. J. Wildland Fire* **2007**, *16*, 414–425. [CrossRef]
13. Baeza, M.; De Luıs, M.; Raventós, J.; Escarré, A. Factors influencing fire behaviour in shrublands of different stand ages and the implications for using prescribed burning to reduce wildfire risk. *J. Environ. Manag.* **2002**, *65*, 199–208. [CrossRef] [PubMed]
14. Miller, C.; Hilton, J.; Sullivan, A.; Prakash, M. SPARK—A bushfire spread prediction tool. In *Proceedings of the International Symposium on Environmental Software Systems, 2015*; Springer: Berlin/Heidelberg, Germany; 2015; pp. 262–271.
15. Finney, M.A. FARSITE: Fire Area Simulator-model development and evaluation. In *Res. Pap. RMRS-RP-4, Revised 2004*. Ogden, UT; US Department of Agriculture, Forest Service, Rocky Mountain Research Station: Fort Collins, CO, USA, 1998; Volume 4, 47p.
16. Rothermel, R.C. A mathematical model for predicting fire spread in wildland fuels. In *Res. Pap. INT-115*; US Department of Agriculture, Intermountain Forest and Range Experiment Station: Ogden, UT, USA, 1972; Volume 115, 40p.
17. Vakalis, D.; Sarimveis, H.; Kiranoudis, C.; Alexandridis, A.; Bafas, G. A GIS based operational system for wildland fire crisis management I. Mathematical modelling and simulation. *Appl. Math. Model.* **2004**, *28*, 389–410. [CrossRef]
18. Nahmias, J.; Téphany, H.; Duarte, J.; Letaconnoux, S. Fire spreading experiments on heterogeneous fuel beds. Applications of percolation theory. *Can. J. For. Res.* **2000**, *30*, 1318–1328. [CrossRef]
19. Méndez, V.; Llebot, J.E. Hyperbolic reaction-diffusion equations for a forest fire model. *Phys. Rev. E* **1997**, *56*, 6557. [CrossRef]
20. Noble, I.; Gill, A.; Bary, G. McArthur's fire-danger meters expressed as equations. *Aust. J. Ecol.* **1980**, *5*, 201–203. [CrossRef]
21. Trucchia, A.; Egorova, V.; Pagnini, G.; Rochoux, M.C. On the merits of sparse surrogates for global sensitivity analysis of multi-scale nonlinear problems: Application to turbulence and fire-spotting model in wildland fire simulators. *Commun. Nonlinear Sci. Numer. Simul.* **2019**, *73*, 120–145. [CrossRef]
22. Hilton, J.E.; Stephenson, A.G.; Huston, C.; Swedosh, W. Polynomial Chaos for sensitivity analysis in wildfire modelling. In Proceedings of the International Congress on Modelling and Simulation, Hobart, Australia, 3–8 December 2017.
23. Tolhurst, K.; Shields, B.; Chong, D. Phoenix: development and application of a bushfire risk management tool. *Aust. J. Emerg. Manag.* **2008**, *23*, 47.
24. Ujjwal, K.; Garg, S.; Hilton, J. An efficient framework for ensemble of natural disaster simulations as a service. *Geosci. Front.* **2020**, *11*, 1859–1873.
25. Riley, K.; Thompson, M.; Riley, K.; Webley, P.; Thompson, M. An uncertainty analysis of wildfire modeling. *Nat. Hazard Uncertain. Assess. Model. Decis. Support. Monogr.* **2017**, *223*, 193–213.

26. Kaschek, D.; Mader, W.; Fehling-Kaschek, M.; Rosenblatt, M.; Timmer, J. Dynamic modeling, parameter estimation and uncertainty analysis in R. *bioRxiv* **2016**, 085001
27. Nossent, J.; Elsen, P.; Bauwens, W. Sobol'sensitivity analysis of a complex environmental model. *Environ. Model. Softw.* **2011**, *26*, 1515–1525. [CrossRef]
28. Saltelli, A.; Ratto, M.; Andres, T.; Campolongo, F.; Cariboni, J.; Gatelli, D.; Saisana, M.; Tarantola, S. *Global Sensitivity Analysis: The Primer*; John Wiley & Sons: Hoboken, NJ, USA, 2008.
29. Cai, L.; He, H.S.; Liang, Y.; Wu, Z.; Huang, C. Analysis of the uncertainty of fuel model parameters in wildland fire modelling of a boreal forest in north-east China. *Int. J. Wildland Fire* **2019**, *28*, 205–215. [CrossRef]
30. Brohus, H.; Nielsen, P.V.; Petersen, A.J.; Sommerlund-Larsen, K. Sensitivity analysis of fire dynamics simulation. In Proceedings of the Roomvent 2007, Helsinki, Finland, 13–15 June 2007.
31. Salvador, R.; Pinol, J.; Tarantola, S.; Pla, E. Global sensitivity analysis and scale effects of a fire propagation model used over Mediterranean shrublands. *Ecol. Model.* **2001**, *136*, 175–189. [CrossRef]
32. Li, X.; Hadjisophocleous, G.; Sun, X.Q. Sensitivity and Uncertainty Analysis of a Fire Spread Model with Correlated Inputs. *Procedia Eng.* **2018**, *211*, 403–414. [CrossRef]
33. Yuan, X.; Liu, N.; Xie, X.; Viegas, D.X. Physical model of wildland fire spread: Parametric uncertainty analysis. *Combust. Flame* **2020**, *217*, 285–293. [CrossRef]
34. Marcot, B.G.; Thompson, M.P.; Runge, M.C.; Thompson, F.R.; McNulty, S.; Cleaves, D.; Tomosy, M.; Fisher, L.A.; Bliss, A. Recent advances in applying decision science to managing national forests. *For. Ecol. Manag.* **2012**, *285*, 123–132. [CrossRef]
35. Ujjwal, K.; Garg, S.; Hilton, J.; Aryal, J.; Forbes-Smith, N. Cloud Computing in natural hazard modeling systems: Current research trends and future directions. *Int. J. Disaster Risk Reduct.* **2019**, *38*, 101188.
36. Bass, B.; Bedient, P. Surrogate modeling of joint flood risk across coastal watersheds. *J. Hydrol.* **2018**, *558*, 159–173. [CrossRef]
37. Bermúdez, M.; Ntegeka, V.; Wolfs, V.; Willems, P. Development and comparison of two fast surrogate models for urban pluvial flood simulations. *Water Resour. Manag.* **2018**, *32*, 2801–2815. [CrossRef]
38. Kim, S.W.; Melby, J.A.; Nadal-Caraballo, N.C.; Ratcliff, J. A time-dependent surrogate model for storm surge prediction based on an artificial neural network using high-fidelity synthetic hurricane modeling. *Nat. Hazards* **2015**, *76*, 565–585. [CrossRef]
39. Jia, G.; Taflanidis, A.A. Kriging metamodeling for approximation of high-dimensional wave and surge responses in real-time storm/hurricane risk assessment. *Comput. Methods Appl. Mech. Eng.* **2013**, *261*, 24–38. [CrossRef]
40. Easum, J.A.; Nagar, J.; Werner, P.L.; Werner, D.H. Efficient multiobjective antenna optimization with tolerance analysis through the use of surrogate models. *IEEE Trans. Antennas Propag.* **2018**, *66*, 6706–6715. [CrossRef]
41. Sanchez, F.; Budinger, M.; Hazyuk, I. Dimensional analysis and surrogate models for the thermal modeling of Multiphysics systems. *Appl. Therm. Eng.* **2017**, *110*, 758–771. [CrossRef]
42. Meng, D.; Yang, S.; Zhang, Y.; Zhu, S.P. Structural reliability analysis and uncertainties-based collaborative design and optimization of turbine blades using surrogate model. *Fatigue Fract. Eng. Mater. Struct.* **2019**, *42*, 1219–1227. [CrossRef]
43. Ökten, G.; Liu, Y. Randomized quasi-Monte Carlo methods in global sensitivity analysis. *Reliab. Eng. Syst. Saf.* **2021**, *210*,107520. [CrossRef]
44. Sharples, J.J.; Bahri, M.F.; Huntley, S. A universal rate of spread index for Australian fuel types. In Proceedings of the International Conference on Forest Fire Research, Coimbra, Portugal, 10–16 November 2018;Available online: https://www.environment.gov.au/system/files/pages/5b3d2d31-2355-4b60-820c-e370572b2520/files/bioregions-new.pdf (accessed on 12 March 2020).
45. IBRA7. Available online: http://www.environment.gov.au/system/files/pages/5b3d2d31-2355-4b60-820c-e370572b2520/files/bioregions-new.pdf (accessed on 12 March 2020).
46. Tasmania's Bioregions. Available online: https://dpipwe.tas.gov.au/conservation/flora-of-tasmania/tasmanias-wetlands (accessed on 12 April 2021).
47. Tasmanian Department of Primary Industries, Parks, W.; Monitoring, E.T.V.; Program, M. TasVeg 3.0. 2013. Available online: https://www.threatenedspecieslink.tas.gov.au/Pages/tasveg-3.aspx (accessed on 12 December 2020).
48. McArthur, A.G. Fire Behaviour in Eucalypt Forests Forestry and Timber Bureau. Canberra, 1967. Available online: https://catalogue.nla.gov.au/Record/2275488 (accessed on 12 March 2021).
49. Cheney, N.P.; Gould, J.S.; McCaw, W.L.; Anderson, W.R. Predicting fire behaviour in dry eucalypt forest in southern Australia. *For. Ecol. Manag.* **2012**, *280*, 120–131. [CrossRef]
50. Marsden-Smedley, J.; Catchpole, W.R. Fire behaviour modelling in Tasmanian buttongrass moorlands. II. Fire behaviour. *Int. J. Wildland Fire* **1995**, *5*, 215–228. [CrossRef]
51. Anderson, W.R.; Cruz, M.G.; Fernandes, P.M.; McCaw, L.; Vega, J.A.; Bradstock, R.A.; Fogarty, L.; Gould, J.; McCarthy, G.; Marsden-Smedley, J.B.; et al. A generic, empirical-based model for predicting rate of fire spread in shrublands. *Int. J. Wildland Fire* **2015**, *24*, 443–460. [CrossRef]
52. Cheney, N.; Gould, J.; Catchpole, W.R. Prediction of fire spread in grasslands. *Int. J. Wildland Fire* **1998**, *8*, 1–13. [CrossRef]
53. List Data. https://listdata.thelist.tas.gov.au/opendata/ (accessed on 12 March 2021).
54. Olson, J.S. Energy storage and the balance of producers and decomposers in ecological systems. *Ecology* **1963**, *44*, 322–331. [CrossRef]
55. Birk, E.M.; Simpson, R. Steady state and the continuous input model of litter accumulation and decompostion in Australian eucalypt forests. *Ecology* **1980**, *61*, 481–485. [CrossRef]

56. *Fire Prediction Services, PHOENIXRapidFire: Technical Reference Guide*; A Technical Guide to the PHOENIX RapidFire Bushfire Characterisation ModelVersion 4; Australasian Fire and Emergency Service Authorities Council: Melbourne, Victoria, 2019; Volume 1.
57. Abatzoglou, J.T.; Hatchett, B.J.; Fox-Hughes, P.; Gershunov, A.; Nauslar, N.J. Global climatology of synoptically-forced downslope winds. *Int. J. Climatol.* **2021**, *41*, 31–50. [CrossRef]
58. Fox-Hughes, P. A fire danger climatology for Tasmania. *Aust. Meteorol. Mag.* **2008**, *57*, 109–120.
59. KC, U.; Garg, S.; Hilton, J.; Aryal, J. *Fire Simulation Data Set for Tasmania*; CSIRO: Hobart, Australia; University of Tasmania: Hobart, Australia, 2021; [CrossRef]
60. Hamby, D. A review of techniques for parameter sensitivity analysis of environmental models. *Environ. Monit. Assess.* **1994**, *32*, 135–154. [CrossRef] [PubMed]
61. Wagner, C.V. A simple fire-growth model. *For. Chron.* **1969**, *45*, 103–104. [CrossRef]
62. Levenberg, K. A method for the solution of certain non-linear problems in least squares. *Q. Appl. Math.* **1944**, *2*, 164–168. [CrossRef]
63. Kohavi, R. A study of cross-validation and bootstrap for accuracy estimation and model selection. *InIjcai* **1995**, *14*, 1137–1145.
64. Benesty, J.; Chen, J.; Huang, Y.; Cohen, I. Pearson correlation coefficient. In *Noise Reduction in Speech Processing*; Springer: Berlin/Heidelberg, Germany, 2009; pp. 1–4.
65. Penman, T.D.; Ababei, D.A.; Cawson, J.G.; Cirulis, B.A.; Duff, T.J.; Swedosh, W.; Hilton, J.E. Effect of weather forecast errors on fire growth model projections. *Int. J. Wildland Fire* **2020**, *29*, 983–994. [CrossRef]
66. Gould, J.S.; McCaw, W.L.; Cheney, N.P. Quantifying fine fuel dynamics and structure in dry eucalypt forest (Eucalyptus marginata) in Western Australia for fire management. *For. Ecol. Manag.* **2011**, *262*, 531–546. [CrossRef]
67. Gould, J.S.; McCaw, W.; Cheney, N.; Ellis, P.; Matthews, S. *Field Guide: Fire in Dry Eucalypt Forest: Fuel Assessment and Fire Behaviour Prediction in Dry Eucalypt Forest*; CSIRO Publishing: Collingwood, Australia, 2008.
68. Matthews, S.; Gould, J.; McCaw, L. Simple models for predicting dead fuel moisture in eucalyptus forests. *Int. J. Wildland Fire* **2010**, *19*, 459–467. [CrossRef]
69. Catchpole, W.; Bradstock, R.; Choate, J.; Fogarty, L.; Gellie, N.; McCarthy, G.; McCaw, W.; Marsden-Smedley, J.; Pearce, G. Cooperative development of equations for heathland fire behaviour. In Proceedings of the 3rd International Conference on Forest Fire Research and 14th Conference on Fire and Forest Meteorology, Coimbra, Portugal, 16–20 November 1998; Volume 2, pp. 16–20.

Article

Predicting Fire Propagation across Heterogeneous Landscapes Using WyoFire: A Monte Carlo-Driven Wildfire Model

Cory W. Ott [1], Bishrant Adhikari [2], Simon P. Alexander [3], Paddington Hodza [4], Chen Xu [4] and Thomas A. Minckley [1,*]

1. Department of Geology & Geophysics, University of Wyoming, Laramie, WY 82071, USA; ottcory3@gmail.com
2. Texas A & M Forest Service, College Station, TX 77845, USA; bishrnt@gmail.com
3. Advanced Research Computing Center, University of Wyoming Laramie, WY 82071, USA; Simon.Alexander@uwyo.edu
4. Wyoming Geographic Information Science Center, University of Wyoming, Laramie, WY 82071, USA; phodza@uwyo.edu (P.H.); cxu3@uwyo.edu (C.X.)
* Correspondence: Minckley@uwyo.edu

Received: 29 October 2020; Accepted: 8 December 2020; Published: 11 December 2020

Abstract: The scope of wildfires over the previous decade has brought these natural hazards to the forefront of risk management. Wildfires threaten human health, safety, and property, and there is a need for comprehensive and readily usable wildfire simulation platforms that can be applied effectively by wildfire experts to help preserve physical infrastructure, biodiversity, and landscape integrity. Evaluating such platforms is important, particularly in determining the platforms' reliability in forecasting the spatiotemporal trajectories of wildfire events. This study evaluated the predictive performance of a wildfire simulation platform that implements a Monte Carlo-based wildfire model called WyoFire. WyoFire was used to predict the growth of 10 wildfires that occurred in Wyoming, USA, in 2017 and 2019. The predictive quality of this model was determined by comparing disagreement and agreement areas between the observed and simulated wildfire boundaries. Overestimation–underestimation was greatest in grassland fires (>32) and lowest in mixed-forest, woodland, and shrub-steppe fires (<−2.5). Spatial and statistical analyses of observed and predicted fire perimeters were conducted to measure the accuracy of the predicated outputs. The results indicate that simulations of wildfires that occurred in shrubland- and grassland-dominated environments had the tendency to over-predict, while simulations of fires that took place within forested and woodland-dominated environments displayed the tendency to under-predict.

Keywords: wildfire; predictive modeling; fire spread model; Monte Carlo; spatial modeling; area difference index; statistics; precision; recall; principal components analysis

1. Introduction

Wildfires have increased in size, frequency, and severity in the past decades as global temperatures have continued to warm, leading to an elevated concern about the health and safety of individuals who inhabit areas prone to wildfire activity [1–4]. Climatic changes have triggered ecosystem alterations in the form of significant vegetation shifts, which have ultimately led to more acreage being burned by wildfires [1,5]. The effects of changes in wildland fire regimes have warranted the development of dynamic wildfire propagation models amongst scientific modelling communities [6].

Wildfire modelling has evolved from the initial deterministic fire models based on the fundamental equations proposed by Rothermel [7]. These models typically generate empirical results that do not account for variability in the input measurements. Wildfires are impelled by dynamic variations in weather and fuel conditions that can produce a chain reaction in local environmental conditions such as fuel moisture, vapour deficits, and wind patterns. Deterministic fire models can use snapshot observations as inputs, and these data are statically accurate but limited to the moments of the observations. Moving from a "static" to a dynamic wildfire modelling environment can follow two solution paths. The first path is to automate the collection of observational data that could be used to initialize the wildfire spread models. This reduces the time required to gather the required datasets and initiate a model run. Rapid initialization is desirable when fire models are used as risk assessment tools during active fires, but model run times might limit the utility of the results as real-world conditions change [8]. The second path is for models that operate in near real time. Near-real-time modelling is a computational approach, which dynamically captures new observations (such as remotely sensed fuel moisture, wind direction, and temperature) while simultaneously forward-projecting results between input data updates. If a model is run during an actual fire event the output should represent a close approximation of fire trajectory based on current conditions and potential behaviour until the next input data update. Regardless of the modelling pathway, the approach used for any model needs to be validated to make it useful in the field or to explore simulated fire behaviour.

While recent increases in model development have provided wildfire scientists with multiple tools for fire research, there has been limited consistent use of appropriate evaluation procedures and performance metrics to effectively quantify the performance of spatially explicit fire spread models [6]. Despite the availability of a framework that can account for varying levels of stochasticity present within meteorological data, fuel-bed conditions, and the overall burnable environment, it remains challenging to predict the propagation of wildfire in near real time due to each event being so unique and transient [6]. The utilization of an effective evaluation process to assess predictive performance requires comprehensive knowledge of the specific model type used [9]. The most effective method for assessing model accuracy and reliability is to test the level of agreement between simulated and observed wildfire perimeters [10]. In order to accomplish this, a set of performance metrics deemed most appropriate for determining the accuracy of model predictions were implemented to quantify performance.

The purpose of this research is to evaluate the predictive performance of a wildfire simulation model called WyoFire, developed at the University of Wyoming as part of a risk assessment tool within the architecture of the Wyoming Wildfire Risk Portal (https://wywrap.wyo.gov/app.html) [8,11]. WyoFire employs a probabilistic approach by implementing a Monte Carlo-driven structure using Gaussian distributions of meteorological and fuel moisture data to account for stochasticity within environments possessing a diverse range of characteristics [8,11,12]. The results of each model run yield a potential or predicted perimeter for individual wildfire simulation that can then be validated against observed fire perimeters. WyoFire overlays the predicted wildfire perimeters from each Monte Carlo simulation and counts the number of times a specific area is predicted as burned to estimate the probability of wildfire front spreading over the study area [8,11]. WyoFire employs a probabilistic approach for generating a range of inputs centred on the observed weather and other environmental data. The input data are variated in line with the degree of randomness apparent in environmental datasets. WyoFire is able to simulate crown fire spread to a certain extent by checking crown fire spread potential for each pixel and the availability of appropriate fuel load [8,11]. WyoFire is able to utilize the updated weather and fuel information during the model execution [8]. However, for this study all the weather and fuel data were downloaded in advance to minimize their effect on the model execution time.

To better understand the strengths and weaknesses of the WyoFire model and assess the predictive accuracy of individual wildfire simulations, we conducted an evaluation that quantifies levels of predictive performance within multiple burnable environments based on Wyoming wildfires that occurred in 2017 and 2019. Our goal was to determine the extent to which WyoFire accurately represents the natural world. We hypothesized that the variance in the predictive performance of the model was the same among different environments based on fuel loading model and terrain complexity.

We validated the predictive performance of WyoFire based on statistical indices first utilized by Adhikari et al. [8] throughout the initial developmental phase of our wildfire simulation system.

2. Materials and Methods

2.1. Overview

Wyoming has many favorable characteristics for studying wildfire, including a wide variety of topographies, vegetation types-steppe to alpine, and low population densities so fire can propagate naturally in many cases. We studied nine and one wildfire that occurred in Wyoming and Montana, respectively during the 2017 and 2019 fire seasons for simulation and analysis (Figure 1). The Montana wildfire occurred within a 25-mile buffer outside of the Wyoming border was included because of its potential to cross into Wyoming. The different wildfire events were simulated within their respective environments composed of unique assemblages of vegetation types, degrees of terrain complexity, fuel loadings, and meteorological conditions (i.e., burnable environment). We used the following performance metrics: Overestimation, Underestimation, Intersection, Area Difference Index (ADI), Area Difference Index for Overestimation (ADI_{oe}), Area Difference Index for Underestimation (ADI_{ue}), F1 Score, Precision, and Recall. Duff, Chong, and Tolhurst [6] concluded that ADI, ADI_{oe}, and ADI_{ue} are the performance indices best suited to assess and portray the specific types of modelling error (e.g., Overestimation or Underestimation). The procedural structure for our performance evaluation was adapted from Duff, Chong, and Tolhurst [6], as their study serves as the foundation for research involving what we considered to be the most widely accepted process for the evaluation of wildfire simulation models.

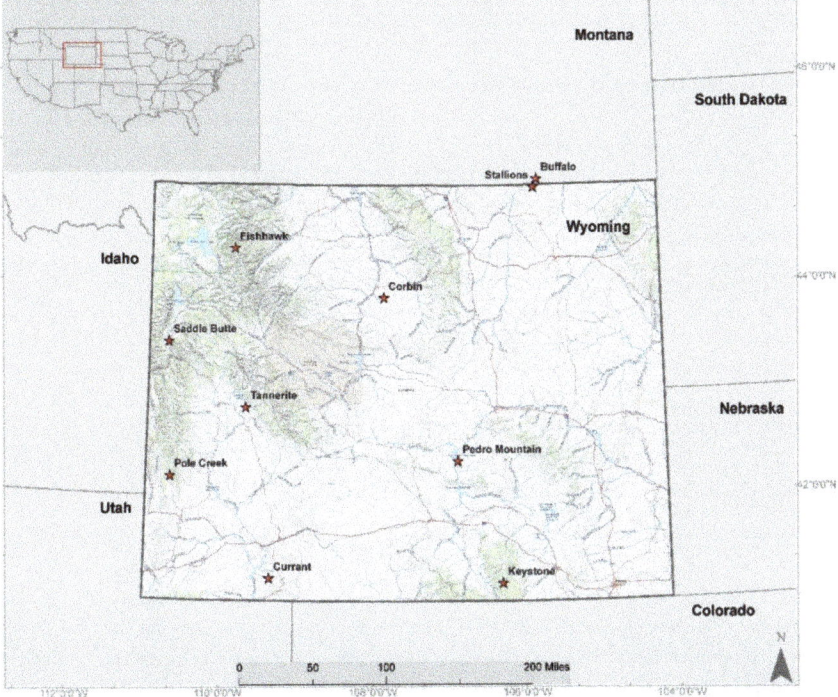

Figure 1. Distribution of 2017 and 2019 Wyoming and Montana fire events used in this study.

2.2. Model Description

WyoFire was developed by Adhikari et al. [8] using Python programming language. The model employs mathematical functions developed by Rothermel [7], Wagner [13], and Finney [14] to create elliptical wildfire propagation across different landscapes [8]. WyoFire accounts for natural stochasticity of independent variables within the burnable environment using Gaussian distributions for (1) fuel moisture and (2) High-Resolution Rapid-Refresh (HRRR) meteorological forecast datasets (i.e., relative humidity, temperature, wind direction, and wind speed), which are automatically created by the Monte Carlo structure. Gaussian distributions are then used for model runs to simulate wildfire propagation and estimate natural stochasticity inherent in environmental datasets. The wildfire simulation uses random points of fuel moisture and meteorological HRRR data from the generated Gaussian distributions. WyoFire employs mathematical functions for wildfire spread developed by Rothermel [7], Wagner [13], and Finney [14] to achieve elliptical wildfire propagation across a given landscape [8]. By applying the Huygens Wavelet principle [15], the model created ellipses around each ignition point at the end of each iteration. The ellipses define the extent of each fire propagation which was then buffered by a convex hull using a minimum bounding geometry function [8]. Ignition points can be randomly generated to initiate wildfire propagation; however, we use polygons rendered from observed VIIRS and MODIS hot spot data to identify fire origination of the 10 wildfires used in this study. Ignition points are generated along the active flaming perimeter of the original ignition polygon.

2.3. Data

Table 1 lists the datasets for the wildfire simulations performed in this study. Existing Vegetation Type and Fuel Loading Model datasets were acquired from the United States Geological Survey (USGS) LANDFIRE database. For 2019 wildfire simulations, two datasets were downloaded daily using Python scripts that were scheduled to run automatically using cron. These datasets consisted of HRRR meteorological forecast data from the National Oceanic and Atmospheric Administration (NOAA) and fuel moisture data from the Wyoming State Forestry Division (WSFD). Downloaded HRRR datasets were coded to only index four meteorological variables of wind direction, wind speed, relative humidity, and temperature. The 2017 wildfire datasets consisted of previously archived HRRR raster data accessed from archives at the University of Utah. Previously archived fuel moisture datasets were also integrated into this study to replicate the simulation environments for the 2017 wildfires as previously performed by Adhikari et al. [8].

Table 1. Metadata of datasets used to simulate wildfire events, adapted from Adhikari et al. [8].

Dataset	Source	Data Volume (Gigabyte/Iteration)	Resolution Spatial (m)	Temporal (h)
High-Resolution Rapid Refresh (HRRR)	NOAA	11.5	3000	1
VIIRS Active Fire Hotspot	NASA	0.02	375	12
Dead Fuel Moisture	WSFD/USFS	1	2500	24
Digital Elevation Model	USGS	26	10	Updated: January 2017
Vegetation and Fuels	LANDFIRE	3.5	30	Updated: January 2014
Historical Wildfire Perimeter (Observed)	USGS GeoMAC	0.1	N/A	24
Historical HRRR Data	Uni. Of Utah	9	3000	24

Observed wildfire perimeters were obtained from the Geospatial Multi-Agency Coordination (GeoMAC) data archives, maintained by the USGS. Observed perimeters were used as control layers to evaluate predictive model performance against the resulting simulated perimeters. Although the new on-site measurements of wildfire perimeter and weather conditions might provide more exact representation of the simulation environment, they were not included in this study as these simulations were performed on the wildfires that already occurred and all the required datasets were downloaded as well as processed beforehand by the python script. For this study, the term *burnable environment* is defined as the combination of existing vegetation type, dominant fuel loading model, and mean level of terrain complexity within an observed wildfire perimeter. The terrain complexity index value for each wildfire was calculated by running Slope and Focal Statistics functions on Digital Elevation Model data in ArcMap. Time and date stamps attributed to the observed fire perimeters represented in spatial data shapefiles did not always align with the initial time of ignition and propagation of each fire, thus requiring the use of supplemental observed datasets in the form of active hot-spot point data from the VIIRS fire detection satellites. Shapefiles of active hot-spot point data were obtained from the Visible Infrared Imaging Radiometer Suite (VIIRS) data archive published by NASA and were used to interpolate active perimeters that were not available within the GeoMAC database for the date and time of the simulation.

2.4. Wildfire Simulation

Two parameters are coded into the simulation configuration module that can be manually adjusted by individuals operating the model: (1) centroid distances (CD) of generated ellipses surrounding ignition points along the propagative front and (2) time step values in minutes for each iteration completed within a given Monte Carlo run. CD can be defined as the radius of each elliptical polygon generated from individual ignition points established along the flaming front. All simulations were configured to sixty-minute time steps, which is equal to one iteration of one Monte Carlo simulation. For this study, active wildfire perimeters were simulated from fire origination to ~eighteen hours due to availability constraints of the HRRR datasets and availability of their observed perimeter data (Table 2).

Table 2. Simulation parameters for the ten wildfires analyzed within this study.

Wildfire	Total Size (Acres)	Simulation Duration (h)	Observed Perimeter Source	Total Simulations (n)
Keystone, 2017	1102	13	VIIRS	600
Pole Creek, 2017	2139	12	GeoMAC and VIIRS	600
Buffalo, 2017	3515	12	GeoMAC and VIIRS	600
Stallions, 2017	1111	12	VIIRS	600
Tannerite, 2019	1349	18	GeoMAC	600
Pedro Mountain, 2019	9388	18	GeoMAC and VIIRS	600
Currant, 2019	381	12	GeoMAC	600
Corbin, 2019	164	8	GeoMAC	600
Fishhawk, 2019	2359	18	GeoMAC and VIIRS	600
Saddle Butte, 2019	252	12	GeoMAC	600
			Total Simulations	6000

An idealized analysis was conducted to train the model and identify which CD value yielded the best performance results across all simulation environments [15]. A direct relationship between CD, dominant fuel loading model, and predictive accuracy was observed throughout this study. As CD increases, a subsequent decrease in predictive accuracy will occur in simulations of wildfires burning in higher fuel loads dominated by canopy fuels. In contrast, predictive accuracy increased when the CD was decreased for simulations of wildfires occurring within those higher canopy fuel loads. Through an iterative analysis, a mean CD of 5 m was identified as optimal across all fires and was used to achieve reported results for the rest of this study.

Simulations were run in Coordinated Universal Time (UTC) to align with the HRRR data format. For terminology purposes, one simulation predicts fire spread for the next x-number of hours. Intermediate or iterative predictions are generated on an hourly basis. Therefore, within a single simulation, there are x-number of iterations. One simulation consists of a simultaneous run of y-number of sample model configurations. Each model execution composed of a unique sample configuration is the equivalent to one Monte Carlo run. Simulated perimeters were evaluated against concurrent observed perimeters to assess variation in model performance for each target fire event. Performance of the WyoFire model was tested across a range of existing vegetation types, terrain complexity values, and fuel loading models in order to identify variables within each burnable environment that induce the greatest variance in the predictive performance of the model.

2.5. Assessing Model Performance

Simulation code was run across the High Performance Computing cluster (Teton), managed by the University of Wyoming's Advanced Research Computing Center. Teton allowed multiple independent simulations to be run concurrently, distributed across multiple nodes, as well as scaling up the number of individual Monte Carlo simulations that could be run in parallel on individual nodes. Where a standard desktop could run 10 simulation queued sequentially, utilizing 4 or 8 cores at one time, the cluster enabled 10 (or more) simulations to be run concurrently, with up to 32 cores (i.e., 32 Monte Carlo simulations) running simultaneously on each node. Utilization of the cluster enabled both vertical (time to run a simulation) and horizontal (number of concurrent simulations) scaling, reducing the computational time from days down to hours. We used simultaneous batch processing of numerous wildfire simulations. Following the conclusion of all wildfire simulation jobs, logged results were transferred from the Linux system to a single-node workstation. Statistical and spatial analysis scripts were written in R-Studio. All data were graphed using R-Studio factoextra and ggplot2 packages. Simulated perimeter data were then analyzed to assess predictive performance by employing a series of spatial and statistical analyses using the aforementioned scripts. In order to calculate performance indices, a spatial intersection was conducted first to identify the critical areas of model Overestimation, Underestimation, and Intersection (Figure 2). The final burned area prediction was calculated using spatial overlay of all predicted wildfire perimeters obtained from each individual Monte Carlo simulation. The multipart polygons and individual polygons that were less than 1sq. meters were removed to generate the final predicted wildfire perimeter. The areas that were predicted to be burned only once were not included in the final perimeter. The results for each simulation were not compared amongst each other in this study due to the inherent randomness of the input weather and fuel conditions generated by the simulation. The resulting area of each predictive zone from the spatial intersection can then be integrated into a series of algebraic formulas to calculate performance indices that are indicative of overall model performance, e.g., Area Difference Index (ADI), Precision, Recall, and F1 Score.

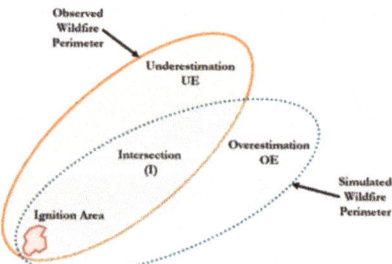

Figure 2. Graphical representation of the observed and simulated wildfire perimeters, as well as the three predictive areas, underprediction, intersection and overprediction, that are a product of each simulation, adapted from Duff, Chong, and Tolhurst [6].

ADI uses an index of incorrect estimation as a ratio of the correctly predicted area of intersection between the simulated and observed wildfire perimeters [6,8,16,17]. All performance indices calculations were conducted in R-Studio. *ADI* is calculated as:

$$ADI\ (t) = (OE\ (t) + UE\ (t))/(I\ (t)) \tag{1}$$

ADI can also be decomposed into partial metrics which attempt to explain whether the source of the modelling error is a result of net Overestimation or Underestimation being the *ADI* of Overestimation

(ADI_{oe}) and the ADI of Underestimation (ADI_{ue}) [6,16]. The partial indices of ADI_{oe} and ADI_{ue} are calculated as:

$$ADIu_e = UE(t)/(I(t)) \quad (2)$$

$$ADIo_e = OE(t)/(I(t)) \quad (3)$$

Precision and Recall are also considered in this study as partial metrics that combine to compose the F1 Score statistic [6,16]. Precision is functionally a measure of over-prediction, and Recall is functionally a measure of under-prediction [6,16]. Precision and Recall are calculated as:

$$Precision = I(t)/I(t) + OE(t) \quad (4)$$

$$Recall = I(t)/I(t) + UE(t) \quad (5)$$

F1 Score is a measure of the overall state of agreement between the over- and underestimation of each simulated perimeter and is functionally equivalent to Sorensen's Familiarity Index, which has been applied in pattern research but not in the discipline of fire science [6,8,16]. F1 Score functions as an evaluative index that essentially combines Precision and Recall values to assess the overall level of predictive agreement between simulated and observed perimeters [6,8,16]. F1 Score is calculated as:

$$F1 = (2 * I(t))/(I(t) + UE(t) + I(t) + OE(t)) \quad (6)$$

Applying an appropriate combination of evaluative indices to assess the predictive performance of the model provides a foundation of computational results to conduct further analyses. Adhikari et al. [8] evaluated three 2017 wildfire events: (1) Keystone, (2) Pole Creek, and (3) Buffalo fires. Here, we apply a paralleled evaluation approach to an additional seven wildfires to increase the sample size and range in diversity of burnable environments tested. Computational results were analyzed in congruence with empirically derived modelling results, e.g., the morphology of predicted perimeters within a GIS platform to determine whether the accuracy of the fire spread model is sufficient for the application of wildfire education. Single-value performance metrics are useful when conducting rapid assessments of model performance for a particular simulation event, but they do not provide a sufficient level of detail regarding the sources of error within each simulation [6,18].

2.6. Principle Components Analysis

A Principal Components Analysis (PCA) was conducted on the simulation results to identify and understand the particular variables that might have induced the most variance on model performance and to identify emergent environmental properties of modeled fire events. Bi-plot visualizations were rendered using the factoextra package within R-Studio.

3. Results

3.1. Statistical Performance

We implemented a series of single-value performance metrics to evaluate how well WyoFire simulations performed across a range of landscapes (Figure 3). Each metric is a unitless index that represent specific facets of simulative model performance and can be formulated to determine rates of Overestimation, Underestimation, and Intersection for each series of the wildfire simulations. Mean performance indices for all simulations are found in Table 3. Performance outcomes for all simulations varied considerably between the 10 wildfire events. The Area Difference Index was designed as a simple metric to describe wildfire model performance, while the closer a value is to being equal to one suggests a less predictive error in simulation results in contrast to values much greater than one, which suggests more significant amounts of predictive modelling error [6,16]. Variables likely driving model over- or under-prediction in different environments are shown in Table 3.

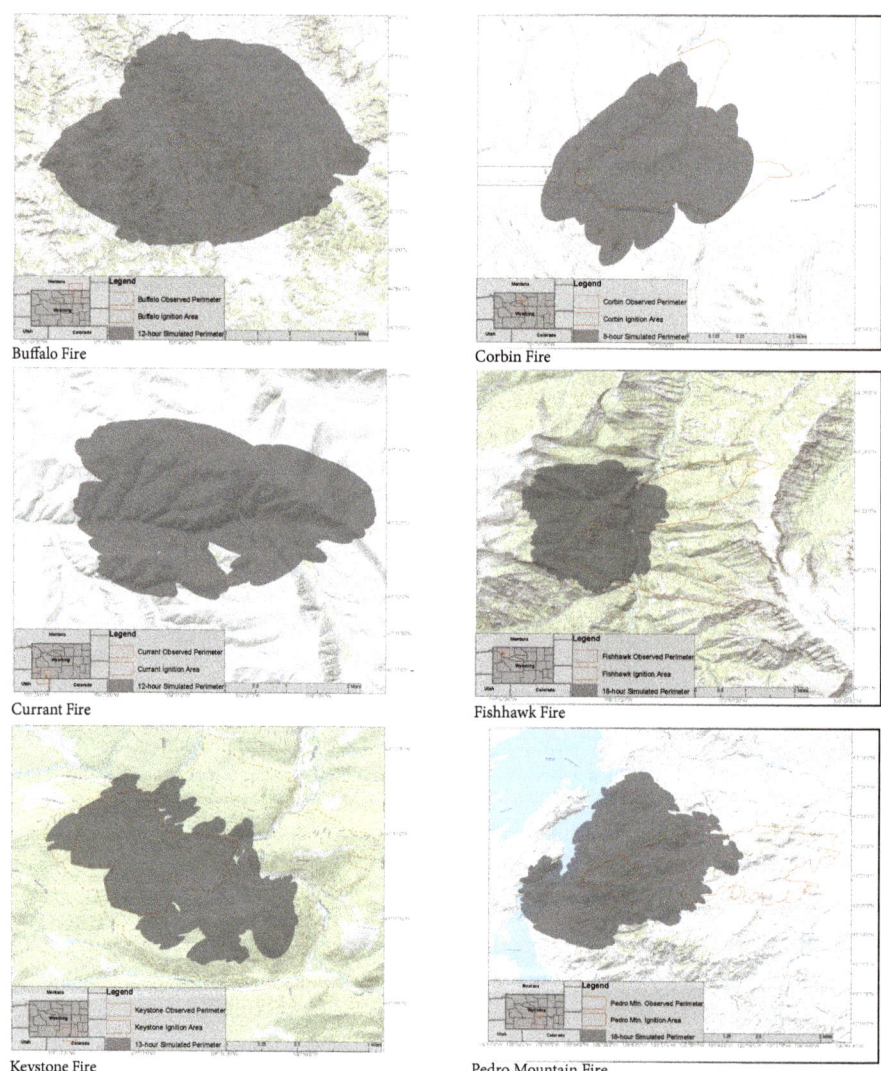

Figure 3. *Cont.*

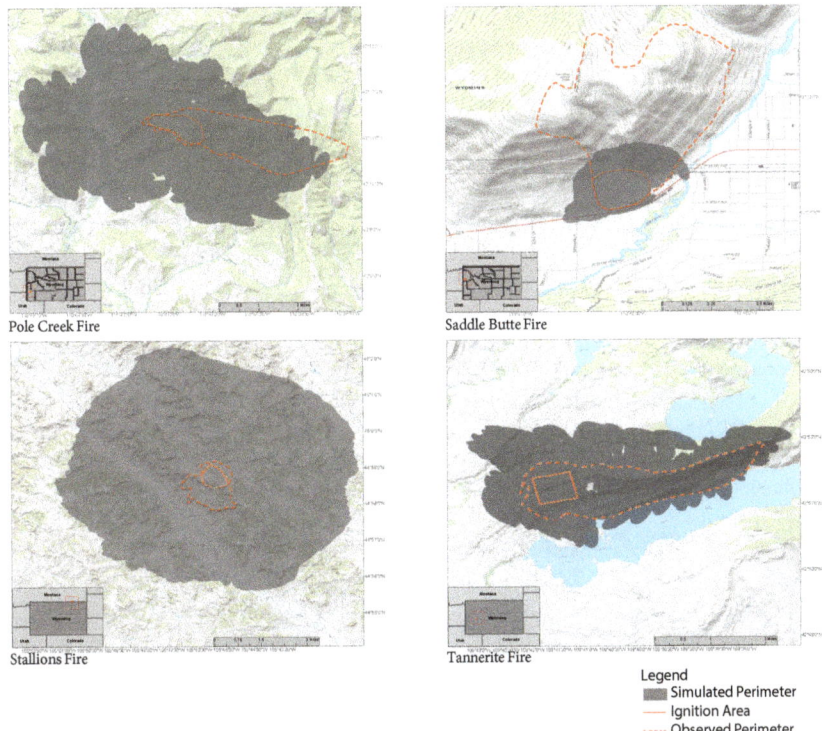

Figure 3. Maps of showing ignition polygon, fire perimeter (actual) and 18-h simulated fire perimeter for each of 10 fires modelled in this study.

Table 3. Mean performance index values derived from all wildfire simulations based on the optimal Centroid Distance of five meters.

Wildfire	Existing Vegetation Type(s)	Terrain Roughness	ADI_{oe}–ADI_{ue}	Fuel Loading
Currant	Grassland, Shrubland	3	37.940	Med–High
Stallions	Northwestern Great Plains Grassland, Ponderosa-Pine Forest	6	32.430	Med–High
Buffalo	Northwestern Great Plains Grassland, Sage Brush Steppe	3	8.479	Low–Med
Tannerite	Sage Brush Steppe, Montane–Foothill–Valley Grassland	5	1.481	Low–High
Pole Creek	Spruce-Fir Woodland, Aspen and Mixed-Conifers	7	1.031	Low–Med
Pedro Mountain	Sage Brush Shrubland, Limber Pine-Juniper Forest	6	−0.093	Low–Med
Keystone	Lodgepole Forest, Spruce-Fire Woodland, Aspen and Mixed Conifers	1	−1.733	Low–Med
Fishhawk	Subalpine Woodland, Spruce-Fir Woodland, Douglas-Fir Forest	5	−2.450	Low
Saddle Butte	Sage Brush Shrubland, Sage Brush Steppe, Montane Meadow	4	−2.789	Low
Corbin	Sage Brush Steppe, Sage Brush Shrubland, Semi-Desert Shrub Steppe	2	−4.248	Low

Simulation of wildfire events occurring in environments with medium-to-high total fuel loadings dominated by shrubland and grassland vegetation types, such as the Currant, Stallions, and Buffalo fires, produced the highest rates of overestimation–underestimation (ADI_{oe}–ADI_{ue}) (Table 3). In contrast, simulations of wildfire events occurring in environments with lower fuel loadings, dominated by mixed-forest, woodland, and shrub-steppe vegetation types, such as the Fishhawk, Saddle Butte, and Corbin fires, yielded the lowest rates of overestimation–underestimation. Wildfires that lie in mixed fuel types, Tannerite, Pole Creek, Pedro Mountain, and Keystone fires, displayed the most balanced performance in terms of overestimation and underestimation rates.

3.2. Principle Components Analysis

PCA was conducted on fuel characteristics and balance of predictive performance indices within each respective burnable environment. The first two principle components account for 78 percent of the total variance in model performance (Figure 4). The position of each fire relative to one another in the Bi-Plot (Figure 4) indicates the relative similarity of the models' performance in predicting actual wildfire perimeters. Examining the burnable environment within each group also helps reveal what vegetation conditions yield over- and under-predictions by WyoFire. PCA yielded three distinct groups of wildfire events (Figure 4). Group 1 consists of the Corbin, Saddle Butte, Keystone, and Fishhawk fires, which can be observed in Quadrant IV of the Principle Components Analysis Bi-Plot. Group 2 is composed of the Currant, Stallions, and Buffalo fires, which can be seen in Quadrants II and III. Lastly, Group 3 consists of the Pole Creek, Tannerite, and Pedro Mountain fires, which can be found in Quadrant I.

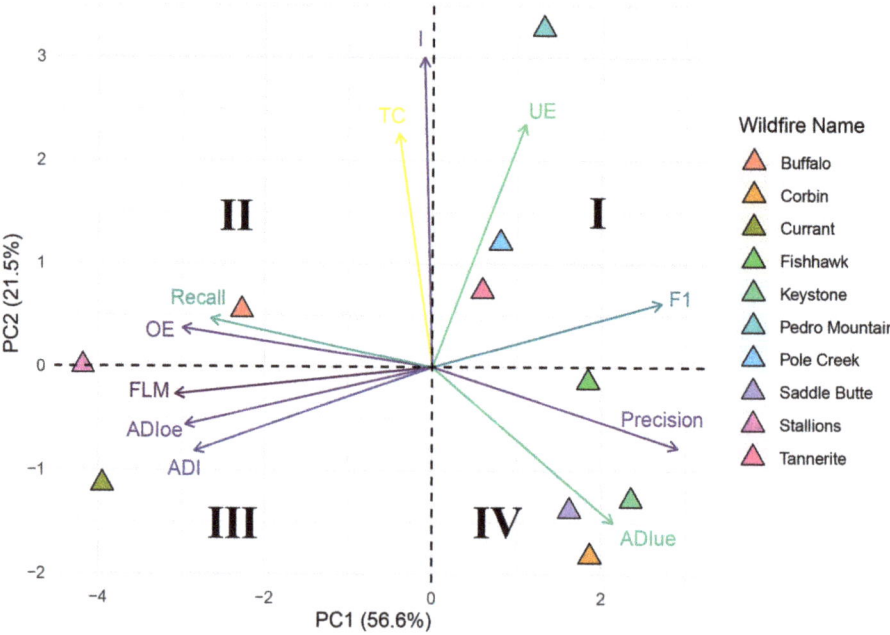

Figure 4. Principal Components Analysis (PCA) Bi-Plot of variable weightings and coordinate locations of individual wildfire events. The events are color-coded and symbolized by triangles within the Bi-Plot. FLM = Fuel Loading Model, TC = Terrain Complexity, ADI = Area Difference Index, OE = Overestimation, UE = Underestimation, ADI_{oe} = ADI of Overestimation, ADI_{ue} = ADI of Underestimation. These variables are represented by individual arrow vectors of varying lengths and colors, in which a lengthier vector signifies that the variable is better represented within the analysis and a darker color fill indicates a better Quality of Representation ($\cos^2$).

The first Principal Component (PC-1) explains 56.6 percent of the variance in model performance across all simulations. Under-prediction indices were prominent along this axis, showing how the model performed within environments containing forested elements. Fuel loading became an emergent variable along this axis, while each group of individuals is ordered along the continuum by existing vegetation type and fuel-bed characteristics. The Fishhawk and Stallions fire simulations form the end members along the PC-1 axis. PC-1 appears to be primarily described by the existing vegetation type and dominant fuel loading models present within each respective burnable environment. WyoFire tended to under-predict in situations where the fuel load was low to medium with poor fuel-bed continuity. The Fishhawk Fire was characterized by low surface fuel loads but higher canopy fuel loads, as it occurred primarily in Rocky Mountain Subalpine Forest and Woodland vegetation types with a low total fuel loading. The Stallions Fire was primarily in North Western Great Plains Mixed Grass Prairie and Inter-Mountain Basin Big Sage Brush Steppe communities, possessing a low total fuel load and poor fuel-bed continuity.

Performance results for Fishhawk fire simulations displayed a much higher rate of underestimation than overestimation, while results for Stallions simulations show a relatively higher rate of overestimation than underestimation. The Fishhawk fire simulations had a greater rate of under-prediction in congruence with its high ratio of the canopy to surface fuels, further suggesting that this model may struggle to accurately model transitions of a flaming front propagating from the surface to canopy fuel types. Both wildfires burned across landscapes with average terrain complexity, as this variable did not prove to have a significant effect on the outcomes for these fire simulations.

The second Principle Component (PC-2) explains 21.5 percent of the variance in the predictive performance of the wildfire simulations. Certain burnable environments for wildfire simulations that possessed greater terrain complexity, meaning that the landscape has a greater degree of localized elevational variance, displayed overestimation rates similar to those with relatively low levels of terrain complexity, such as Pedro Mountain (6) and Corbin (2) fire simulations. Overestimation indices are pointed toward the Stallions, Currant, and Buffalo fire simulations as a result of simulating wildfire in higher fuel load with increased continuity. The Pedro Mountain and Corbin fires burned through similar environments dominated by Inter Mountain Basin Big Sage Brush Steppes and *Artemisia tridentata ssp. vaseyana* Shrubland Alliances, yet these two fires appear as opposing end members along the PC-2 axis. The significant difference between these two burnable environments is the influence of grassland vegetation present throughout the Corbin Fire but not the Pedro Mountain Fire.

The Pedro Mountain Fire had a strong influence of limber pine and juniper woodland vegetation types interwoven on the burnable landscape. This disparity may reflect that overall predictive performance results can be a product of the general fuel load and the specific fuel type present within the burnable environment. In this case, the presence-or-absence of canopy fuels may have been a driving factor of model error for these two wildfires. Heterogeneity of fuel loading models and the configuration of existing vegetation types across the landscape were the variables that induced the most significant amount of variance on performance results for each series of wildfire simulations.

Fires in Group 1 are characterized by low total fuel load and low degree of fuel-bed continuity, which resulted in higher rates of under-prediction. Simulation results for these four fires displayed the highest rates of under-prediction out of all model runs. The Corbin and Saddle Butte fires burned through sagebrush steppe and semi-arid shrubland vegetation types, while the Keystone and Fishhawk fires burned predominantly through Subalpine forest and woodland vegetation types such as spruce-fir, lodgepole pine, Douglas-fir, aspen, and other mixed conifers. Existing vegetation type(s) is the only significant difference amongst this grouping of individual wildfire environments, as variations in surface fuel loading appear to be the primary driver of model performance. It can be inferred that the most significant rates of under-prediction are yielded when simulating wildfire events in environments with high ratios of canopy-to-surface fuels. Given the discontinuous nature of sagebrush and semi-arid shrubland vegetation communities across the landscape, the interstitial spacing between clusters of burnable vegetation may result in simulations to under-predicting wildfire activity.

Group 2 is characterized by medium-to-high total fuel loadings and a high level of fuel-bed continuity, which resulted in higher rates of over-prediction. In contrast to Group 1, results for the simulations within Group 2 displayed the highest rates of over-prediction across all model simulations. The Buffalo, Currant, and Stallions fires burned in predominantly of Northwestern Great Plains mixed grassland and sagebrush steppe vegetation types. These landscapes are more homogeneous than landscapes present in Groups 1 and 3. Higher rates of overestimation are achieved when modelling wildfire propagation in herb- and grassland-dominated environments. The Currant and Stallions fire simulations yielded substantially higher rates of overestimation than did model runs for Buffalo Fire, which is likely attributable to the relatively lower level of surface fuel loading within the Buffalo Fire. We infer that higher rates of over-prediction are associated with simulative environments dominated by herb and grassland vegetation types with medium-to-high surface fuel loads. Simulations for wildfires that have burned on landscapes dominated by herb and grassland vegetation types possess a more continuous fuel bed, which results in a more uniform propagation pattern. The grassland vegetation types inherent to these simulation environments create a more continuous fuel bed than shrubland, forest, and woodland vegetation types do.

Group 3 is associated with low-to-medium fuel loads with varying levels of fuel-bed continuity, which resulted in a more accurate prediction with minimal over- or under-prediction. It can be inferred that these landscapes possessed a relatively higher degree of heterogeneity among existing vegetation types and fuel loadings due to the diversification of surface and canopy fuel types within these environments. All fires in this group were burned environments consisting of an increased mixture of canopy and surface fuel types. Tannerite Fire simulations yielded a higher rate of over-prediction than simulations of the Pole Creek and Pedro Mountain fires, as this is likely attributable to the presence of Montane–Foothill–Valley Grassland vegetation types across the burnable landscape for the Tannerite Fire. An increased level of heterogeneity among fuel types in these environments allows the model to simulate more transitionary events of the flaming front propagating from surface to canopy. Simulative results for the Pole Creek and Pedro Mountain fires displayed slightly lower rates of overestimation than the simulations for the Tannerite Fire.

4. Discussion

As the scope of wildfire activity and severity continues to increase [1], it has become increasingly important to identify and employ accurate predictive wildfire models to ensure timely interventions and protect property, lives, and biodiversity. In the Central Rocky Mountain Region, forest and woodland areas are typically characterized by steep and highly variable terrain elevation, which poses a series of complex challenges for wildfire models to resolve while integrating datasets with limited spatial resolutions. Our research quantified and evaluated the predictive performance of WyoFire, a Monte Carlo-based wildfire simulation model using a set of evaluative indices and corresponding PCA results to assess how the model performs over a range of diverse landscapes.

The different environments of the 10 wildfire events had unique vegetation type assemblages which helped bring about statistically significant variations in model performance. The significant variations in model performance were supported by statistically testing the accuracy results by using the equation ADI_{oe}—ADI_{ue} to determine ranges in over- and underestimation within particular fuel types and loadings. Fuel loading was found to induce the most variance in model performance of all variables present within the wildfire simulations, while terrain complexity appeared to be the second most import factor on performance.

Simulations of wildfires occurring in shrubland- and grassland-dominated environments displayed the tendency to over-predict, while fire simulations of forested and woodland dominated-environments displayed the tendency to under-predict. In part, these performance differences may reflect interactions of the models used for wildfire spread in WyoFire [7,8,13,14], particularly during fuel condition changes (grassland to forest) or actual historic wind speeds used in our simulations as compared to modelled wind speed restrictions based on the source model assumptions [18–20]. This information is pertinent to researchers examining the processes of wildfire propagation across heterogeneous landscapes, as assumptions can be made about expected model performance across heterogeneous environments, as is evident by WyoFires' output.

The results of this study reveal that, relative to vegetation type and fuel loading, terrain complexity has a minimal effect on predictive modelling performance when employing a Monte Carlo simulation approach. Rates of model overestimation and underestimation can be primarily attributed to fuel loading models and vegetation types within burnable environments. WyoFire displayed its highest rates of underestimation when simulating fires in environments with low surficial fuel loads that also have a low degree of fuel-bed continuity. These results are reasonable as wildfires burning within environments possessing lower degrees of fuel-bed continuity will invoke higher rates of under-prediction due to interstitial spaces between burnable vegetation [21]. In contrast, the highest rates of overestimation occurred when simulating wildfires in environments with medium-to-high total fuel loads that have a higher degree of fuel-bed continuity. This increased rate of overestimation likely occurred due to the increased connectivity of vegetation across the heterogeneous landscape [22]. These results will help us understand the environmental characteristics that lead to the tendencies for wildfire simulation models to over- or-under-predict wildfire behaviour. It should be noted that a range of centre distances should always be tested for each wildfire simulation environment in order to determine which value will yield the most accurate results. After analysing each of the ten fires within this study, we observed that parameterizing WyoFire with a higher CD yielded more favourable results in homogenous grasslands and shrubland landscapes. When parameterizing the model with a lower CD, we observed more favourable prediction results in forested and transitionary fuel types.

In its current state, WyoFire performs exceptionally well across a range of burnable environments as defined by fuel load, vegetation type and terrain complexity characteristics. The most adverse challenge faced in wildfire modelling is accounting for the stochastic nature of natural wildfire caused by internal dynamics, the ability to locally modify winds and fuels, coupled with mesoscale weather conditions that can shift rapidly. By employing a probabilistic approach that uses local historic variability to model wildfire propagation across heterogeneous landscapes, WyoFire incorporates natural stochasticity within each burnable environment. This approach begins to deal with the problem of wildfire simulation models coupled with meteorological forecast datasets maintaining accuracy due to the growth in error with each hour of prediction [23]. Monte Carlo simulation models that account for physical stochasticity within the natural environment are invaluable tools for a better understanding of how wildfires propagate across heterogeneous landscapes. Researchers leveraging deterministic wildfire prediction models for training purposes would benefit from implementing probabilistic Monte Carlo simulation models such as WyoFire.

WyoFire allows the user to parameterize specific model inputs to optimize the quality of predictive performance results regarding characteristics present within the desired burnable environment. If the characteristics associated with the burnable environment are known at the time of simulation, then a general hypothesis can be developed to address whether the simulation will result in over- or under-prediction. This model serves as a practical educational tool that can improve our understanding of wildfire behaviour in the lab and in the classroom. Future improvements to WyoFire largely hinge upon interpretations made from the results of this study, as conducting a comprehensive performance evaluation of model simulation results is pertinent for understanding its strengths and limitations [6,24].

5. Conclusions

In this study, we assessed the predictive performance of a Monte Carlo-driven wildfire simulation model, WyoFire, employing a set of single-value performance metrics and results from a Principal Components Analysis. Ten wildfire events in and around the state of Wyoming were simulated to assess the causes of variance in model performance which were mainly explained by existing vegetation types, fuel loadings, and degrees of terrain complexity variables. The PCA yielded three apparent groups of individual wildfire events based upon a measure of similarity between their resulting performance metric values and the physical characteristics that comprise each burnable environment. The fuel loading model emerged as the variable that induces the most substantial amount of variance on model performance when simulating wildfire events on a particular landscape, while terrain complexity was found to be relatively less significant in altering model performance.

Results from this research further confirm Adhikari et al.'s [8] finding that WyoFire is reliably effective and efficient across various heterogeneous landscapes. The model tends to over-predict fire spread in environments with a higher total fuel load in contrast to under-predicting wildfire activity in environments possessing lower total fuel loads. Results from all optimized simulations suggest that WyoFire performs exceptionally well, as accentuated rates of over- and under-prediction align with the fuel loading model and fuel-bed continuity present within the burnable environment. This model also displays the tendency to over-predict at higher rates when simulating wildfire events occurring on relatively smoother landscapes dominated by grassland vegetation types, in contrast to yielding substantially higher rates of under-prediction in environments dominated by shrubland and woodland vegetation types with a higher level of terrain complexity. WyoFires' predictive ability across various fuel loading models and vegetation types may prove to be an effective tool to understand potential fire risk and potential processes that affect wildfire behaviour.

Author Contributions: Conceptualization, T.A.M., P.H., C.X.; methodology, B.A., C.W.O., T.A.M., P.H., C.X.; software, C.X., B.A., C.W.O., S.P.A.; formal analysis, C.W.O., S.P.A.; writing—original draft preparation, C.W.O., T.A.M., C.X., P.H., B.A., S.P.A. All authors have read and agreed to the published version of the manuscript.

Funding: This research received no external funding.

Conflicts of Interest: Authors declare no conflicts of interest.

References

1. Garfin, G.; Gonzalez, P.; Breshears, D.; Brooks, K.; Brown, H.; Elias, E.; Gunasekara, A.; Huntly, N.; Maldonado, J.; Mantua, N. Southwest. In *Impacts, Risks, and Adaptation in the United States: Fourth National Climate Assessment*; Reidmiller, D.R., Avery, C.W., Easterling, D.R., Kunkel, K.E., Lewis, K.L.M., Maycock, T.K., Stewart, B.C., Eds.; U.S. Global Change Research Program: Washington, DC, USA, 2018; Volume II, pp. 1101–1184.
2. Westerling, A.L. Increasing western US forest wildfire activity: Sensitivity to changes in the timing of spring. *Philos. Trans. R. Soc. B Biol. Sci.* **2016**, *371*, 20150178. [CrossRef] [PubMed]
3. Stavros, E.N.; Abatzoglou, J.T.; McKenzie, D.; Larkin, N.K. Regional projections of the likelihood of very large wildland fires under a changing climate in the contiguous Western United States. *Clim. Chang.* **2014**, *126*, 455–468. [CrossRef]
4. Higuera, P.E.; Abatzoglou, J.T. Record-setting climate enabled the extraordinary 2020 fire season in the western United States. *Glob. Chang. Biol.* **2020**. [CrossRef] [PubMed]
5. McWethy, D.B.; Schoennagel, T.; Higuera, P.E.; Krawchuk, M.; Harvey, B.J.; Metcalf, E.C.; Schultz, C.; Miller, C.; Metcalf, A.L.; Buma, B. Rethinking resilience to wildfire. *Nat. Sustain.* **2019**, *2*, 797–804. [CrossRef]
6. Duff, T.J.; Chong, D.M.; Tolhurst, K.G. Indices for the evaluation of wildfire spread simulations using contemporaneous predictions and observations of burnt area. *Environ. Model. Softw.* **2016**, *83*, 276–285. [CrossRef]
7. Rothermal, R. *A Mathematical Model for Predicting Fire Spread in Wildland Fuels*; Intermountain Forest & Range Experiment Station, Forest Service, US Department of Agriculture: Ogden, UT, USA, 1972; 40p.
8. Adhikari, B.; Xu, C.; Hodza, P.; Minckley, T.A. Developing a geospatial data-driven solution for rapid natural wildfire risk assessment. *J. Appl. Geogr.* In Revision.
9. Bennett, N.D.; Croke, B.F.; Guariso, G.; Guillaume, J.H.; Hamilton, S.H.; Jakeman, A.J.; Marsili-Libelli, S.; Newham, L.T.; Norton, J.P.; Perrin, C. Characterising performance of environmental models. *Environ. Model. Softw.* **2013**, *40*, 1–20. [CrossRef]
10. Kelso, J.K.; Mellor, D.; Murphy, M.E.; Milne, G.J. Techniques for evaluating wildfire simulators via the simulation of historical fires using the Australis simulator. *Int. J. Wildland Fire* **2015**, *24*, 784–797. [CrossRef]
11. Adhikari, B. *A Web GIS Portal for Modeling Wildfire Spread in Near Realtime and Assessing Associated Risk*; University of Wyoming: Laramie, WY, USA, 2018.

12. Filippi, J.B.; Mallet, V.; Nader, B. Representation and evaluation of wildfire propagation simulations. *Int. J. Wildland Fire* **2014**, *23*, 46–57. [CrossRef]
13. Wagner, C.V. Conditions for the start and spread of crown fire. *Can. J. For. Res.* **1977**, *7*, 23–34. [CrossRef]
14. Finney, M.A. *FARSITE: Fire Area Simulator—Model Development and Evaluation*; Research Paper RMRS-RP-4 Revised; USDA Forest Service, Rocky Mountain Research Station: Ogden, UT, USA, 2004; 47p.
15. Ott, C.W. *Performance Evaluation of a Monte Carlo Driven Wildfire Simulation Model: Assessing Model Performance to Advance Education and Improve Human Safety*; University of Wyoming: Laramie, WY, USA, 2020.
16. Fawcett, T. An introduction to ROC analysis. *Pattern Recognit. Lett.* **2006**, *27*, 861–874. [CrossRef]
17. Pugnet, L.; Chong, D.; Duff, T.; Tolhurst, K. Wildland–urban interface (WUI) fire modelling using PHOENIX Rapidfire: A case study in Cavaillon, France. In Proceedings of the 20th International Congress on Modelling and Simulation, Adelaide, Australia, 1–6 December 2013; pp. 1–6.
18. Duff, T.J.; Cawson, J.G.; Cirulis, B.; Nyman, P.; Sheridan, G.J.; Tolhurst, K.G. Conditional performance evaluation: Using wildfire observations for systematic fire simulator development. *Forests* **2018**, *9*, 189. [CrossRef]
19. Moon, K.; Duff, T.; Tolhurst, K. Sub-canopy forest winds: Understanding wind profiles for fire behaviour simulation. *Fire Saf. J.* **2019**, *105*, 320–329. [CrossRef]
20. Andrews, P.L.; Cruz, M.G.; Rothermel, R.C. Examination of the wind speed limit function in the Rothermel surface fire spread model. *Int. J. Wildland Fire* **2013**, *22*, 959–969. [CrossRef]
21. Loehle, C. Applying landscape principles to fire hazard reduction. *Forest Ecol. Manag.* **2004**, *198*, 261–267. [CrossRef]
22. Kerby, J.D.; Fuhlendorf, S.D.; Engle, D.M. Landscape heterogeneity and fire behavior: Scale-dependent feedback between fire and grazing processes. *Landsc. Ecol.* **2007**, *22*, 507–516. [CrossRef]
23. Coen, J.L.; Schroeder, W. Use of spatially refined satellite remote sensing fire detection data to initialize and evaluate coupled weather-wildfire growth model simulations. *Geophys. Res. Lett.* **2013**, *40*, 5536–5541. [CrossRef]
24. Salis, M.; Arca, B.; Alcasena, F.; Arianoutsou, M.; Bacciu, V.; Duce, P.; Duguy, B.; Koutsias, N.; Mallinis, G.; Mitsopoulos, I. Predicting wildfire spread and behaviour in Mediterranean landscapes. *Int. J. Wildland Fire* **2016**, *25*, 1015–1032. [CrossRef]

Publisher's Note: MDPI stays neutral with regard to jurisdictional claims in published maps and institutional affiliations.

© 2020 by the authors. Licensee MDPI, Basel, Switzerland. This article is an open access article distributed under the terms and conditions of the Creative Commons Attribution (CC BY) license (http://creativecommons.org/licenses/by/4.0/).

Article

Integrating Remote Sensing Methods and Fire Simulation Models to Estimate Fire Hazard in a South-East Mediterranean Protected Area

Panteleimon Xofis [1,*], Pavlos Konstantinidis [2], Iakovos Papadopoulos [3] and Georgios Tsiourlis [2]

1. Department of Forestry and Natural Environment, International Hellenic University, 1st km Drama-Mikrohori, GR66100 Drama, Greece
2. ELGO-DEMETER, Hellenic Agricultural Organization "Demeter" Forest Research Institute, GR57006 Vassilika Thessaloniki, Greece; pavkon@fri.gr (P.K.); gmtsiou@fri.gr (G.T.)
3. Hellenic Forest Service, Decentralized Administration of Macedonia and Thrace, Poligiros, GR63100 Chalkidiki, Greece; iakpap@damt.gov.gr
* Correspondence: pxofis@for.ihu.gr; Tel.: +30-6973035416

Received: 5 June 2020; Accepted: 17 July 2020; Published: 19 July 2020

Abstract: Unlike low intensity fire which promotes landscape heterogeneity and important ecosystem services, large high-intensity wildfires constitute a significant destructive factor despite the increased amount of resources allocated to fire suppression and the improvement of firefighting tactics and levels of organization. Wildfires also affect properties, while an increasing number of fatalities are also associated with wildfires. It is now widely accepted that an effective wildfire management strategy can no longer rely on fire suppression alone. Scientific advances on fire behavior simulation and the increasing availability of remote sensing data, along with advanced systems of fire detection can significantly reduce fire hazards. In the current study remote sensing data and methods, and fire behavior simulation models are integrated to assess the fire hazard in a protected area of the southeast Mediterranean region and its surroundings. A spatially explicit fire hazard index was generated by combining fire intensity estimations and proxies of fire ignition probability. The results suggest that more than 50% of the study area, and the great majority of the protected area, is facing an extremely high hazard for a high-intensity fire. Pine forest formations, characterized by high flammability, low canopy base height and a dense shrub understory are facing the most critical hazard. The results are discussed in relation to the need for adopting an alternative wildfire management strategy.

Keywords: wildfire risk; object-oriented image analysis; Sentinel-2; fire behavior; flammap; wildfire management

---

1. Introduction

Fire is an environmental factor with a long history on earth, dating back to the first appearance of terrestrial vegetation, and as a result it has played a vital role in the ecology and evolution of species, ecosystems and humans [1–4]. While low intensity fires, resembling prehuman fire regimes, are important drivers for maintaining landscape and habitat heterogeneity and promote a number of important ecosystem services to humans [5], large high-intensity wildfires constitute a significant destructive factor for many biomes and ecosystems across the world. Man has dramatically changed fire behavior and regime, converting it into a primarily anthropogenic factor [6], with more than 464 million ha affected annually by wildfires [7]. In the Mediterranean Europe, in particular, fire affects an average area of 350.000 ha annually, with an interannual variation ranging from a little more than 100 thousand ha to almost one million ha, following fluctuations in weather patterns [8]. Wildfires do not only threaten the ecosystem's ecological integrity, but in recent years they have been responsible for significant damages to infrastructures and properties and a high number of human fatalities. The wildfires of 2007 in Peloponnese had a high cost in human lives with 78 deaths, while more than 100 villages were affected, resulting in a significant loss of property [9]. More recently, in 2018, almost 100 people lost their lives in the fire occurred in the Wildland Urban Interface (WUI) of Attica, Greece. According to the data reported by Molina-Terren et al., [10] there has been an important increase in the mean annual number of fatalities in Mediterranean countries in the period 1979–2016 compared to the period 1945–1979.

The decade of 1970s seems to coincide with an important turning point in the behavior of wildland fires and their related consequences, with an observed increase in the total number of fires and area burned [11]. Furthermore, the relevant importance of weather conditions and fuel availability in determining the fire regime and behavior appear to have shifted, in favor of the former, turning wildfires from "fuel-driven" to "weather-driven" [12–15]. The availability of fuel has increased dramatically since the second half of the 20th century as a result of major socioeconomic changes which resulted in agricultural abandonment in mountainous and semi-mountainous regions and a decrease in agropastoral activities [16,17]. Both trends resulted in an expansion of woody semi-natural vegetation into formerly cultivated lands and rangelands, increasing the continuity of fuel and its total load. An analysis performed by Koutsias et al. [18] revealed the significant contribution of transitional woodland scrub vegetation, associated with a reduction in pastoralism activities, and abandoned former agricultural land in the fuel consumed during the megafires in Peloponnese in 2007. Another important shift in the wildfire pattern, observed in several studies, is that recently they tend to occur in higher altitudes and more humid regions affecting communities that are not considered fire prone [19,20]. Meanwhile, species growing in fire prone environments have developed adaptive or exaptive traits to overcome the detrimental effects of fire [21–24]; species and communities growing in regions where fires are rare do not possess such traits and their long term presence, at least at a local scale, is threatened [19]. The proven difficulties in controlling the impacts of wildfires, and especially megafires, impose the need to reconsider wildfire management strategy and take advantage of the technological and methodological advances and the wide range of available data, which allow for the accurate estimation of wildfire hazards.

A wildfire hazard is related to the possibility of a fire event turning into a high-intensity fire with a high potential for loss [25]. A prior and spatially explicit knowledge of areas and ecosystems with the high potential to host a high-intensity fire allows not only targeted management towards decreasing the hazard, but also the increase of the level of preparedness and alertness for a quick detection and initial attack [26]. Initial attack can have a vital impact on reducing a wildfire hazard under certain circumstances regarding fuel load and structure [27]. Over the last two decades, spatially explicit fire behavior simulation models have been integrated into wildfire risk assessment protocols in several fire prone regions of the world with promising results. They calculate a number of different components of fire behavior, including fireline intensity, rate of spread, flame height etc., in a spatially explicit manner, and they integrate fuel characteristics, topographic data and weather scenarios. In eastern Mediterranean, in particular, Mitsopoulos et al. [28] employed the Minimum Travel Time (MTT) algorithm [29] embedded in Flammap [30] to assess fire behavior in a WUI. The results obtained demonstrate the elevated fire risk, especially under dry conditions, and constitute an important contribution in the spatial planning and management of WUI, to reduce the risk of loss of property and human lives [31]. Fire simulation using Flammap has also been employed to assess the fire hazard under different scenarios of climate change [32,33] or in areas of high conservation importance [34]. These studies demonstrate the importance of fire hazard assessment, using advanced simulation methods, for the effective planning of wildfire management and better allocation of resources.

The performance and accuracy of fire behavior simulation models, and as a result of fire hazard estimation protocols, depend largely on the quality of input data, with the most important and complex being the fuel data. The Mediterranean region is characterized by a complex relief which results in complex vegetation patterns and structures. The wide availability of satellite remote sensing data at high spatial, spectral and temporal resolution offer a great tool to, thematically and spatially, accurately map the complex vegetation mosaics observed in the Mediterranean landscape [35–37]. While moderate resolution Landsat data have been widely used for vegetation and fuel mapping, as have various very high resolution data, the suitability of Sentinel-2 data for fuel mapping in the complex landscape of eastern Mediterranean region remains largely unexplored. A recent study by Stefanidou et al. [38], where various landcover types have been identified using Sentinel-2 data in Greece, achieved promising results and suggested that these data could significantly improve the accuracy of fuel mapping at a regional and national scale. Remote sensing data and methods can serve other aspects of wildfire management also. Sifakis et al. [39] presented an automatic fire detection algorithm based on data obtained by the sensor Spinning Enhanced Visible and Infrared Imager (SEVIRI) on board the Meteosat Second Generation (MSG) geostationary satellite (henceforth: MSG-SEVIRI), with a spatial resolution of approximately 4 km and a temporal resolution of 15 min. The algorithm was tested during the megafires of 2007 in Greece. The results obtained show an 82% efficiency in detecting fire spots using the Short-Wave Infrared (SWIR) channel 4 and the Thermal Infrared (TIR) channel 9. A similar work has been developed and tested in Australia by Xu and Zhong [40] employing multispectral Himawari-8 imagery. The results reported demonstrate the high accuracy and time efficiency of the method in detecting hot spots. Such tools are of great value for the monitoring of fire evolution, spread direction and dynamics, especially if one takes into account the very high temporal resolution of the data. The only drawbacks of the presented methods lie on the coarse spatial resolutions of the imagery, which exceeds 2 km and restricts the applicability of the method in directing the suppression forces and support real-time firefighting tactics. The integration of such methods with higher resolution data could be a step forward in utilizing remote sensing technologies for time efficient and spatially accurate fire detection [39,40]. Remote sensing data and methods are also widely employed for post-fire mapping of the burned area, as well as for the post-fire characterization of wildfires in terms of fire severity [41–43]. Furthermore, the integration of data acquired by active sensors allows for the mapping of the vertical characteristics of the complex forest structure and can assist the process of characterizing and mapping forest fuels and post-fire damage assessment [44].

In the current study we employ remote sensing data, fire behavior simulation modeling and other fire hazard related components to generate a spatially explicit estimation of fire hazard in a protected area in Cyprus using a Fire Danger Index (FDI) estimation formula. The resulting product is expected to assist in the identification of vulnerable areas and promote the adoption of appropriate management measures for hazard reduction. Furthermore, it will be integrated in an automatic fire detection system to assist the process of decision making and mobilization of resources, once a potential ignition is detected.

2. Materials and Methods

2.1. Study Area

The study site covers the Western part of the Troodos mountains which constitutes the largest mountain range in Cyprus with geographic coordinates 34°53′29.06″ N and 32°51′54.24″ E. Mount Troodos is characterized by its high biodiversity and diversity of endemic plants which resulted in its inclusion in the list of 13 "plant diversity hotspots" in the Mediterranean region. The central part of Mount Troodos has been a designated National Forest Park (NFP) since 1992 and it is included in the NATURA 2000 Network of protected areas (CY5000004). The altitudinal range is from 455 to 1963 m a.s.l. Administratively, it falls within the Prefecture of Limasol and the management of the forested part is conducted by the Forestry Directorate of Cyprus.

Precipitation at the NFP of Troodos ranges from 660 mm at the lowest parts and exceeds 1100 mm at its peak altitude. The first snowfall usually occurs at the beginning of December and the last at the middle of April. According to the weather data of the nearby weather station of Saitas, at an altitude of

640 m, the hottest month is July, with an average daily temperature of 26.5 °C and a monthly average maximum temperature of 38.6 °C. The coldest month is February with an average daily temperature of 8.4 °C and an average monthly minimum temperature of −1.5 °C. The wettest month is December with an average precipitation of 130.1 mm, and the driest is July with 6.6 mm precipitation [45].

The area of the NFP of Troodos is dominated by forest ecosystems of *Pinus brutia, P. nigra, Juniperus foetidissima* and *Quercus alnifolia*, in pure or mixed formations [46]. The *P. brutia* forest starts from the lowest altitudinal zones and extends locally up to c. 1.400 m. The shrub understory consists of *Q. alnifolia, Arbutus andrachne, Pistacia terebinthus, Styrax officinalis, Genista fasselata. P. nigra*, starts at altitudes of 1200 m in mixed formations with *P. brutia* and extends up to the top altitude, often in mixture with *J. foetidissima*. Pure stands with *J. foetidissima* are mainly restricted to the central part, from about 1.300 m up to the highest peak. On rocky slopes the stands are usually open whereas on better sites they are dense. Riparian broadleaved forests occur from the lowest parts up to an altitude of 1.600 m, and consist of *Alnus orientalis, Platanus orientalis, Laurus nobilis* and *Salix alba*, while *Nerium oleander* is sometimes present. A large part of the study area consists of evergreen broadleaved shrubland formations at varying densities, from very dense and 2–3 m heigh t formations to low density short formations.

2.2. Remote Sensing Data and Analysis

For the identification of vegetation formations and the generation of a landcover map of the study area a time series of Sentinel-2 images were employed, and analyzed in an object-oriented image analysis (OBIA) environment, using the software eCognition [47]. The time series compilation of Sentinel-2 images consisted of four images acquired on 31/10/2018, 03/02/2019, 25/03/2019 and 03/07/2019, respectively (Figure 1). The images were selected to cover an entire annual vegetation cycle because this allows for the capture of the vegetation, both natural and anthropogenic, in the different phenological stages. This is of particular importance for mapping dynamic land cover types, such as low-density shrubs, which have a significant grassland cover, low density pines as well as agricultural land with annual crops. All selected images were cloud free.

Sentinel-2 mission consists of two polar-orbiting satellites launched by European Space Agency [48], carrying the Multispectral Instrument (MSI) sensor and delivering data at spatial resolutions between 10 and 60 m, covering a wide part of the electromagnetic spectrum, from visual to infrared (Table 1), and a temporal resolution of 5 days. The images were processed at Level 2A, so they already provided geometrically and atmospherically corrected at Bottom of Atmosphere (BoA) reflectance. For the image analysis, only the spectral bands at 10 and 20 m spatial resolution were employed. Furthermore, the Aster Global Digital Elevation Model (GDEM) was employed, downloaded by the NASA EarthData database. The sharp relief of the study area results in a zonation in the distribution of vegetation types and the use of a DEM was indispensable to achieve maximum classification accuracy. The landcover mapping was done using the software eCognition v. 9.4, which allows the integration in the same classification project data from different sources and different spatial resolutions, without the need for resampling prior to the classification. However, the spectral bands with a spatial resolution of 20 m were resampled at 10 m, and a band composite of 10 bands for each image date was created at a spatial resolution of 10 m. The resampling does not, of course, improve the spatial resolution of the 20 m resolution bands, but it was done simply for practical purposes and for the better organization of the project. OBIA, as implemented in the software ecognition, has several advantages compared to traditional pixel-based approaches when applied on High Spatial Resolution data. It effectively deals with within-class variability, and since the classification is done not on the pixel unit but on the object unit, various shape, texture and context characteristics can be employed in the classification process [49,50]. Furthermore, OBIA avoids the pixelized (salt and pepper) representation of landcover classes which is often observed in pixel-based approaches, while the final classification product can be integrated directly into a vector-GIS for further analysis [51].

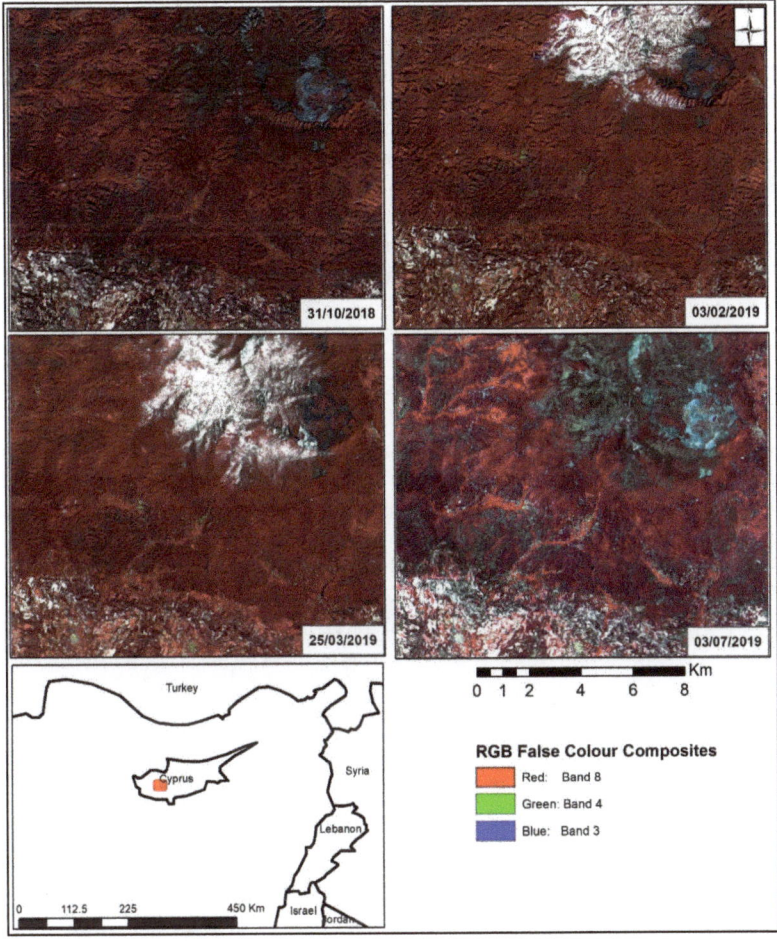

Figure 1. Sentinel 2 images used for the land cover mapping of the study area.

Table 1. Spectral and spatial resolution of the multispectral instrument (MSI) instrument onboard Sentinel-2 satellites.

Spectral Band	Color Description	Wavelength Range (nm)	Spatial Resolution
Band 1	Coastal aerosol	433–453	60
Band 2	Blue	458–523	10
Band 3	Green	543–578	10
Band 4	Red	650–680	10
Band 5	Red Edge 1	698–713	20
Band 6	Red Edge 2	733–748	20
Band 7	Red Edge 3	773–793	20
Band 8	Near Infrared	785–900	10
Band 8A	Narrow Near Infrared	855–875	20
Band 9	Water vapour	395–955	60
Band 10	Shortwave infrared-Cirrus	1360–1390	60
Band 11	Shortwave infrared 1	1565–1655	20
Band 12	Shortwave infrared 2	2100–2280	20

Sentinel-2 mission consists of two polar-orbiting satellites launched by European Space Agency [48], carrying the Multispectral Instrument (MSI) sensor and delivering data at spatial resolutions between 10 and 60 m, covering a wide part of the electromagnetic spectrum, from visual to infrared (Table 1), and a temporal resolution of 5 days. The images were processed at Level 2A, so they already provided geometrically and atmospherically corrected at Bottom of Atmosphere (BoA) reflectance. For the image analysis, only the spectral bands at 10 and 20 m spatial resolution were employed. Furthermore, the Aster Global Digital Elevation Model (GDEM) was employed, downloaded by the NASA EarthData database. The sharp relief of the study area results in a zonation in the distribution of vegetation types and the use of a DEM was indispensable to achieve maximum classification accuracy. The landcover mapping was done using the software eCognition v. 9.4, which allows the integration in the same classification project data from different sources and different spatial resolutions, without the need for resampling prior to the classification. However, the spectral bands with a spatial resolution of 20 m were resampled at 10 m, and a band composite of 10 bands for each image date was created at a spatial resolution of 10 m. The resampling does not, of course, improve the spatial resolution of the 20 m resolution bands, but it was done simply for practical purposes and for the better organization of the project. OBIA, as implemented in the software ecognition, has several advantages compared to traditional pixel-based approaches when applied on High Spatial Resolution data. It effectively deals with within-class variability, and since the classification is done not on the pixel unit but on the object unit, various shape, texture and context characteristics can be employed in the classification process [49,50]. Furthermore, OBIA avoids the pixelized (salt and pepper) representation of landcover classes which is often observed in pixel-based approaches, while the final classification product can be integrated directly into a vector-GIS for further analysis [51].

Apart from the original spectral bands the following spectral indices were calculated and used in the analysis:

$$Normalised\ Difference\ Vegetation\ Index\ (NDVI) = \frac{Band\ 8 - Band\ 4}{Band\ 8 + Band\ 4} \quad (1)$$

$$Normalised\ Difference\ Snow\ Index\ (NDSI) = \frac{Band\ 3 - Band\ 11}{Band\ 3 + Band\ 11} \quad (2)$$

$$Bare\ Soil\ Index\ (BSI) = \frac{(Band\ 11 + Band\ 4) - (Band\ 8 + Band\ 2)}{(Band\ 11 + Band\ 4) + (Band\ 8 + Band\ 2)} \quad (3)$$

$$Band\ Ratio\ for\ Built\text{-}up\ Area;\ (BRBA) = \frac{Band\ 3}{Band\ 8} \quad (4)$$

The classification process was assisted by ground truth data, collected in May 2019, in situ, at 50, 20 × 20 m plots, where the land cover type was recorded together with various vegetation structural characteristics, which were later used to allocate fuel models to the various vegetation types. Training data were also collected by visual inspection of the Very High-Resolution Images available on Google Earth. A scale parameter of 50 was selected, using a trial and error approach and visually inspecting the results, for the creation of homogenous objects based on the spectral characteristics included in the segmentation process. Only the spectral bands of 10 m spatial resolution, of all four images, were used for the segmentation. The first step of the analysis was to identify the areas covered by snow in the two images of February and March 2019, respectively. This was achieved by using the NDSI and an altitudinal threshold of 1200 m. The objects identified as snow were merged together and re-segmented with a scale parameter of 50, using only the spectral bands of 10 m resolution of the July 2019 and October 2018 images, and were classified separately from the rest of the scene using spectral bands and indices from the July 2019 and October 2018 images. The total number of created objects was 23,155 with an average size of 0.86 ha, minimum of 0.01 ha, maximum of 13.85 ha and Standard Deviation of 0.86 ha. These objects represent homogenous areas and constitute the basic unit for classification. Classification Trees (CART), Support Vector Machines (SVM) and Random Forests (RF) classifiers, all embedded in eCognition, were tested for their effectiveness in identifying the selected classes. The evaluation of each classifier's performance was assessed by the efficiency in reproducing the training set. Random Forests [52] classification algorithm was found to be the most accurate and was employed for the classification. The RF classifier is a nonparametric machine learning technique which consists of an ensemble of classification trees. Each tree is built based on a random subset of training data and a random subset of predictor variables. Variables that appear more often in the ensemble are those with the highest predictive power. Similarly, each tree predicts inclusion of a training sample in one class. The class predicted by the final model is the one predicted by the highest number of individual trees. The minimum mapping unit was set at 0.25 ha and various classification refinements were made in order to avoid isolated small polygons across the final classification map. The final classification accuracy was estimated using an error matrix on an independent set of 365 points, located randomly across the study area, where the landcover type was verified on the ground.

2.3. Fire Behavior Simulation and Estimation of Fire Hazard

As already mentioned, a survey was conducted on 50 plots across the study area, following a stratified random sampling design. The stratification was done using the Corine Land Cover data, which was the only available land cover dataset prior the commence of the study and the development of the landcover map with the method described above. On each plot the relevant abundance of all tree and shrub species was recorded and used to estimate the dominant and accompanying species. Various structural characteristics were also recorded including breast height diameter (BHD), tree height (used to calculate canopy height), live and dead crown base height, crown diameter and canopy cover. For shrubs, instead of BHD, the diameter at the ground level was recorded. Furthermore, litter depth was measured, and the existence of dead branches and their average diameter was also recorded.

The above measurements were used in allometric equations developed for vegetation types similar to the ones observed in the study area [53–57] to obtain an estimate of the fuel load present on each plot, split into diameter categories as required for the description of fuel models. Based on the fuel estimates and the morphological characteristics of each vegetation type, a fuel model (FM) was assigned to each plot. Two custom FM were built, for the high- and low-density pines, respectively, and five standard FMs, described by Scott and Burgan [58], were employed to capture the variation on the vegetation formations present in the study area. Each landcover type was represented by a single FM and the fuel model raster dataset was created.

Normally, at the final stage of any OBIA classification process, adjacent objects of the same landcover type are merged together before exporting. In this case, this was not done, and the original 23,155 objects were kept separate, since each one of them represented a homogenous area in terms of vegetation compositional and structural characteristics. On each of these objects (polygons) the structural characteristics of Canopy Height (CH), Canopy Cover (CC) and Canopy Base Height (CBH) of the nearest plot of the same landcover type and aspect were assigned, using the point-to-polygon spatial join function in ArcGIS 10.7. This method was preferred to calculating the average values of plots falling on the same landcover type because it allowed for the better representation in the relevant mapping products of the high landscape complexity observed in the study area. As a result, three additional raster datasets were created representing CH, CC and CBH, respectively, for the entire study area, which were necessary to perform fire simulation. Three additional raster datasets were created using the Aster GDEM, a Digital Elevation Model, a Digital Aspect Model (DAM) and a Digital Slope Model (DSM). The latter were resampled at 10 m spatial resolution and spatially matched perfectly to the FM raster dataset, because simulation cannot be executed if the spatial data do not have the exact same number of columns and rows. The resampling of the Aster GDEM derived products does not of course lead to any actual improvement in their spatial resolution, but unfortunately there was not a better DTM available for the study area.

The model Flammap [30] was employed to simulate fire behavior. Flammap is a two-dimensional fire simulation model which estimates fire behavior under uniform weather conditions. It calculates important components of fire behavior including fireline intensity, rate of spread, flame height and crown fire activity. Although flammap does not take into account the ignition points or the variation of weather conditions during a fire event, it is a powerful method typically used to estimate fire risk based on the vegetation and topographic characteristics [59]. In that sense, it is more appropriate for a study like this, compared to other simulators, such as Farsite [60], because it allows for the identification of areas with a high potential of giving a high-intensity fire, once appropriate weather conditions exist.

The aim of the study is to assess the fire hazard in the area, in a spatially explicit manner, and under conditions that favor a high-intensity fire. The largest fires that have recently occurred in the Eastern Mediterranean region (Peloponnese Greece) occurred in the summer of 2007, which was the hottest recorded in the area for almost a century, with three consecutive heatwaves from June to August and wind patterns that favored the spread and intensity of fires [61,62]. These extreme weather conditions are more likely to occur more often in the future than to be an exceptional case of extremely rare occurrence [61,62]. For this reason, the fire behavior simulation was performed under the extreme scenario described by Mitsopoulos et al. [32] in the Eastern Mediterranean. The general orientation of the area is west so a west wind was assumed, which promotes upslope fire, while the fuel moisture parameters were set at 5%, 7%, 9%, 80% and 100% for 1-h, 10-h, 100-h, live-woody and live-herbaceous fuel, respectively. In the absence of in situ measurements of crown bulk density (CBD) it was decided to be kept constant at 0.3 kg/m^3 for all fuels, which is the average value observed in an area of similar orientation, continentality and vegetation composition (forest of profitis Ilias) on the Island of Lesvos by Palaiologou [63].

Fire behavior simulation using flammap generates a number of different parameters of fire behavior including Fireline intensity (often called Byram's intensity), rate of spread, flame length etc. Fireline intensity describes the heat release per meter of the fire front and is calculated by the following equation [64] which is an adjustment of the original equation developed by Byram [65]:

$$I = 0.007\, H * W * R \tag{5}$$

where:

I = Fireline intensity in Kw/m
H = Heat yield in cal/g
W = Fuel loading in tonnes/ha
R = Rate of spread in m/min

Fireline intensity estimations were rescaled to a scale between 0 and 1 to form the Fire Intensity (FI) component of the Fire Danger Index (FDI) formula (6) presented by Xofis et al. [66]. Rate of Spread (ROS) is also an important component of fire behavior, since it determines to a great extent the time required for an ignition to turn into a large size and hard to suppress wildfire. A preliminary analysis in two study areas showed that Fireline intensity and rate of spread are significantly and positively correlated to a degree of around 60%. However, various landcover types, such as grasslands and meadows, have a low amount of released energy, due primarily to the low fuel load, but still have a high ROS [66]. In a complex landscape mosaic, such as the one in the study area, and in the wider Mediterranean region, these landcover types impose a high fire hazard due to the increased possibility of a wildfire to quickly spread into a nearby area of high fuel load and high potential for a high-intensity fire. Given that the purpose of the FDI is not only to identify the most vulnerable to high-intensity wildfire areas, but also to increase the level of organization of suppression forces, the ROS is included as a separate component in the FDI formula, with a relatively low weight, despite the fact that it constitutes a component of the Fireline intensity equation, and in a sense it is already included in the FDI formula. ROS values were also rescaled to a scale between 0 and 1 and formed the ROS component of the FDI formula (6)

$$FDI = 0.5 * FI + 0.2 * ROS + 0.2 * HI + 0.1 * PH \tag{6}$$

The above formula integrates two more important components of fire risk, the Human Index (HI) and the Pyric History Index (PH), which serve as proxies of fire ignition probability. The HI attempts to capture the relevant risk for a fire event associated with anthropogenic activities. As a proxy of anthropogenic activities, the distance to roads was adopted, with values ranging from 0, for areas at a distance of 200 m or more from roads, to 1 for areas at immediate proximity to them. Distance to roads has been reported to be positively associated with ignition frequency [67,68]. The Pyric History does not have a direct relationship with fire hazard, especially if one takes into account that a past wildfire may reduce the fuel load and subsequently the intensity of a fire event. However, it provides an indication of the fire pattern of an area which might indicate trends related to specific land uses and spatial locations. For instance, in the Eastern Mediterranean region, where free range pastoralism is still practiced in the mountainous and semi-mountainous regions, it is very common for deliberate fires to be set by shepherds for the improvement of grazing conditions. This trend results in a clustering of past fire episodes indicating the higher risk for fire in such areas. Although these fires rarely become high-intensity ones, because they primarily occur in autumn and in areas with a low fuel load, the possibility to spread into nearby more flammable ecosystems and vegetation formations always exists. The inclusion of the PH index in the above formula will allow for the identification of areas where a specific spatial pattern of past fires does exist. The PH was calculated using data on ignition events over the last 25 years provided by the local Forest Service. A Kernel Density Estimation Function was applied to convert the point ignition data to a raster file with values ranging from 0, for areas away from a past ignition point, to 1, for areas at immediate proximity to a past ignition. The FDI calculated with the above formula varies between 0 and 1 with higher values indicating a higher fire hazard.

2.4. Validation of Fire Hazard Estimation

To validate the results of the estimated fire hazard using the FDI formula, we compared them with the risk estimated using the Conditional Flame Length (CFL) index, which has been widely used as an estimate of fire hazard [28,35,69]. For the CFL estimation a fire simulation under the Minimum Travel Time (MTT) algorithm, embedded in Flammap, with 10.000 randomly distributed fire ignitions and a fire duration of 8 hours was performed. The simulation was done under the exact same burning conditions as the first simulation, performed for the estimation of FI and ROS. The MTT algorithm [29] employs the Huygen's principle to replicate fire growth, where the growth of the fire edge is a vector or wave front. Since the burning conditions are constant, the MTT algorithm reveals the effect of fuel conditions and topography on fire growth [70]. The CFL was estimated using the ArcFuels 10 software, which was embedded as an extension in ArcGIS 10.7, using as input data the flame-length probabilities (FLP) metrics file generated in Flammap. CFL is an estimate of the mean flame length (FL) of all the ignitions (10.000) that burned a particular pixel.

The calculated FDI values were classified into four classes which were defined as very low hazard ($0 < FDI \leq 0.1$), low hazard ($0.1 < FDI \leq 0.35$), high hazard ($0.35 < FDI \leq 0.7$) and extremely high hazard ($FDI > 0.7$). The lowest class incorporates unburnable land cover classes and areas where some vegetations exist but where it is impossible to sustain a high-intensity fire. The upper threshold of the second class was chosen based on the relevant weight of FI and ROS in the FDI. Both these components account cumulatively for the 70% of the FDI. Since areas that score bellow 0.5 in these components carry a lower risk of a high-intensity fire and the weight of those components in the FDI is 0.7, the resulted threshold was set at 0.35. This class incorporates areas where the intensity and rate of spread of a possible fire is low, so even in the case of a fire event the probability of ending up in a high-intensity fire is low. The third class incorporates all these areas that score highly in the components of FI and ROS. The upper class represents the most vulnerable to wildfire areas, since in addition to the favorable fuel characteristics, these areas are also close to roads and/or past fire spots.

The validation was made on 388 points distributed across the study area, under a stratified random sampling design and with a minimum distance of 300 m between them, using the Sampling

Design Tool extension in the ArcGIS 10.7 platform. The request was for 500 points, but only 388 could be located under the restrictions of the sampling design and the minimum distance. On each generated point the FDI class was assigned as well as the CFL value that corresponded to the particular location. One-Way Analysis of Variance (ANOVA) and Post-hoc Tukey HSD (Honestly Significant Difference) tests were employed to test for statistical differences in the CFL among the four classes of fire hazard.

2.5. Assesment of between Landcover Types Differences in Estimated Fire Hazard

From the 388 points selected for the validation of the estimated fire hazard, 353 points corresponded to burnable landcover types. On each one of these 353 points the estimated FDI value was assigned together with the landcover type, the vegetation structural characteristics (CC, CBH, CH) and topographic data. One-Way ANOVA and Post-hoc Tukey HSD tests were employed to test for significant differences in the estimated fire hazard between landcover types.

For the points corresponding to high and low-density pines landcover types a second step of exploratory analysis was done using Regression Trees (RT), in order to assess the effect of topography and vegetation structural characteristics on the estimated fire hazard. Regression Trees is a data mining method which handles simultaneously continuous and categorical independent variables and they are very effective in handling non-linear and non-hierarchical relationships. Regression trees repeatedly divide the data into two mutually excluded groups using one independent variable at a time, until homogenous groups, in terms of the dependent variable, are achieved or the data cannot be divided any further [71].

3. Results

Seven classes were identified in the study area, shown in Figure 2 and Table 2. The overall accuracy of the final product was estimated at 93% and the Kappa statistics was calculated at 0.9, indicating an excellent performance by the classifier (Table A1 in Appendix A).

Low- and high-density pines collectively account for more than 50% of the total area, consisting of pure and mixed stands of *P. brutia* and *P. nigra*, often with a dense understory of shrubs. Figure 2 demonstrates the zonation in the distribution of landcover types observed in the study area. The lowest altitudes are occupied by agricultural land, and low and mid altitudes by scarce and low shrublands, which locally turn into dense shrublands, and as the altitude increases the pines dominate. The highest altitudes are dominated by *J. foetidissima* formations often in mixture with pines. Broadleaved forests tend to occur only along the various streams present in the study area.

Following the process described above, on each polygon generated in OBIA which represents a homogenous, in terms of vegetation structure and composition area, one fuel model was assigned, together with the additional structural characteristics (CH, CC, CBH) of the nearest sample plot of the same land cover class and aspect. Two custom fuel models and five standard fuel models were employed, and the corresponding fuel loads are shown in Table 3. For the custom fuel models the properties of FM type, Surface-Area-to-Volume (SAV) ratio, fuel depth, dead fuel extinction moisture and heat content were adopted by Palaiologou [63].

Table 2. Landcover classes identified in the study area.

Landcover Class	Area (ha)	Cover (%)
Bare ground -Agricultural Land.	1737.8	8.7
Scarce and low shrublands	4923.1	24.6
Dense shrublands	1369.9	6.9
Broadleaved Forests	407.9	2.0
Juniperus Foetidissima	1537.3	7.7
Low Density Pines	6032.9	30.2
High Density Pines	3976.7	19.9
Total	19,985.6	100

Table 3. Fuels models and corresponding land cover types used in the study.

FM Code	Landcover Class	Fuel Load (t/ha)				
		1-h	10-h	100-h	Live Herb	Live Woody
NB9	Bare ground -Agricultural Land	N/A	N/A	N/A	N/A	N/A
SH1	Scarce and low shrublands	0.6	0.6	0.0	0.4	3.2
SH7	Dense shrublands	8.6	13.1	5.4	0.0	8.4
TL9	Broadleaved Forests	16.4	8.2	10.3	0.0	0.0
SH4	*Juniperus Foetidissima*	2.1	2.8	0.5	0.0	6.3
CFM01	High Density Pines	10.9	13.7	8.4	0.1	15.8
CFM02	Low Density Pines	7.6	7.7	5.0	0.1	10.9

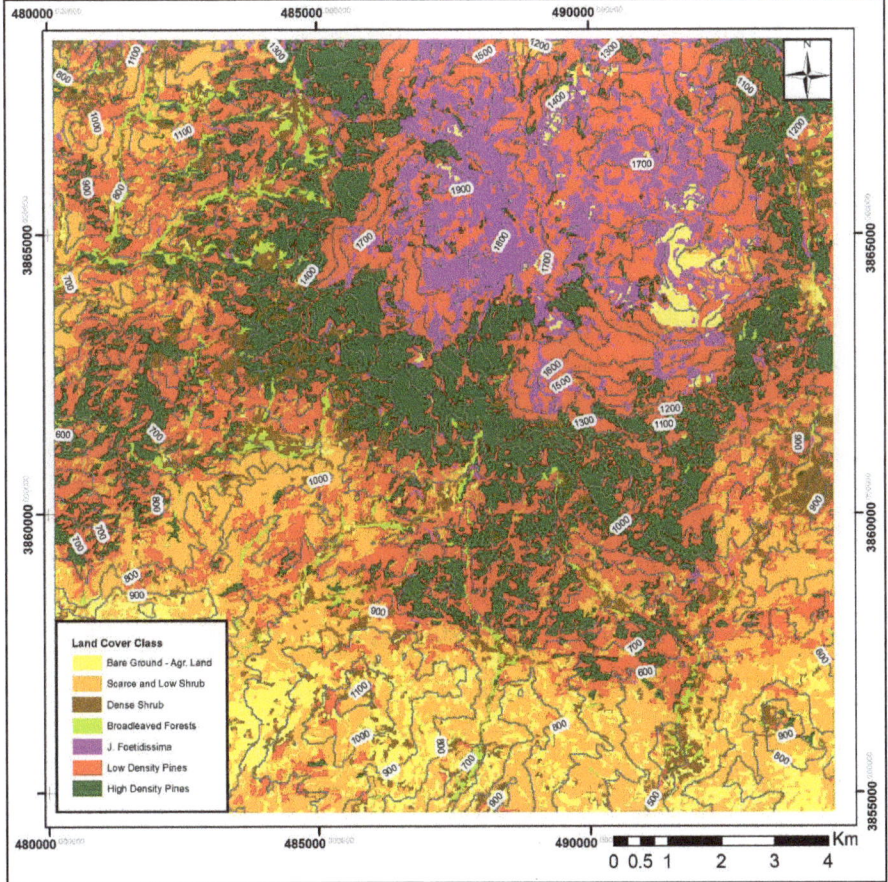

Figure 2. Distribution of land cover types in the study area according to the OBIA classification. *Coordinate System: WGS_1984_UTM_Zone_36N.*

Flammap simulation and the resulted layers of Fireline intensity and rate of spread revealed the diversity of burning conditions in the area (Figure 3a,b). FI exceeds the threshold of 0.5 in more than 42% of the area, while the ROS exceeds the value of 0.5 in more than 53% of the study area. The highest values of FI seem to coincide with the distribution of pine forest formations. High ROS are also observed in the distribution zone of pines, but the *J. foetidissima* formations at the highest altitudes also appear to score high in the ROS index. The other two components of the FDI (Figure 3d,c), which are mainly related to the possibility of a fire event, refer to a much smaller proportion of the study area.

The integration of these four components using the FDI formula resulted in the final, spatially explicit, index of fire hazard in the study area (Figure 4).

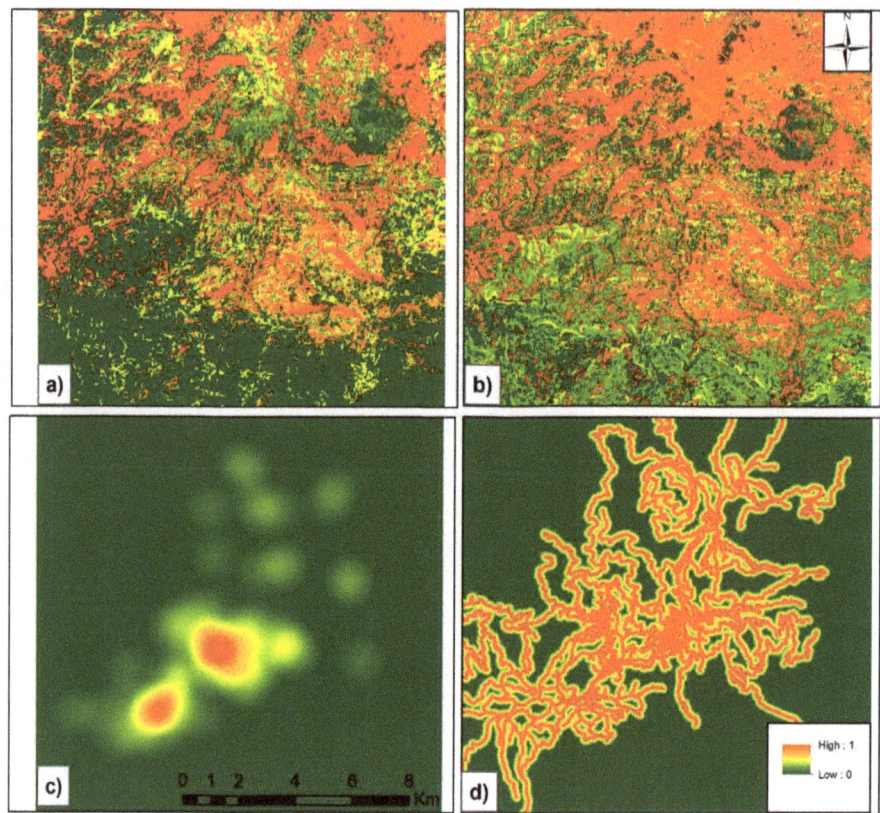

Figure 3. Components of the Fire Danger Index (FDI) used in the study. FI (Fireline Intensity) (**a**), ROS (Rate of Spread) (**b**), PH (Pyric History) (**c**), HI (Distance to Roads) (**d**).

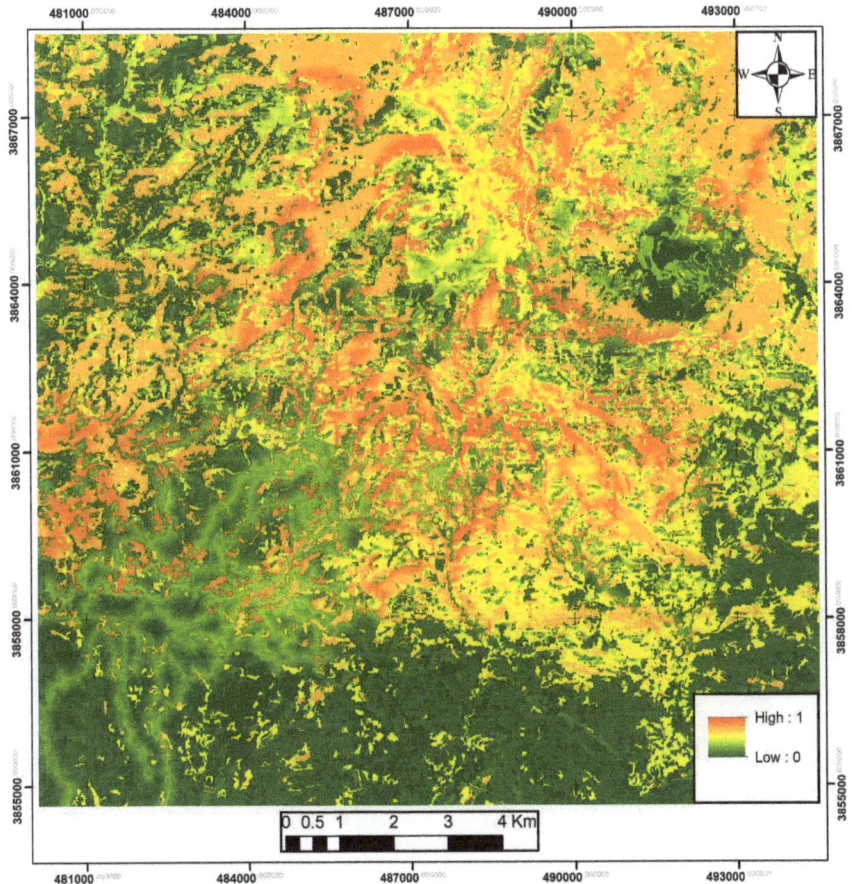

Figure 4. Spatially explicit index of fire hazard in the study area.

The value of FDI ranges between 0 in the lowland parts of the study area, occupied primarily by non-burnable classes, to 1 in the uplands, where forest formations exist, often in close proximity to roads or past fire spots. The FDI reveals a complex mosaic of areas with different flammability properties and different degree of fire hazard, reflecting the compositional and structural vegetational diversity, as well as the complex relief of the study area. The classification of the FDI into four classes of fire hazard (Figure 5) indicates that 36% of the study area faces a high fire hazard and an additional 14% an extremely high hazard. In total, 50% of the study area faces a low or very low fire hazard.

The ANOVA results for the validation of the FDI (Figure 6) show how the four classes of fire hazard, as estimated using the FDI formula have significantly different means in the estimated CFL, which is adopted in the current study as an independent index of fire hazard. All fire hazard classes differ significantly between them, as tested using the post-hoc Tukey HSD test. CFL increases dramatically between the two classes of low and high fire hazard while the increase between the high and extremely high hazard is smaller but still significantly different. Although CFL is impossible to have captured the increased risk of fire hazard imposed by the anthropogenic activities and past fire occurrence, the observed difference between the two high hazard classes is probably the result of the highest consistency, in terms of estimated fire behavior parameters, on the areas belonging to the extremely high hazard class.

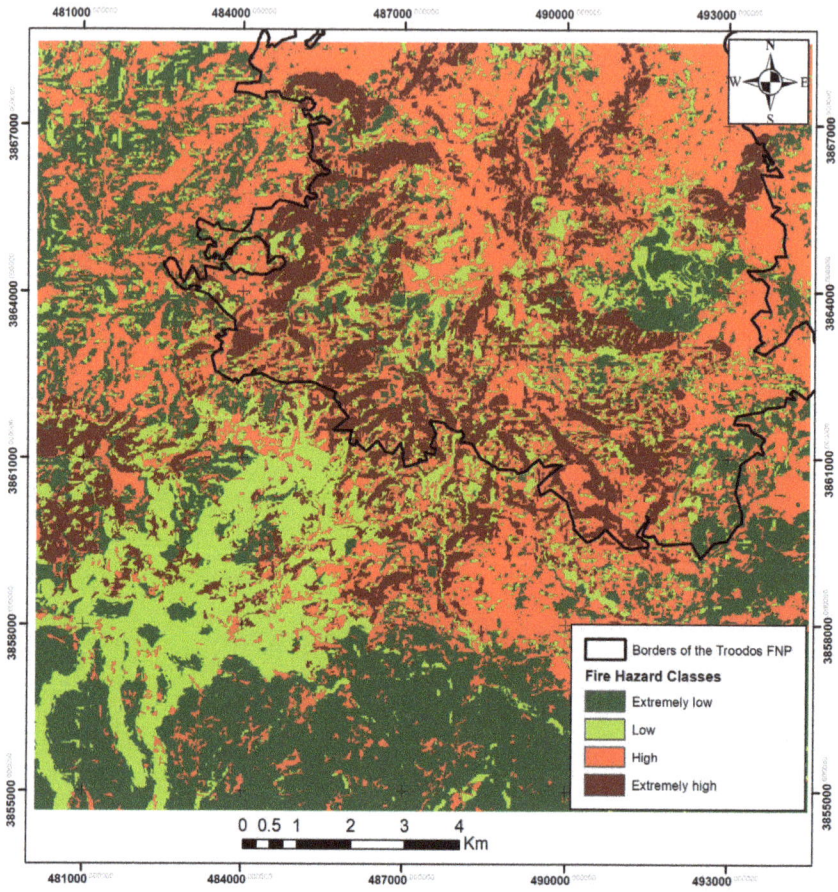

Figure 5. Fire hazard map of the study area. *Coordinate System: WGS_1984_UTM_Zone_36N.*

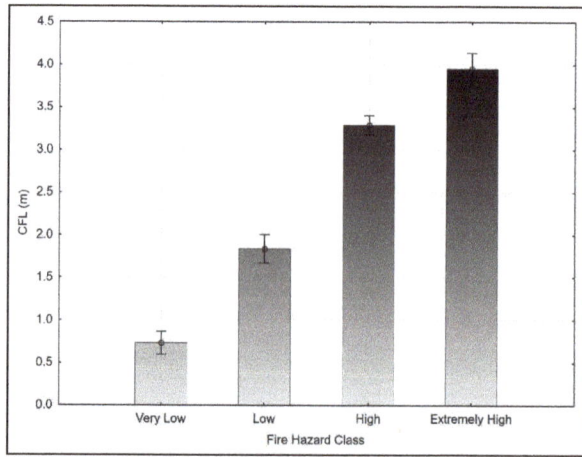

Figure 6. Conditional Flame Length (CFL) of the four identified fire hazard classes estimated using the FDI formula. Vertical bars denote ± one standard error.

The ANOVA followed by a Tukey test revealed, as it was expected, that high density pine formations face the highest fire hazard among the land cover types present in the study area (Figure 7). High fire hazard is also faced by low density pine and dense shrubs, while surprisingly high fire hazard is also faced by the *J. foetidissima* formations. Scarce and low shrubs on the other hand, are the least prone to high-intensity fires by land cover type, along with broadleaved forests, where no significant difference was found between them, despite the higher FDI value observed in broadleaved forests.

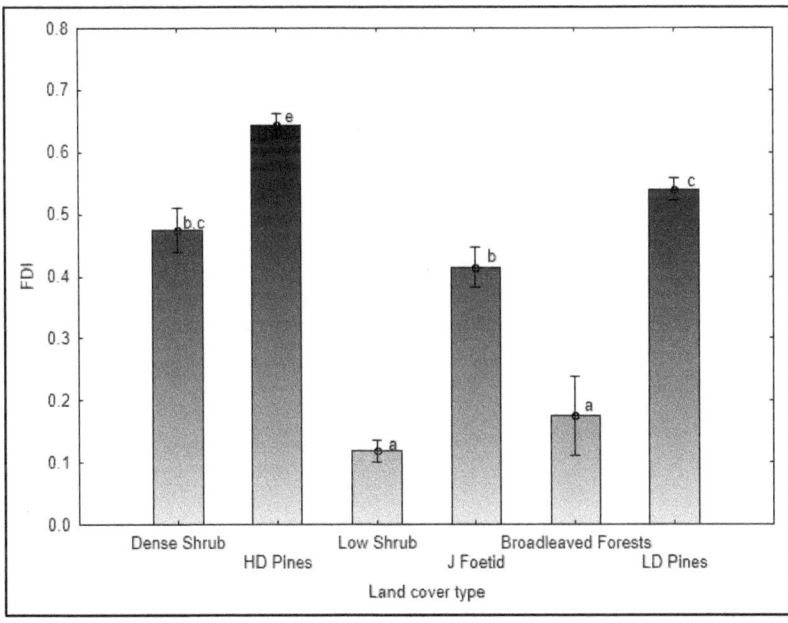

Figure 7. FDI mean for the burnable landcover classes in the study area. Different letters indicate statistically significant differences. Vertical bars denote ± one standard error.

The results of the Regression Trees analysis indicate that both, topographic and vegetation structural characteristics affect the degree of fire hazard (Figure 8). Canopy cover, as expected, determined to a great extent the fire behavior, with pine formation of very low density facing a low fire hazard. Aspect and slope are the two topographic characteristics with the greatest effect on fire hazard. Southeast, south and southwest slopes generally result in lower values compared to the rest of the possible aspects. A positive relationship between slope and fire hazard is also observed under every possible aspect. Canopy base height is a very important vegetation structural characteristic in determining the possibility of a high-intensity fire. Even on relatively gentler slopes, a lower CBH, which does not exceed 3.35 m, can lead to a higher fire hazard in pine forests.

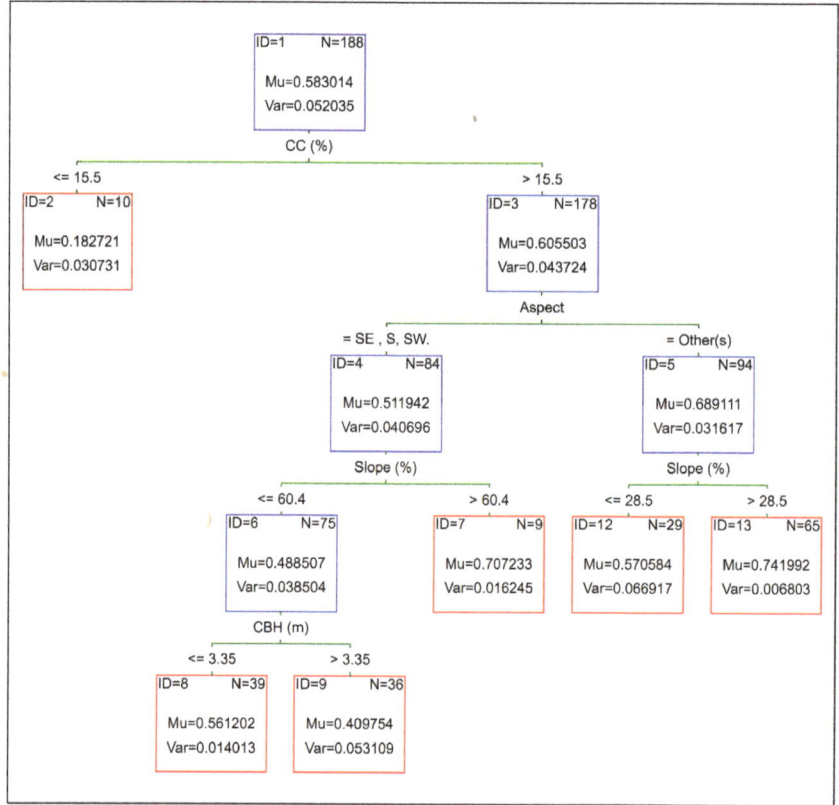

Figure 8. Regression tree for low- and high-density pine forests with FDI as the response variable and topographic and vegetation structural characteristics as independent variable. (N = number of cases in the node).

4. Discussion

In this study we integrated remote sensing data and methods, fire behavior simulation, and proxies of ignition risk to assess the hazard imposed by wildfires in a protected area and its surroundings in the South-East Mediterranean. The results and the approach presented here can provide a significant tool towards a more effective fire management strategy. The spatially explicit fire hazard classification allows its integration into an automatic fire detection system, consisting of land optical cameras and thermal cameras loaded on Unmanned Aerial Vehicles (UAVs) and it is expected to improve its detection accuracy. It can also be integrated in a decision support system to direct the fire fighting forces in a manner that will reduce the time between ignition and first announcement, increasing the efficiency of the initial attack. Satellite-based automatic fire detection systems have already been developed and tested for their efficiency in fire prone areas [39,40]. These systems, despite the coarse spatial resolution of the imagery (i.e., MSG-Seviri and Himawari-8), are of high operational value in wildfire management [72] since they can detect fires as small as 0.1 ha [73]. The only objection to those systems is the difficulty in the precise positioning of the fire event within the large sized pixel. The integration of the spatially explicit fire hazard index developed here into the raw data processing chain may allow for the identification and allocation of new fire spots with higher spatial accuracy. The identification of the vegetation formation that faces the highest fire hazard is also an important outcome of the study and the adopted approach.

Starting from the lowlands to the uplands, the fire hazard in the study area increases. The low altitude areas are dominated by the low hazard formations of scarce and low shrubs, which are often intermixed with agricultural land and other anthropogenic activities, including pastoralism. The latter is often associated with frequent fires of a small area and low heat yield, burnt for the improvement of grazing conditions. The intensive human presence in this area has resulted in the degradation of vegetation which currently consists of phryganic-type formations with a scarce presence of grazing resistant shrubs. The fuel load is kept low, and, as a result, the possibility of a high-intensity fire in these areas is low. In the absence of grazing or any other degradation factors the vegetation turns into dense shrublands, dominated by *Q. alnifolia* which constitute one of the most hazardous vegetation formations in the study area. Dense shrublands have been reported to be among the most fire prone vegetation types and the main carriers of fires in the Eastern Mediterranean, with a high contribution in the annually burned area [18,19]. Furthermore, Moreira et al. [74] reported a higher probability of an ignition to turn into a big fire when it occurs in shrubland forest formations.

Broadleaved forests face a relatively low hazard by wildfires. In the study area these forests are mainly confined across the streams, from high altitudes to the lowlands. These forests, despite the relatively high dead fuel load, which is generated by the annual drop of leaves and the relatively slow decomposition rate of the Mediterranean region, have humid conditions that locally prevail in these microhabitats, which increase the fuel moisture content and prevent an ignition from turning into a high-intensity fire.

Low- and high-density pine formations are the most vulnerable to fires and are those with the highest possibility to result in a high-intensity fire, and subsequently face the highest fire hazard. The largest and the most catastrophic wildfires that have occurred in the Eastern Mediterranean region over the last 15 years have primarily burned pine forests, dominated by *P. halepensis* and *P. brutia* [18,19]. All conifers and especially the Mediterranean pines *P. halepensis* and *P. brutia* have a high resin content, both in the woody parts and in the needles, which result in high flammability compared to other vegetation types. At the same time, the shrub understory, rich in flammable essential oils, promotes flammability and increases the fire hazard [75]. Furthermore, the canopy height of pine forests, which often exceeds 20 m, allows fire brands to disperse in longer distances promoting the advancement of fire front by spotting.

Another important observation in the current study is the significant role of vegetation structural characteristics in determining the fire hazard in the vegetation formations of pine forests most prone to high-intensity fires. CC is a major determinant of fire hazard with low tree densities being associated with low fire hazard. CBH is also important in determining the fire hazard in pine forests, with a CBH that exceeds 3.35 m resulting in a lower hazard on relatively gentler slopes. The trend of Mediterranean pines to retain dead branches in the crown and along the trunk has been thought of and reported to be a fire adaptive trait which promotes fire as a mechanism to create a favorable regeneration environment (niche construction) for the seeds retained in serotinous cones in the crown [76,77]. However, it is more likely that this trend is related to the slow decomposition rate of the Mediterranean pine forests rather than a niche construction mechanism. Nevertheless, the negative correlation between CBH and fire hazard provides an indication on the management measures that could prevent large high-intensity fires. Sheltered fuelbreaks on both sides of roads and, in some cases, in the cores of forest patches, where the CC would be kept low and the CBH high, have been proposed by Xanthopoulos et al. [78] as a method for the horizontal disruption of fuel continuity in fire-prone ecosystems. Such a practice would prevent the evolvement of an ignition into a high-intensity crown fire without exposing the soil to conditions of high soil erosion and degradation, often associated with clear cut firebreaks. This aspect was also found to have a significant role in determining the fire hazard with southern slopes scoring lower values. This was a little surprising since southern slopes are generally drier which should favor high-intensity fires. However, the dry conditions of southern slopes most likely prevent the accumulation of high fuel loads and, as a result, reduces the risk of a high-intensity fire.

Estimating fire hazard and identifying the most vulnerable areas to high-intensity fires is probably the necessary first step towards developing a more effective fire management strategy. Traditional approaches of wildfire management which rely almost exclusively on fire suppression have reached their limits of success [79], despite the improvement in fire-fighting tactics and methods and the increased amount of resources allocated to it [80]. An alternative strategy is suggested by many experts, which should be based on a better balance between fire suppression, an increased level of preparedness involving fuel management and affordable cost in the burned area [81–83]. Fire-smart forest management has been proposed as a more effective approach of fire management, which employs forest management practices, including fuel reduction, conversion and isolation to decrease the probability of an ignition to turn into a high-intensity fire and increase the resistance of forests stands to fires [79,84]. Despite the uncertainties behind the effect of projected climate change on fire regimes, particularly with regards to the changes in the fire landscape ("firescape"), it seems that wildfires will become wilder in the future. Prescribed burning, tested primarily under fire simulation scenarios, seems to be reducing the intensity of future wildfires and, with the exception of the WUI, it deserves higher attention as a fuel reduction measure [84,85]. The same purpose of reducing fuel load could also be served by biomass extraction for bioenergy, especially when the intensity of extraction is high, and it is coupled with an intensive suppression effort [86].

5. Conclusions

The current study reveals the critical fire hazard faced by the most valuable protected area of Cyprus and one of the most valuable areas in the Mediterranean region. In total, 50% of the entire study area and the great majority of the designated FNP are in the high and extremely high class of fire hazard. Its protection relies on preventing megafires that could result in a significant loss of habitats, possible properties and infrastructure. The FDI employed in the study for the estimation of fire hazard is composed by components that can be easily interpreted, in relation to fire behavior and fire risk. It also integrates components related to local peculiarities in terms of ignition risk, so it also integrates local knowledge. The data required (satellite images and topographic data) are freely available while the software used are well documented and widely used. As a result, the proposed approach can easily be transferred to other fire-prone regions of Europe and elsewhere to support an effective wildfire management strategy. Increasing the level of preparedness through the appropriate allocation of resources and the adoption of management measures which could reduce the fire intensity and rate of spread, are steps in the right direction. When fuel management measures are adopted it is essential to ensure the ecological integrity of the ecosystems and avoid measures that could lead to significant degradation. Furthermore, the integration of the resulted mapping products in automatic fire detection systems could decrease significantly the time required for the initial attack. When wildfire management strategy is developed, one should always bear in mind that the main objective should not be the complete elimination of fires, which is both impossible and ecologically dubious, but rather the establishment of a fire regime with low intensity fires and, most importantly, a low socioeconomic and ecological cost [87].

Author Contributions: Conceptualization, P.X.; G.T. and P.K.; Methodology, P.X. and G.T.; Validation, P.X., G.T. and P.K.; Formal analysis, P.X. and I.P.; Resources, P.X., G.T., I.P.; Data curation, P.X., G.T., and I.P.; writing—original draft preparation, P.X.; writing—review and editing, P.X.; Visualization, P.X.; Project administration G.T.; funding acquisition, G.T., P.K. All authors have read and agreed to the published version of the manuscript.

Funding: The study was funded by the European Union, under the program "Interreg V-B Balkan-Mediterranean 2014–2020" and National Funds.

Acknowledgments: The authors wish to acknowledge the significant contribution of the Forestry Directorate of Cyprus for their support in completing this study through the provision of extremely valuable data of the study area, and the in-situ validation of the achieved results.

Conflicts of Interest: The authors declare no conflict of interest. The funders had no role in the design of the study; in the collection, analyses or interpretation of data; in the writing of the manuscript, or in the decision to publish the results.

Appendix A

Table A1. Confusion Matrix and accuracy assessment of the landcover mapping.

Landcover Type	(1)	(2)	(3)	(4)	(5)	(6)	(7)	Total	Users Accuracy
Broadleaved Forests (1)	13	0	0	0	0	0	0	13	1.00
Bare ground-Agr Land (2)	0	37	0	0	0	0	0	37	1.00
Scarse and Low shrublands (3)	0	2	35	0	0	6	0	43	0.81
Low Density Pines (4)	0	0	0	79	5	6	0	90	0.88
Juniperus foetidisima (5)	0	0	0	2	78		0	80	0.98
High Density Pines (6)	0	0	0	6	0	80	0	86	0.93
Dense shrublands (7)	0	0	0	0	0	0	16	16	1.00
Total	13	39	35	87	83	92	16	365	
P_Accuracy	1.00	0.95	1.00	0.91	0.94	0.87	1.00		
Kappa	0.90								
Overall Accuracy	0.93 (93%)								

References

1. Pausas, J.G.; Keeley, J.E. A burning story: The role of fire in the history of life. *BioScience* **2009**, *59*, 593–601. [CrossRef]
2. Bond, W.J.; Keeley, J.E. Fire as a global "herbivore": The ecology and evolution of flammable ecosystems. *Trends Ecol. Evol.* **2005**, *20*, 387–394. [CrossRef]
3. Bond, W.J.; Woodward, F.I.; Midgley, G.F. The global distribution of ecosystems in a world without fire. *New Phytol.* **2005**, *165*, 525–538. [CrossRef]
4. Keeley, J.E.; Rundel, P.W. Fire and the Miocene expansion of C4 grasslands. *Ecol. Lett.* **2005**, *8*, 683–690. [CrossRef]
5. Pausas, J.G.; Keeley, J.E. Wildfires as an ecosystem service. *Front. Ecol. Environ.* **2019**, *17*, 289–295. [CrossRef]
6. Trabaud, L.V.; Christensen, N.L.; Gill, A.M. Historical biogeography of fire in temperate and mediterranean ecosystems. In *Fire in the Environment: Its Ecological and Atmospheric Importance*; Grutzen, P.J., Goldamer, J.G., Eds.; John Wiley: New York, NY, USA, 1993; pp. 277–295.
7. Randerson, J.T.; Chen, Y.; van der Werf, G.R.; Rogers, B.M.; Morton, D.C. Global burned area and biomass burning emissions from small fires. *J. Geophys. Res. Biogeosci.* **2012**, *117*, 1–23. [CrossRef]
8. San-Miguel-Ayanz, J.; Durrant, T.; Boca, R.; Libertà, G.; Branco, A.; de Rigo, D.; Ferrari, D.; Maianti, P.; Artés Vivancos, T.; Oom, D.; et al. *Forest Fires in Europe, Middle East and North Africa 2018*; Publications Office of the European Union: Brussels, Belgium, 2019.
9. Xanthopoulos, G. Wildland fires: Mediterranean. *Crisis Response J.* **2009**, *5*, 50–51.
10. Molina-Terren, D.M.; Xanthopoulos, G.; Diakakis, M.; Ribeiro, L.; Caballero, D.; Delogu, G.M.; Viegas, D.X.; Silva, C.A.; Cardil, A. Analysis of forest fire fatalities in Southern Europe: Spain, Portugal, Greece and Sardinia (Italy). *Int. J. Wildland Fire* **2019**, *28*, 85–98. [CrossRef]
11. Pausas, J.G. Changes in fire and climate in the Eastern Iberian Peninsula (Mediterranean Basin). *Clim. Chang.* **2004**, *63*, 337–350. [CrossRef]
12. Dimitrakopoulos, A.P.; Vlahou, M.; Anagnostopoulou, C.G.; Mitsopoulos, I.D. Impact of drought on wildland fires in Greece: Implications of climate change? *Clim. Chang.* **2011**, *109*, 331–347. [CrossRef]
13. Pausas, J.G.; Fernandez-Munoz, S. Fire regime changes in the Western Mediterranean Basin: From fuel limited to draught-driven fire regime. *Clim. Chang.* **2012**, *110*, 215–226. [CrossRef]
14. Koutsias, N.; Xanthopoulos, G.; Founda, D.; Xystrakis, F.; Nioti, F.; Pleniou, M.; Mallinis, G.; Arianoutsou, M. On the relationships between forest fires and weather conditions in Greece from long-term national observations (1894–2010). *Int. J. Wildland Fire* **2013**, *22*, 493–507. [CrossRef]

15. Turco, M.; Bedia, J.; Di Liberto, F.; Fiorucci, P.; von Hardenberg, J.; Koutsias, N.; Llasat, M.C.; Xystrakis, F.; Provenzale, A. Decreasing fires in Mediterranean Europe. *PLoS ONE* **2016**, *11*, e0150663. [CrossRef] [PubMed]
16. Moreira, F.; Viedma, O.; Arianoutsou, M.; Curt, T.; Koutsias, N.; Rigolot, E.; Barbati, A.; Corona, P.; Vaz, P.; Xanthopoulos, G.; et al. Landscape and wildfire interactions in southern Europe: Implications for landscape management. *J. Environ. Manag.* **2011**, *92*, 2389–2402. [CrossRef] [PubMed]
17. Vacchiano, G.; Garbarino, M.; Lingua, E.; Motta, R. Forest dynamics and disturbance regimes in the Italian Apennines. *For. Ecol. Manag.* **2017**, *388*, 57–66. [CrossRef]
18. Koutsias, N.; Arianoutsou, M.; Kallimanis, A.S.; Mallinis, G.; Halley, J.M.; Dimopoulos, P. Where did the fires burn in Peloponnisos, Greece the summer of 2007? Evidence for a synergy of fuel and weather. *Agric. For. Meteorol.* **2012**, *156*, 41–53. [CrossRef]
19. Kontoes, C.; Keramitsoglou, I.; Papoutsis, I.; Sifakis, N.I.; Xofis, P. National scale operational mapping of burnt areas as a tool for the better understanding of contemporary wildfire patterns and regimes. *Sensors* **2013**, *13*, 11146–11166. [CrossRef] [PubMed]
20. Xofis, P.; Poirazidis, K. Combining different spatio-temporal resolution images to depict landscape dynamics and guide wildlife management. *Biol. Conserv.* **2018**, *218*, 10–17. [CrossRef]
21. Keeley, J.E.; Pausas, J.G.; Rundel, P.W.; Bond, W.J.; Bradstock, R.A. Fire as an evolutionary pressure shaping plant traits. *Trends Plant Sci.* **2011**, *16*, 406–411. [CrossRef] [PubMed]
22. Bradshaw, S.D.; Dixon, K.W.; Hopper, S.D.; Lambers, H.; Turner, S.R. Little evidence for fire-adapted plant traits in Mediterranean climate regions. *Trends Plant Sci.* **2011**, *16*, 69–76. [CrossRef]
23. Gill, A.M. Adaptive Responses of Australian Vascular Species to Fires. In *Fire and the Australian Biota*; Gill, A.M., Groves, R.H., Noble, R.I., Eds.; Australian Academy of Science: Canberra, Australia, 1981; pp. 243–272.
24. Mooney, H.A.; Hobbs, R.J. Resilience at the Individual Plant Level. In *Resilience in Mediterranean-Type Ecosystems*; Dell, B.B., Lamont, D., Hopkins, A.J., Eds.; Springer: The Hague, The Netherlands, 1986; pp. 65–82.
25. Miller, C.; Ager, A.A. A review of recent advances in risk analysis for wildfire management. *Int. J. Wildland Fire* **2013**, *22*, 1–14. [CrossRef]
26. Jolly, W.M.; Freeborn, P.H.; Page, W.G.; Butler, B.W. Severe Fire Danger Index: A Forecastable Metric to Inform Firefighter and Community Wildfire Risk Management. *Fire* **2019**, *2*, 47. [CrossRef]
27. Reimer, J.; Thompson, D.K.; Povak, N. Measuring Initial Attack Suppression Effectiveness through Burn Probability. *Fire* **2019**, *2*, 60. [CrossRef]
28. Mitsopoulos, I.; Mallinis, G.; Arianoutsou, M. Wildfire Risk Assessment in a Typical Mediterranean Wildland-Urban Interface of Greece. *Environ. Manag.* **2015**, *55*, 900–915. [CrossRef]
29. Finney, M. Fire growth using minimum travel time methods. *Can. J. For. Res.* **2002**, *32*, 1420–1424. [CrossRef]
30. Finney, M. An overview of FlamMap modelling capabilities. In *Fuels Management—How to Measure Success: Conference Proceedings*; Andrews, P., Butler, B., Eds.; USDA, Forest Service, Rocky Mountain Research Station: Fort Collins, CO, USA, 2006; pp. 213–219.
31. Calkin, D.E.; Cohen, J.D.; Finney, M.A.; Thomson, P. How risk management can prevent future wildfire disasters in the wildland-urban interface. *Proc. Natl. Acad. Sci. USA* **2014**, *111*, 746–751. [CrossRef] [PubMed]
32. Mitsopoulos, I.; Mallinis, G.; Karali, A.; Giannakopoulos, C.; Arianoutsou, M. Mapping fire behaviour under changing climate in a Mediterranean landscape in Greece. *Reg. Environ. Chang.* **2016**, *16*, 1929–1940. [CrossRef]
33. Vilà-Vilardell, L.; Keeton, W.S.; Thom, D.; Gyeltshen, C.; Tshering, K.; Gratzer, G. Climate change effects on wildfire hazards in the wildland-urban-interface-Blue pine forests of Bhutan. *For. Ecol. Manag.* **2020**, *461*, 117927. [CrossRef]
34. Mangeas, M.; André, J.; Gomez, C.; Despinoy, M.; Wattelez, G.; Touraivane, T. A spatially explicit integrative model for estimating the risk of wildfire impacts in New-Caledonia. *Int. J. Parallel Emergent Distrib. Syst.* **2019**, *34*, 37–52. [CrossRef]
35. Mallinis, G.; Mitsopoulos, I.; Beltran, E.; Goldammer, J. Assessing Wildfire Risk in Cultural Heritage Properties Using High Spatial and Temporal Resolution Satellite Imagery and Spatially Explicit Fire Simulations: The Case of Holy Mount Athos, Greece. *Forests* **2016**, *7*, 46. [CrossRef]
36. Stefanidou, A.; Dragozi, E.; Stavrakoudis, D.; Gitas, I.Z. Fuel type mapping using object-based image analysis of DMC and Landsat-8 OLI imagery. *Geocarto Int.* **2018**, *33*, 1064–1083. [CrossRef]

37. Mallinis, G.; Galidaki, G.; Gitas, I. A Comparative Analysis of EO-1 Hyperion, Quickbird and Landsat TM Imagery for Fuel Type Mapping of a Typical Mediterranean Landscape. *Remote Sens.* **2014**, *6*, 1684–1704. [CrossRef]
38. Stefanidou, A.; Gitas, I.; Katagis, T. A national fuel type mapping method improvement using sentinel-2 satellite data. *Geocarto Int.* **2020**. [CrossRef]
39. Sifakis, N.I.; Iossifidis, C.; Kontoes, C.; Keramitsoglou, I. Wildfire Detection and Tracking over Greece Using MSG-SEVIRI Satellite Data. *Remote Sens.* **2011**, *3*, 524–538. [CrossRef]
40. Xu, G.; Zhong, X. Real-time wildfire detection and tracking in Australia using geostationary satellite: Himawari-8. *Remote Sens. Lett.* **2017**, *8*, 1052–1061. [CrossRef]
41. Amos, C.; Petropoulos, G.P.; Ferentinos, K.P. Determining the use of Sentinel-2A MSI for wildfire burning & severity detection. *Int. J. Remote Sens.* **2019**, *40*, 905–930.
42. Colson, D.; Petropoulos, G.P.; Ferentinos, K.P. Exploring the Potential of Sentinels-1 & 2 of the Copernicus Mission in Support of Rapid and Cost-effective Wildfire Assessment. *Int. J. Appl. Earth Obs. Geoinf.* **2018**, *73*, 262–276.
43. Brown, A.; Petropoulos, G.P.; Ferentinos, K.P. Appraisal of the Sentinel-1 & 2 use in a large-scale wildfire assessment: A case study from Portugal's fires of 2017. *Appl. Geogr.* **2017**, *100*, 78–89.
44. Skowronski, N.S.; Gallagher, M.R.; Warner, T.A. Decomposing the Interactions between Fire Severity and Canopy Fuel Structure Using Multi-Temporal, Active, and Passive Remote Sensing Approaches. *Fire* **2020**, *3*, 7. [CrossRef]
45. Sotiriou, A.R. Phytosociological Research of the National Forest Park of Mountain Troodos of Cyprus. Ph.D. Thesis, Aristotelean University of Thessaloniki, Thessaloniki, Greece, 2010. (In Greek) [CrossRef]
46. Kaufmann, S.; Berg, C. Bryophyte Ecology and Conservation in the Troodos Mountains, Cyprus. *Herzogia* **2014**, *27*, 165–187. [CrossRef]
47. Trimple. *Ecognition Developer Reference Book*; Trimble Documentation: Munich, Germany, 2014.
48. Drusch, M.; Del Bello, U.; Carlier, S.; Colin, O.; Fernandez, V.; Gascon, F.; Hoersch, B.; Isola, C.; Laberinti, P.; Martimort, P.; et al. Sentinel-2: ESA's Optical High-Resolution Mission for GMES Operational Services. *Remote Sens. Environ.* **2012**, *120*, 25–36. [CrossRef]
49. Kim, M.; Madden, M.; Warner, T. Forest type mapping using object-specific texture measures from multispectral Ikonos imagery: Segmentation quality and image classification issues. *Photogramm. Eng. Remote Sens.* **2009**, *75*, 819–829. [CrossRef]
50. Rittl, T.; Cooper, M.; Heck, R.J.; Ballester, M.V.R. Object-based method outperforms per-pixel method for land cover classification in a protected area of the Brazilian Atlantic rainforest region. *Pedosphere* **2013**, *23*, 290–297. [CrossRef]
51. Bock, M.; Xofis, P.; Rossner, G.; Wissen, M.; Mitchley, J. Object oriented methods for habitat mapping in multiple scales: Case studies from Northern Germany and North Downs, GB. *J. Nat. Conserv.* **2005**, *13*, 75–89. [CrossRef]
52. Breiman, L. Random forests. *Mach. Learn.* **2001**, *45*, 5–32. [CrossRef]
53. Tsiourlis, G.M. Etude d'un écosystème de maquis à *Juniperus phoenicea* L. (Naxos, Cyclades, Grèce): Phytomasse et nécromasse épigées. *Ecologie* **1992**, *23*, 59–69.
54. Tsiourlis, G.M. Phytomasse et productivité primaire d'une phytocénose peupleraie (Populus cultivar robusta) d'âges différents (Hainaut, Belgique). *Belg. J. Bot.* **1994**, *127*, 134–144.
55. Smiris, P.; Maris, F.; Vitoris, K.; Stamou, N.; Kalambokidis, K. Aboveground biomass of Pinus halepensis Mill. forests in the Kassandra Peninsula-Chalkidiki. *Silva Gandav.* **2000**, *65*, 173–187. [CrossRef]
56. Ruiz-Peinado, R.; del Rio, M.; Montero, G. New models for estimating the carbon sink capacity of Spanish softwood species. *For. Syst.* **2011**, *20*, 176–188. [CrossRef]
57. Nunes, L.; Lopes, D.; Castro Rego, F.; Gower, S. Aboveground biomass and net primary production of pine, oak and mixed pine-oak forests on the Vila Real district, Portugal. *Ecol. Manag.* **2013**, *305*, 38–47. [CrossRef]
58. Scott, J.H.; Burgan, R.E. *Standard Fire Behaviour Fuel Models: A Comprehensive Set for Use with Rothermel's Surface Fire Spread Model*; USDA, Forest Service, Rocky Mountain Research Station: Fort Collins, CO, USA, 2005; 72p.
59. Calkin, D.E.; Ager, A.A.; Gilbertson-Day, J. *Wildfire Risk and Hazard: Procedures for the First Approximation*; USDA, Forest Service, Rocky Mountain Research Station: Fort Collins, CO, USA, 2010; 62p.

60. Finney, M.A. *FARSITE: Fire Area Simulator-Model Development and Evaluation*; USDA, Forest Service, Rocky Mountain Research Station: Fort Collins, CO, USA, 2004; 47p.
61. Tolika, K.; Maheras, P.; Tegoulias, I. Extreme temperatures in Greece during 2007: Could this be a "return to the future"? *Geophys. Res. Lett.* **2009**, *36*, 10. [CrossRef]
62. Founda, D.; Giannakopoulos, C. The exceptionally hot summer of 2007 in Athens, Greece—A typical summer in the future climate? *Glob. Planet. Chang.* **2009**, *67*, 227–236. [CrossRef]
63. Palaiologou, P. Design of Fire Behaviour Prediction and Assessment with the Use of Geoinformation. Ph.D. Thesis, University of the Aegean, Izmir, Turkey, 2015. (In Greek)
64. Chandler, C.; Cheney, P.; Thomas, P.; Trabaud, L.; Williams, D. *Fire in Forestry. Volume I. Forest Fire Behavior and Effects*; John Wiley & Sons: New York, NY, USA, 1983; pp. 1–30.
65. Byram, G.M. Combustion of forest fuels. In *Forest Fire: Control and Use*; Davis, K.P., Ed.; McGraw-Hill Book Company: New York, NY, USA, 1959; pp. 61–89.
66. Xofis, P.; Tsiourlis, G.; Konstantinidis, P. A Fire Danger Index for the early detection of areas vulnerable to wildfires in the Eastern Mediterranean region. *Euro-Mediterr. J. Environ. Integr.* **2020**. [CrossRef]
67. Ricotta, C.; Bajocco, S.; Guglietta, D.; Conedera, M. Assessing the Influence of Roads on Fire Ignition: Does Land Cover Matter? *Fire* **2018**, *1*, 24. [CrossRef]
68. Catry, F.X.; Rego, F.C.; Bacao, F.; Moreira, F. Modelling and mapping wildfire ignition risk in Portugal. *Int. J. Wildland Fire* **2009**, *18*, 921–931. [CrossRef]
69. Scott, J.H.; Thompson, M.P.; Calkin, D.E. *A Wildfire Risk Assessment Framework for Land and Resource Management*; USDA, Forest Service, Rocky Mountain Research Station: Fort Collins, CO, USA, 2013; 92p.
70. Kalabokidis, K.; Athanasis, N.; Palaiologou, P.; Vasilakos, C.; Finney, M.; Ager, A. Minimum travel time algorithm for fire behavior and burn probability in a parallel computing environment. In *Advances in Forest Fire Research*; Viegas, D.X., Ed.; Imprensa da Universidade de Coimbra: Coibra, Portugal, 2014; pp. 882–891.
71. De'ath, G.; Fabricious, K.E. Classification and regression trees: A powerful yet simple technique for ecological data analysis. *Ecology* **2000**, *81*, 3178–3192. [CrossRef]
72. Filizzola, C.; Corrado, R.; Marchese, F.; Mazzeo, G.; Paciello, R.; Pergola, N.; Tramutoli, V. RST-FIRES, an exportable algorithm for early-fire detection and monitoring: Description, implementation, and field validation in the case of the MSG-SEVIRI sensor. *Remote Sens. Environ.* **2016**, *186*, 196–216. [CrossRef]
73. Laneve, G.; Castronuovo, M.M.; Cadau, E.G. Continuous Monitoring of Forest Fires in the Mediterranean Area Using MSG. *IEEE Trans. Geosci. Remote Sens.* **2006**, *44*, 2761–2768. [CrossRef]
74. Moreira, F.; Catry, F.X.; Rego, F.; Bacao, F. Size-dependent pattern of wildfire ignitions in Portugal: When do ignitions turn into big fires? *Landsc. Ecol.* **2010**, *45*, 1405–1417. [CrossRef]
75. Ne'eman, G.; Goubitz, S.; Nathan, R. Reproductive traits o? *Pinus halepensis* in the light of fire—A critical review. *Plant Ecol.* **2004**, *171*, 69–79.
76. Schwilk, D.W. Flammability is a Niche construction trait: Canopy archtecture affects fire intensity. *Am. Nat.* **2003**, *162*, 725–733. [CrossRef] [PubMed]
77. Schwilk, D.W.; Ackerly, D.D. Flammability and serotiny as strategies: Corelated evolution in pines. *Oikos* **2001**, *94*, 326–336. [CrossRef]
78. Xanthopoulos, G.; Caballero, D.; Galante, M.; Alexandrian, D.; Rigolot, E.; Marzano, R. Forest fuels management in Europe. In *Fuels Management—How to Measure Success: Conference proceedings*; USDA, Forest Service, Rocky Mountain Research Station: Fort Collins, CO, USA, 2006; pp. 29–46.
79. Hirsch, K.; Kafka, V.; Tymstra, C.; McAlpine, R.; Hawkes, B.; Stegehuis, H.; Quintilio, S.; Gauthier, S.; Peck, K. Fire-smart forest management: A pragmatic approach to sustainable forest management in fire-dominated ecosystems. *For. Chron.* **2001**, *77*, 357–363. [CrossRef]
80. Raftoyannis, Y.; Nocentini, S.; Marchi, E.; Calama, R.; Garcia Guemes, C.; Pilas, I.; Peric, S.; Paulo, J.A.; Moreira, A.C.; Costa-Ferreira, M.; et al. Perceptions of forest experts on climate change and fire management in European Mediterranean forests. *iFor. Biogeosci. For.* **2014**, *7*, 33–41. [CrossRef]
81. Donovan, G.H.; Brown, T.C. Be careful what you wish for: The legacy of Smokey Bear. *Front. Ecol.* **2007**, *5*, 73–79. [CrossRef]
82. Doerr, S.H.; Santı́n, C. Global trends in wildfire and its impacts: Perceptions versus realities in a changing world. *Philos. Trans. R. Soc. B* **2016**, *371*, 1696. [CrossRef]
83. San-Miguel-Ayanz, J.; Moreno, J.M.; Camia, A. Analysis of large fires in European Mediterranean landscapes: Lessons learned and perspectives. *For. Ecol. Manag.* **2013**, *294*, 11–22. [CrossRef]

84. Fernandes, P.M. Fire-smart management of forest landscapes in the Mediterranean basin under global change. *Landsc. Urban Plan.* **2013**, *110*, 175–182. [CrossRef]
85. Duane, A.; Aquilué, N.; Canelles, Q.; Morán-Ordoñez, A.; De Cáceres, M.; Brotons, L. Adapting prescribed burns to future climate change in Mediterranean landscapes. *Sci. Total Environ.* **2019**, *677*, 68–83. [CrossRef]
86. Regos, A.; Aquilué, N.; López, I.; Codina, M.; Retana, J.; Brotons, L. Synergies between forest biomass extraction for bioenergy and fire suppression in Mediterranean ecosystems: Insights from a storyline-and-simulation approach. *Ecosystems* **2016**, *19*, 786–802. [CrossRef]
87. Moreira, F.; Ascoli, D.; Safford, H.; Adams, M.A.; Moreno, J.M.; Pereira, J.M.; Fernandez, P.M. Wildfire management in Mediterranean-type regions: Paradigm change needed. *Environ. Res. Lett.* **2020**, *15*, 1. [CrossRef]

© 2020 by the authors. Licensee MDPI, Basel, Switzerland. This article is an open access article distributed under the terms and conditions of the Creative Commons Attribution (CC BY) license (http://creativecommons.org/licenses/by/4.0/).

MDPI
St. Alban-Anlage 66
4052 Basel
Switzerland
Tel. +41 61 683 77 34
Fax +41 61 302 89 18
www.mdpi.com

Fire Editorial Office
E-mail: fire@mdpi.com
www.mdpi.com/journal/fire

www.ingramcontent.com/pod-product-compliance
Lightning Source LLC
LaVergne TN
LVHW070405100526
838202LV00014B/1395